# Turn the graphing calculator into a powerful tool for your success!

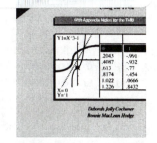

## Explorations in Beginning and Intermediate Algebra Using the TI-82 With Appendix Notes for the TI-85, CASIO fx-7700GE, and HP48G
by Deborah J. Cochener and Bonnie M. Hodge,
both of Austin Peay State University

$25.95 Single Copy Price. 314 pages. Spiral bound. 8 1/2 x 11.
ISBN: 0-534-34091-1. © 1996. Published by Brooks/Cole.

*Deborah Jolly Cochener*
*Bonnie MacLean Hodge*

***Learn to use the graphing calculator to develop problem-solving and critical-thinking skills that will help improve your performance in your beginning or intermediate algebra course!***

Designed to help you succeed in your beginning or intermediate algebra course, this unique and student-friendly workbook improves both your understanding and retention of algebra concepts—using the graphing calculator. By integrating technology into mathematics, the authors help you develop problem-solving and critical-thinking skills.

To guide you in your explorations, you'll find:
* hands-on applications with solutions
* correlation charts that relate course topics to the workbook units
* a key chart that shows which units introduce keys on the calculator
* a Troubleshooting Section to help you avoid common errors

*I found this workbook to be very well written and thought out and to be quite different from calculator supplements on the market. The integration of concept development and applications makes this usable as an ancillary textbook rather than just a calculator supplement. . . . The students are not taught to rely on the calculator, but to use it as another problem solving tool and as reinforcement of algebraic principles."*
Mary Lou Hammond, Spokane Community College

*The authors are clearly aware of important issues in mathematics education today, and have attempted to respond with a constructive, active-oriented approach. . . . A surprisingly valuable introduction to some not so well understood arithmetic concepts.*
Don Shriner, Frostburg State University

### Topics

*This text contains 35 units divided into the following subsections:*
  Basic Calculator Operations
  Graphically Solving Equations and Inequalities
  Graphing and Applications of Equations in Two Variables
  Stat Plots
Also included are:
  Concept Correlation Charts
  Key Introduction Charts
  Troubleshooting

### Order your copy today!

To receive your copy of ***Explorations in Beginning and Intermediate Algebra Using the TI-82: With Appendix Notes for the TI-85, CASIO fx-7700GE, and HP48G,*** simply mail in the order form attached.

# ORDER FORM

To order, simply fill out this coupon and return it to Brooks/Cole along with your check, money order, or credit card information.

_____Yes! I would like to order ***Explorations in Beginning and Intermediate Algebra Using the TI-82: With Appendix Notes for the TI-85, CASIO fx-7700 GE, and HP48G*** , by Deborah J. Cochener and Bonnie M. Hodge, ISBN: 0-534-34227-2 for $25.95.

Residents of: AL, AZ, CA, CT, CO, FL, GA, IL, IN, KS, KY, LA, MA, MD, MI, MN, MO, NC, NJ, NY, OH, PA, RI, SC, TN, TX, UT, VA, WA, WI must add appropriate state sales tax.

Subtotal _____
Tax _____
Handling ___$4.00_
Total _____

## Payment Options

_____ Check or money order enclosed

or bill my ____VISA ____MasterCard ____American Express

Card Number: _____

Expiration Date: _____

Signature: _____

*Note: Credit card billing and shipping addresses must be the same.*

## Please ship my order to: (Please print.)

Name _____

Street Address_____

City _____ State _____ Zip+4_____

Telephone ( )_____ e-mail _____

## Mail to:

**Brooks/Cole Publishing Company**
Dept. 9BCMA005
511 Forest Lodge Road, Pacific Grove, California 93950-5098
Phone: (408) 373-0728; Fax: (408) 375-6414
Internet: http://www.brookscole.com

**I**(**T**)**P** International Thomson Publishing
Education Group

8/97

*P.S. If this book is appropriate for a course that you teach, please send your request for a complimentary review copy, on department letterhead, to the address listed above. Prices subject to change without notice.*

# Mastering Mathematics: How to Be a Great Math Student, Third Edition
## by Richard Manning Smith

**A complete guide to improving your learning and performance in math courses!**

Providing solid tips for every stage of study, *Mastering Mathematics: How to Be a Great Math Student, Third Edition* stresses the importance of a positive attitude and gives you the tools to succeed in your math course. This practical guide will help you:

- avoid mental blocks during math exams
- identify and improve your areas of weakness
- get the most out of class time
- study more effectively
- overcome a perceived "low math ability"
- be successful on math tests
- get back on track when you are feeling "lost"
  …and much more!

To order a copy of *Mastering Mathematics*, please contact your college store or fill out the form on the back and return with your payment to Brooks/Cole. You can also order on-line at http://www.brookscole.com

# ORDER FORM

_____Yes! Send me a copy of *Mastering Mathematics: How to be a Great Math Student*  (ISBN: 0-534-34864-5)

_____Copies x $15.95* = _____

Residents of: AL, AZ, CA, CT, CO, FL, GA, IL, IN, KS, KY, LA, MA, MD, MI, MN, MO, NC, NJ, NY, OH, PA, RI, SC, TN, TX, UT, VA, WA, WI must add appropriate state sales tax.

Subtotal _____
Tax _____
Handling __$4.00__
Total Due _____

## Payment Options

_____ Check or money order enclosed

Bill my  ____VISA ____MasterCard ____American Express

Card Number: _____

Expiration Date: _____

Signature: _____

*Note: Credit card billing and shipping addresses must be the same.*

## Please ship my order to: (Please print)

Name _____

Institution _____

Street Address_____

_____

City _____  State _____  Zip+4_____

Telephone (     )_____  e-mail _____

Your credit card will not be billed until your order is shipped. Prices subject to change without notice. We will refund payment for unshipped out-of-stock titles after 120 days and for not-yet-published titles after 180 days unless an earlier date is requested in writing from you.

## Mail to:

**Brooks/Cole Publishing Company**
Source Code 9BCMA006
511 Forest Lodge Road
Pacific Grove, California 93950-5098
Phone: (408) 373-0728; Fax: (408) 375-6414
e-mail: info@brookscole.com

10/97

# Proven math help where many need it most: Word problems

## IntelliTutor™:The Word Problem Solver
**by Armstrong Laboratory**

*In the home or in the classroom,* IntelliTutor *will help you achieve success in basic arithmetic and prealgebra.*

Learning and understanding mathematics is a challenge for many students, especially when it comes to word problems. Conquer this difficult part of the curriculum and boost both your confidence *and* proficiency in mathematics with *IntelliTutor*, an exciting new software program now available for the first time to the public. Armstrong Laboratory has been involved in conducting research in Intelligent Tutoring Systems for many years, and has even briefed President Clinton about the benefits of the program. Emulating the human tutor who knows *what* to teach and *how* to teach it, "intelligent tutors" provide advanced, individualized computer-based instruction. Much like a human tutor, the *IntelliTutor* adapts to the needs of the individual learner. Plus, *IntelliTutor* grows with you, ensuring that you'll always work at your skill level.

### Unique Features

- Each topic provides a brief review at the beginning of each Computer Based Training (CBT) Module. A required quiz at the end ensures that you understand the topic before you work on word problems.

- An easy-to-use toolbar interface operates with a simple "point-and-click."

- A "Guided Mode" through the program helps less experienced users reap the benefits of *IntelliTutor's* hint structure, while more experienced users can move more quickly in the "Unguided Mode."

- *IntelliTutor* was used by Armstrong Laboratories in one of the largest research studies ever conducted in educational software. Classroom testing has shown that using *IntelliTutor* with regular classroom learning can improve a remedial mathematics student's performance by one whole letter grade.

### Minimum system requirements

Windows®: 486 PC, 8 MB RAM, 32 MB hard disk, color monitor, Windows 3.1 or higher.

To order a copy of *IntelliTutor: The Word Problem Solver*, please contact your college store or fill out the form on the back and return with your payment to Brooks/Cole.

# ORDER FORM

_____Yes! Send me a copy of *Intellitutor™: The Word Problem Solver,*
by Armstrong Laboratory  (Windows® ISBN: 0-534-34864-5)

_____Copies x $10.95* =_____

Residents of: AL, AZ, CA, CT, CO, FL, GA, IL, IN, KS, KY, LA,
MA, MD, MI, MN, MO, NC, NJ, NY, OH, PA, RI, SC, TN, TX,
UT, VA, WA, WI must add appropriate state sales tax.

Subtotal _____
Tax _____$4.00
Handling _____
Total Due _____

* Call after 12/1/97 for current prices: 1-800-487-3575

## Payment Options

_____ Check or money order enclosed

Bill my ____VISA ____MasterCard ____American Express

Card Number: _____ Expiration Date: _____

Signature: _____

*Note: Credit card billing and shipping addresses must be the same.*

## Please ship my order to: (Please print.)

Name _____

Institution _____

Street Address_____

City _____ State _____ Zip+4_____

Telephone (     )_____ e-mail _____

Your credit card will not be billed until your order is shipped. Prices subject to change without
notice. We will refund payment for unshipped out-of-stock titles after 120 days and for not-yet-
published titles after 180 days unless an earlier date is requested in writing from you.

## Mail to:

**Brooks/Cole Publishing Company**
Source Code 9BCMA007
511 Forest Lodge Road
Pacific Grove, California 93950-5098
Phone: (408) 373-0728; Fax: (408) 375-6414
e-mail: info@brookscole.com

10/97

## *Now you can learn the abstract concepts of algebra through concrete modeling!*

## *Computer Interactive Algeblocks, Volumes 1 and 2*

by Anita Johnston, Jackson Community College

### *Taking algebra from frustrating to fun!*

Learning abstract concepts from a lecture or a book is one of the more difficult hurdles many students face when they enter the algebra curriculum. Frustration may lead some to give up before ever building the strong algebraic foundation necessary for them to move on to higher-level courses. Now, with *Computer Interactive Algeblocks*, you actually learn by doing *and* seeing algebra—giving you the skills you need to continue with confidence.

### *Three easy steps to mathematical success!*

1. *Follow the examples:* On-screen examples are automatic and guide you every step of the way. Each example begins with an explanation of how to solve the problem at hand, and then proceeds with an animated demonstration of how to use *Algeblocks* to solve the problem.

2. *Practice the concepts:* Once comfortable with the examples, you're ready to perform practice exercises. The program gives you the option to solve equations with or without the help of *Algeblocks*.

3. *Self-test:* The self-test allows you to review the concepts presented in the *Algeblocks* lab without the guidance of the program. If you have any difficulty, simply click on "Examples" or "Practice" to review, then return to the self-test when you're ready.

### *System Requirements*

**Macintosh**®: System 7.0 or higher, 68030 Processor or higher, 8MB RAM, color monitor.
**Windows**®: Windows 3.1 or higher, VGA, 386/40 MHz processor, 8MB RAM, 640 x 480 screen resolution.

To order a copy of *Computer Interactive Algeblocks, Volume 1* or *2*, please contact your college store or fill out the form on the back and return with your payment to Brooks/Cole.

# ORDER FORM

_____Yes! Send me a copy of *Computer Interactive Algeblocks, Volume 1*
(Macintosh® ISBN: 0-534-95144-9)

_____Yes! Send me a copy of *Computer Interactive Algeblocks, Volume 1*
(Windows® ISBN: 0-534-95146-5)

_____Yes! Send me a copy of *Computer Interactive Algeblocks, Volume 2*
(Macintosh® ISBN: 0-534-95443-X)

_____Yes! Send me a copy of *Computer Interactive Algeblocks, Volume 2*
(Windows® ISBN: 0-534-95443-X)

_____Copies x $20.95* = _____

Residents of: AL, AZ, CA, CT, CO, FL, GA, IL, IN, KS, KY, LA, MA, MD, MI, MN, MO, NC, NJ, NY, OH, PA, RI, SC, TN, TX, UT, VA, WA, WI must add appropriate state sales tax.

Subtotal _____
Tax _____
Handling __$4.00__
Total Due _____

* Call after 12/1/97 for current prices: 1-800-487-3575

## Payment Options

_____ Check or money order enclosed

Bill my ____VISA ____MasterCard ____American Express

Card Number: _____ Expiration Date: _____

Signature: _____

*Note: Credit card billing and shipping addresses must be the same.*

## Please ship my order to: (Please print.)

Name _____

Institution _____

Street Address_____

City _____ State _____ Zip+4_____

Telephone ( )_____ e-mail _____

Your credit card will not be billed until your order is shipped. Prices subject to change without notice. We will refund payment for unshipped out-of-stock titles after 120 days and for not-yet-published titles after 180 days unless an earlier date is requested in writing from you.

## Mail to:

**Brooks/Cole Publishing Company**
Source Code 9BCMA053
511 Forest Lodge Road
Pacific Grove, California 93950-5098
Phone: (408) 373-0728; Fax: (408) 375-6414
e-mail: info@brookscole.com

# STUDENT SOLUTIONS MANUAL AND STUDY GUIDE
for Goodman and Hirsch's

# Understanding Elementary Algebra with Geometry
*A Course for College Students*

## 4th Edition

Steven Kahan
*Queens College of the City University of New York*

BROOKS/COLE PUBLISHING COMPANY

I(T)P® An International Thomson Publishing Company

Pacific Grove • Albany • Belmont • Bonn • Boston • Cincinnati • Detroit • Johannesburg • London
Madrid • Melbourne • Mexico City • New York • Paris • Singapore • Tokyo • Toronto • Washington

Assistant Editor: *Linda Row*
Editorial Associate: *Peggi Rodgers*
Marketing Team: *Jennifer Huber, Laura Caldwell, & Jean Thompson*
Production Editor: *Laurel Jackson*
Cover Design: *E. Kelly Shoemaker*
Printing and Binding: *Webcom Limited*

*For more information, contact:*

BROOKS/COLE PUBLISHING COMPANY
511 Forest Lodge Road
Pacific Grove, CA  93950
USA

International Thomson Publishing Europe
Berkshire House 168-173
High Holborn
London WC1V 7AA
England

Thomas Nelson Australia
102 Dodds Street
South Melbourne, 3205
Victoria, Australia

Nelson Canada
1120 Birchmount Road
Scarborough, Ontario
Canada M1K 5G4

International Thomson Editores
Seneca 53
Col. Polanco
11560 México, D. F., México

International Thomson Publishing Japan
Hirakawacho Kyowa Building, 3F
2-2-1 Hirakawacho
Chiyoda-ku, Tokyo 102
Japan

International Thomson Publishing Asia
221 Henderson Road
#05-10 Henderson Building
Singapore 0315

International Thomson Publishing GmbH
Königswinterer Strasse 418
53227 Bonn
Germany

Printed in Canada

5     4     3

ISBN 0-534-35502-1

# Table of Contents

# Chapter 1
# The Integers

## Exercises 1.1

1. True
3. True
5. True
7. True

9. False
11. $\{1, 2, 3, 4, 5, 6, 7\}$

13. $\{0, 2, 4, 6, 8, 10, 12, 14, 16, 18\}$
15. $\{0, 1, 2, 3, 4, 5, 6\}$

17. $\{0, 1, 2, 3, 4, 5\}$
19. $\{6, 7, 8, 9, \dots\}$

21. $\{7, 8, 9, 10, \dots\}$
23. $\{3, 4, 5\}$

25. $\{1, 2, 3, 4, 6, 8, 12, 24\}$
27. $\{1, 2, 3, 4, 5, 6, 12, 15\}$

29. $\{0, 4, 8, 12, \dots\}$
31. $\{0, 12, 24, 26, \dots\}$

33. $\emptyset$
35. $>$
37. $=$
39. $>$

41. $=$
43. $=$
45. $<, \leq, \neq$
47. $\geq, \leq, =$

49. $>, \geq, \neq$
51. $<, \leq, \neq$
53. $>, \geq, \neq$
55. $2 \cdot 7$

57. $3 \cdot 11$
59. $2 \cdot 3 \cdot 5$
61. Prime
63. $2 \cdot 2 \cdot 2 \cdot 2 \cdot 2 \cdot 2$

65. $2 \cdot 2 \cdot 2 \cdot 2 \cdot 2 \cdot 3$
67. $3 \cdot 29$

69.
```
 -1 0 1 2 3 4 5 6 7 8 9 10
```
71.
```
  1 2 3 4 5 6 7 8 9 10 11 12
```

73. $\mathbb{W}$ contains the number 0; $\mathbb{N}$ does not.

74. In a sum, the numbers to be added are called terms; in a product, the numbers to be multiplied are called factors. In $2 + 5$, 2 is a term; in $2 \cdot 3$, 2 is a factor.

75. A factor of $n$ is a number that divides exactly into $n$; a multiple of $n$ is a number that is exactly divisible by $n$. 3 is a factor of 12; 12 is a multiple of 3.

76. $C$ is the set containing the first three multiples of 6.

77. $C$ is the set containing the first three multiples of 12.

78. (a) Sum of three terms: $x$, $y$, and $z$.
    (b) Product of three factors: $x$, $y$, and $z$.
    (c) Sum of two terms: $xy$ and $z$.
    (d) Product of two factors: $x$ and $(y + z)$.

## Exercises 1.2

1. True (commutative property of addition)
3. True (commutative property of addition)

5. False
7. False

9. True (commutative property of multiplication)

11. True (commutative property of multiplication)

13. False

15. True (associative property of addition)

17. True (associative property of addition)

19. True (associative property of multiplication)

21. True (associative property of multiplication)

23. False

25. False

27. Associative property of addition

29. Associative property of multiplication.

31. Commutative property of addition, followed by associative property of addition.

33. Commutative property of multiplication, followed by associative property of multiplication.

35. $-4$

37. 4

39. 4

41. 4

43. $-4$

45. $-4$

47. 5

49. 5

51. 2

53. 11

55. 5

57. $7 + 3 \cdot 4 = 7 + 12 = 19$

59. $4 \cdot 3 + 6 = 12 + 6 = 18$

61. $11 - 4 \cdot 2 = 11 - 8 = 3$

63. $15 - 5 \cdot 3 = 15 - 15 = 0$

65. $6 + 3 \cdot 5 - 2 = 6 + 15 - 2 = 19$

67. $12 - 4 \cdot 3 + 6 = 12 - 12 + 6 = 6$

69. $6 + 4(3 + 2) = 6 + 4 \cdot 5 = 6 + 20 = 26$

71. $6 + (4 \cdot 3 + 2) = 6 + (12 + 2)$
$= 6 + 14 = 20$

73. $6 + (4 \cdot 3) + 2 = 6 + 12 + 2 = 20$

75. $(6 + 4)(3 + 2) = 10 \cdot 5 = 50$

77. $30 \div 5(2) = 6 \cdot 2 = 12$

79. $30 \div (5 \cdot 2) = 30 \div 10 = 3$

81. $15 + 12 \div 6 - 3 = 15 + 2 - 3 = 17 - 3 = 14$

83. $\dfrac{15 + 2 \cdot 3}{3 + 2 \cdot 2} = \dfrac{15 + 6}{3 + 4} = \dfrac{21}{7} = 3$

85. $3 + 2[3 + 2(3 + 2)] = 3 + 2[3 + 2 \cdot 5]$
$= 3 + 2[3 + 10]$
$= 3 + 2 \cdot 13$
$= 3 + 26 = 29$

87. $9 - 4[6 - 2(3 - 1)] = 9 - 4[6 - 2 \cdot 2]$
$= 9 - 4[6 - 4]$
$= 9 - 4 \cdot 2$
$= 9 - 8 = 1$

89. $4 + \{3 + 5[2 + 2(3 + 1)]\} = 4 + \{3 + 5[2 + 2 \cdot 4]\}$
$= 4 + \{3 + 5[2 + 8]\}$
$= 4 + \{3 + 5 \cdot 10\}$
$= 4 + \{3 + 50\} = 4 + 53 = 57$

91. $\dfrac{10 + 2(5 + 3)}{2 \cdot 5 + 3} = \dfrac{10 + 2 \cdot 8}{10 + 3} = \dfrac{10 + 16}{13} = \dfrac{26}{13} = 2$

93. Jane has a checking account with overdraft privileges. If she has \$283 in her account and she writes a check for \$300, by how much is Jane's account overdrawn?
($283 - $300 = -$17)

94.

## Exercises 1.3

1. $+5 + (-7) = -(7-5) = -2$

3. $-9 + (-3) = -(9+3) = -12$

5. $-4 + (+11) = +(11-4) = +7$

7. $+8 + (-3) = +(8-3) = +5$

9. $+6 + (-2) = +(6-2) = +4$

11. $-6 + (-2) = -(6+2) = -8$

13. $-6 + (+2) = -(6-2) = -4$

15. $+6 + (+2) = 6 + 2 = 8$

17. $-7 + (-8) = -(7+8) = -15$

19. $10 + (-14) = -(14-10) = -4$

21. $8 + (-8) = 0$

23. $9 + (-16) = -(16-9) = -7$

25. $-30 + (-14) = -(30+14) = -44$

27. $-6 + 12 = +(12-6) = +6$

29. $-5 + (-4) + (-6) = -(5+4) + (-6)$
$= -9 + (-6)$
$= -(9+6) = -15$

31. $5 + (-4) + (-6) = +(5-4) + (-6)$
$= +1 + (-6)$
$= -(6-1) = -5$

33. $-5 + 4 + (-6) = -(5-4) + (-6)$
$= -1 + (-6)$
$= -(1+6) = -7$

35. $-5 + (-4) + 6 = -(5+4) + 6$
$= -9 + 6$
$= -(9-6) = -3$

37. $7 + (-10) + (-2) = -(10-7) + (-2)$
$= -3 + (-2)$
$= -(3+2) = -5$

39. $-3 + 9 + (-5) = +(9-3) + (-5)$
$= +6 + (-5)$
$= +(6-5) = 1$

41. $-4 + (-3) + (-8) = -(4+3) + (-8)$
$= -7 + (-8)$
$= -(7+8) = -15$

43. $15 + (-3) + (-6) = +(15-3) + (-6)$
$= +12 + (-6)$
$= +(12-6) = 6$

45. $-8 + (-2) + 6 = -(8+2) + 6$
$= -10 + 6$
$= -(10-6) = -4$

47. $16 + (-5) + (-7) + 2 = +(16-5) + (-7) + 2$
$= +11 + (-7) + 2 = +(11-7) + 2$
$= +4 + 2 = 6$

49. $-8 + 6 + (-5) + 1 = -(8-6) + (-5) + 1$
$= -2 + (-5) + 1 = -(2+5) + 1$
$= -7 + 1 = -(7-1) = -6$

51. $2 + (-9) + (-3) + (-1) + 6 = -(9 - 2) + (-3) + (-1) + 6$
$$= -7 + (-3) + (-1) + 6$$
$$= -(7 + 3) + (-1) + 6 = -10 + (-1) + 6$$
$$= -(10 + 1) + 6 = -11 + 6 = -(11 - 6) = -5$$

53. $27 + (-56) = -(56 - 27) = -29$

55. $-22 + (-45) = -(22 + 45) = -67$

57. $-31 + (-26) + 48 = -(31 + 26) + 48$
$$= -57 + 48$$
$$= -(57 - 48) = -9$$

59. $-5 + [7 + (-3)] = -5 + [+(7 - 3)]$
$$= -5 + (+4)$$
$$= -(5 - 4) = -1$$

61. $-3 + 5 \cdot 2 + (-6) = -3 + 10 + (-6)$
$$= +(10 - 3) + (-6)$$
$$= +7 + (-6)$$
$$= +(7 - 6) = +1$$

63. $|2 + (-6)| + |2| + |-6| = |-(6 - 2)| + |2| + |-6|$
$$= |-4| + |2| + |-6|$$
$$= 4 + 2 + 6 = 12$$

65. $|-7| + (-7) = 7 + (-7) = 0$

67. $+48 + [(-18) + (-22) + (-15)] + (+50) + [(-28) + (-12)] + (+20) + (-17) + (+27)$
$= +48 + (-55) + (+50) + (-40) + (+20) + (-17) + (+27) = +33$
So the final balance in Carla's checking account is $33.

69. $+25 + (+8) + (-14) + (-5) = +(25 + 8) + (-14) + (-5)$
$$= +33 + (-14) + (-5) = +(33 - 14) + (-5)$$
$$= +19 + (-5) = +(19 - 5) = +14$$

So on fourth down, the team is located on its own 14 yard line.

71. $+\$10,432 + (-\$1,678) + (-\$2,046) + \$7,488 = \$8,754 + (-\$2,046) + \$7,488$
$$= +\$6,708 + \$7,488 = \$14,196 \text{ (profit)}$$

73. $10,000 + 380 + 540 + (-275) + (-600) + (-72) = 10,920 + (-947)$
$$= 9,973 \text{ (feet)}$$

75. When we add an integer to its opposite, we get 0 as our result. This occurs because an integer and its opposite have the same absolute value.

76. If the sum of two integers is zero, the integers must be opposites. In all other cases, the sum of the integers would either be positive or negative.

77. $4 - 7 = -3$, because $-3 + 7 = 4$

78. $4 - (-7) = 11$, because $11 + (-7) = 4$

## Exercises 1.4

1. $6 - (+10) = 6 + (-10) = -4$

3. $-7 - (+4) = -7 + (-4) = -11$

5. $3 - (-6) = 3 + (+6) = 9$

7. $-8 - (-2) = -8 + (+2) = -6$

9. $-5 + (+8) = 3$

11. $5 - (+8) = 5 + (-8) = -3$

13. $5 - (-8) = 5 + (+8) = 13$

15. $-5 - (+8) = -5 + (-8) = -13$

17. $-5 - (-8) = -5 + (+8) = 3$

19. $5 + (-8) = -3$

21. $-5 + (-8) = -13$

23. $6 - (-7) = 6 + (+7) = 13$

25. $-6 - (-7) = -6 + (+7) = 1$

27. $2 + (-6) - (+7) = 2 + (-6) + (-7)$
$= -4 + (-7) = -11$

29. $2 - 6 - 7 = 2 + (-6) + (-7)$
$= -4 + (-7) = -11$

31. $2 - (6 - 7) = 2 - (-1)$
$= 2 + (+1) = 3$

33. $7 - 9 - 3 + 2 = 7 + (-9) + (-3) + 2$
$= -2 + (-3) + 2$
$= -5 + 2 = -3$

35. $11 - 5 + 4 - 7 = 11 + (-5) + 4 + (-7)$
$= 6 + 4 + (-7)$
$= 10 + (-7) = 3$

37. $2 - 3 - 6 - 2 = 2 + (-3) + (-6) + (-2)$
$= -1 + (-6) + (-2)$
$= -7 + (-2) = -9$

39. $-10 + 4 - 9 - (-3) = -10 + 4 + (-9) + (+3)$
$= -6 + (-9) + (+3)$
$= -15 + (+3) = -12$

41. $3 - 6 + 1 - (-4) = 3 + (-6) + 1 + (+4)$
$= -3 + 1 + (+4)$
$= -2 + (+4) = 2$

43. $-1 - 4 - 2 - (-5) = -1 + (-4) + (-2) + (+5)$
$= -5 + (-2) + (+5)$
$= -7 + (+5) = -2$

45. $4 - 8 - 6 + 3 = 4 + (-8) + (-6) + 3$
$= -4 + (-6) + 3$
$= -10 + 3 = -7$

47. $4 - 8 - (6 + 3) = 4 - 8 - 9$
$= 4 + (-8) + (-9)$
$= -4 + (-9) = -13$

49. $4 - (8 - 6) + 3 = 4 - 2 + 3$
$= 4 + (-2) + 3$
$= 2 + 3 = 5$

51. $4 - (8 - 6 + 3) = 4 - (8 + (-6) + 3)$
$= 4 - (2 + 3) = 4 - 5$
$= 4 + (-5) = -1$

53. $-8 + 8 = 0$

55. $-8 - 8 = -8 + (-8) = -16$

57. $-8 - (-8) = -8 + (+8) = 0$

59. $9 - 5 \cdot 4 - 2 = 9 - 20 - 2$
$= 9 + (-20) + (-2)$
$= -11 + (-2) = -13$

61. $9 - 5(4 - 2) = 9 - 5 \cdot 2 = 9 - 10$
$= 9 + (-10) = -1$

63. $9 - (5 \cdot 4 - 2) = 9 - (20 - 2)$
$= 9 - (20 + (-2)) = 9 - 18$
$= 9 + (-18) = -9$

65. $|6 - 2| - |2 - 6| = |4| - |-4|$
$= 4 - 4 = 0$

67. $|2 - 6| - (2 - 6) = |-4| - (-4) = 4 - (-4)$
$= 4 + (+4) = 8$

69. $|-4 - 3 + 2| - 4 - 3 + 2 = |-4 + (-3) + 2| + (-4) + (-3) + 2$
$$= |-7 + 2| + (-4) + (-3) + 2$$
$$= |-5| + (-4) + (-3) + 2 = 5 + (-4) + (-3) + 2$$
$$= 1 + (-3) + 2 = -2 + 2 = 0$$

71. $-6 - (-19) = -6 + (+19) = 13°C$

73. $\$643.47 - (-\$82.94) = \$643.47 + (+\$82.94) = \$726.41$

75. $29,028 - (-1,296) = 29,028 + (+1,296) = 30,324 \, (\text{feet})$

77. $-1,296 - (-35,840) = -1,296 + (+35,840) = 34,544 \, (\text{feet})$

79. $-8 - 2 - 7 = (-8) + (-2) + (-7)$. The terms are $-8$, $-2$, and $-7$.

81. $-3 - m - n = (-3) + (-m) + (-n)$. The terms are $-3$, $-m$, and $-n$.

83. $-15,000 - (-6,000) = -15,000 + (+6,000) = -9,000$. The reduced debt is $\$9,000$.

84. The left side is equal to 4, while the right side is equal to $-4$. In general, the absolute value of a difference of two integers is not the same as the difference of the absolute values of those integers.

85. 4 times $-3$ should mean "add $-3$ four times", giving $(-3) + (-3) + (-3) + (-3) = -12$.

86. We can interpret $-4$ times 3 to mean "subtract 3 four times", giving $-3 - 3 - 3 - 3 = -12$. Similarly, we can interpret $-4$ times 3 to mean "subtract $-3$ four times", giving $-(-3) - (-3) - (-3) - (-3) = +3 + (+3) + (+3) + (+3) = +12$.

## Exercises 1.5

1. $(+5)(-3) = -(5 \cdot 3) = -15$

3. $(-5)(+3) = -(5 \cdot 3) = -15$

5. $(-5)(-3) = +(5 \cdot 3) = 15$

7. $-5 - 3 = -8$

9. $-(5 - 3) = -2$

11. $\dfrac{20}{-5} = -\left(\dfrac{20}{5}\right) = -4$

13. $\dfrac{-20}{5} = -\left(\dfrac{20}{5}\right) = -4$

15. $\dfrac{-20}{-5} = +\left(\dfrac{20}{5}\right) = 4$

17. $\dfrac{0}{7} = 0$

19. Undefined (division by 0 is not allowed)

21. $4(-2) - 6 = -8 - 6 = -14$

23. $4 - 2 - 6 = 2 - 6 = -4$

25. $4 - (2 - 6) = 4 - (-4) = 8$

27. $4(-2 - 6) = 4(-8) = -32$

29. $-6(2)(-5)(-1) = -12(-5)(-1)$
$$= 60(-1) = -60$$

31. $9 - 3(1 - 3) = 9 - 3(-2)$
$$= 9 + 6 = 15$$

33. $8 - 3(2 - 5) = 8 - 3(-3)$
$$= 8 + 9 = 17$$

35. $8 - 3 \cdot 2 - 5 = 8 - 6 - 5$
$$= 2 - 5 = -3$$

37. $8 - (3 \cdot 2 - 5) = 8 - (6 - 5)$
$$= 8 - 1 = 7$$

39. $(8-3)(2-5) = (5)(-3) = -15$

41. $\dfrac{-12-3+1}{-7} = \dfrac{-15+1}{-7} = \dfrac{-14}{-7} = 2$

43. $\dfrac{-8-2-4}{-2} = \dfrac{-10-4}{-2} = \dfrac{-14}{-2} = 7$

45. $\dfrac{-8-(2-4)}{-2} = \dfrac{-8-(-2)}{-2} = \dfrac{-8+2}{-2}$
$= \dfrac{-6}{-2} = 3$

47. $\dfrac{-8-2(-4)}{-2} = \dfrac{-8+8}{-2} = \dfrac{0}{-2} = 0$

49. $\dfrac{8}{-2} - \dfrac{12}{-3} = -4 - (-4) = -4 + 4 = 0$

51. $\dfrac{8(-6)}{8-6} = \dfrac{-48}{2} = -24$

53. $\dfrac{-4(-5)(-4)}{-4(-5)-4} = \dfrac{20(-4)}{20-4} = \dfrac{-80}{16} = -5$

55. $\dfrac{3(-2)-4}{-4-1} = \dfrac{-6-4}{-4-1} = \dfrac{-10}{-5} = 2$

57. $\dfrac{6-4}{4-4} = \dfrac{2}{0}$, which is undefined, since division by 0 is not allowed.

59. $\dfrac{4(-6)}{-2-1} - \dfrac{4-6}{-2(-1)} = \dfrac{-24}{-3} - \dfrac{-2}{2}$
$= 8 - (-1) = 9$

61. $-6[-3 - 2(5-3)] = -6[-3 - 2 \cdot 2]$
$= -6[-3 - 4]$
$= -6(-7) = 42$

63. $5 - 2[3 - (4+6)] = 5 - 2[3 - 10]$
$= 5 - 2(-7)$
$= 5 + 14 = 19$

65. $5 - \dfrac{3-5}{-2} = 5 - \dfrac{-2}{-2} = 5 - 1 = 4$

67. $6 - \dfrac{8-2(-3)}{-7} = 6 - \dfrac{8+6}{-7} = 6 - \dfrac{14}{-7}$
$= 6 - (-2) = 6 + 2 = 8$

69. Even though the symbols are the same in both examples, the parentheses lead to different results. Without the parentheses, we have a subtraction example: $5 - 2 = 3$; with parentheses, we have a multiplication example: $5(-2) = -10$.

70. Again, the same symbols are used in both examples, but the location of the parentheses cause the difference this time. Our order of operations requires that we multiply before we subtract. Therefore, $6(-3) - 2 = -18 - 2 = -20$, while $6 - 3(-2) = 6 + 6 = 12$.

71. When we see $-7 - 8$, it must be interpreted as a subtraction, since no parentheses are present. Therefore, (a) and (b) are both wrong, since $-7 - 8 = -15$. (It would be correct to write $-7(-8) = +56$ or $(-7)(-8) = +56$.) Part (c) is wrong also. Since $2 - 4 = -2$, it follows that $-3 - (2 - 4) = -3 - (-2) = -3 + 2 = -1$. In part (d), the order of operations has not been followed. Here, $-4 - 2(5 - 6) = -4 - 2(-1) = -4 + 2 = -2$. (The calculation $-4 - 2 = -6$ never legitimately enters into the computation of the expression.)

72. The statement "two negatives make a positive" is accurate for multiplication and division but **not** for addition and subtraction. For instance, $(-3) + (-2) = -5$ and $(-3) - (-2) = -1$, and neither of these answers is positive.

73. The pattern suggest that $-1(-2) = 2$ and $-2(-2) = 4$. This in turn suggests that the product of two negative numbers is positive.

## Exercises 1.6

1. $4.2 + 5.9 = 10.1$

3. $4(5.1) = 20.4$

5. $\dfrac{12.8}{3.2} = 4$

7. $\dfrac{36.8}{1.5} = 24.5\overline{3}$

9. $\dfrac{8}{0.2} + \dfrac{12}{0.4} = 40 + 30 = 70$

11. $-2 - 5.3 = -7.3$

13. $-2(5.3) = -10.6$

15. $-2(-5.3) = 10.6$

17. $\dfrac{-8.4}{1.2} = -7$

19. $\dfrac{-6}{1.5} + \dfrac{21.6}{-1.2} = -4 - 18 = -22$

21. $-0.209$ 　　　　23. $-24.57$

25. $1.1$ 　　　　27. $0.302$

29. $2.26$

31. Between 4 and 5.

33. Between $-3$ and $-2$.

35. Between 7 and 8.

37. Between $-1$ and 0.

39. 

41. 

43. 

45. 

47. 

49. (a) False. The rational number $\dfrac{2}{3}$ is not an integer.

   (b) True. Every integer can be written as itself over 1. For example, the integer 8 can be written as $\dfrac{8}{1}$.

   (c) False, because $-15$ is to the left of $-10$ on the number line.

   (d) True, because $|-15| = 15$ and $|-10| = 10$.

   (e) False. The rational number $-\dfrac{1}{4}$ is negative, not positive.

   (f) False. The integer $-2$ is negative, not positive.

   (g) False. The real number $\sqrt{5}$ is not a rational number.

   (h) True. Every rational number has a location on the number line and so is also a real number.

50. (a) Use the commutative property of addition to rewrite the problem as $12.9 - 1.7 - 8.3$. Then this equals $12.9 - 10 = 2.9$.

   (b) Use the commutative property of multiplication to rewrite the problem as $7(10)(-2.3)$. Then this becomes $7(-23) = -161$.

   (c) Use the associative property of multiplication to rewrite the problem as $-10\left(\dfrac{3.2}{-1.6}\right)$. Then this equals $-10(-2) = 20$.

(d)  First, use the commutative property of multiplication to rewrite the problem as
$-30(-10)\left(\dfrac{1.6}{3}\right)$, which becomes $300\left(\dfrac{1.6}{3}\right)$. Then use the associative property of
multiplication to write this as $\dfrac{300}{3}(1.6) = 100(1.6) = 160$.

51.  Since $(1)(1) = 1$ and $(2)(2) = 4$, the square root of 3, to the nearest whole number, is 2. Since $(1.7)(1.7) = 2.89$ and $(1.8)(1.8) = 3.24$, the square root of 3, to the nearest tenth, is 1.7. Since $(1.73)(1.73) = 2.9929$ and $(1.74)(1.74) = 3.0276$, the square root of 3, to the nearest hundredth, is 1.73. Since $(1.732)(1.732) = 2.999824$ and $(1.733)(1.733) = 3.003289$ the square root of 3, to the nearest thousandth, is 1.732.

## Chapter 1 Review Exercises

1.  $\{2, 4, 6, 8, 10, 12, 14, 16, 18\}$

3.  $\{2, 3, 5, 7, 11, 13, 17, 19\}$

5.  $\{23, 29\}$

7.  $2 \cdot 3 \cdot 5$

9.  Prime

11.  $2 \cdot 2 \cdot 5 \cdot 5$

13.  $3 - 7 = -4$

15.  $-7 - 5 = -12$

17.  $-2 - (-6) = -2 + 6 = 4$

19.  $-4 - 5 - 6 = -9 - 6 = -15$

21.  $-7 + 12 - 5 = +5 - 5 = 0$

23.  $7 - 4 + 3 - 9 = 3 + 3 - 9 = 6 - 9 = -3$

25.  $8 - 5 - 6 = 3 - 6 = -3$

27.  $8 - (5 - 6) = 8 - (-1) = 8 + 1 = 9$

29.  $8(5 - 6) = 8(-1) = -8$

31.  $8 - 3 - 6 = 5 - 6 = -1$

33.  $8 - 3(-6) = 8 + 18 = 26$

35.  $8 - (3 - 6) = 8 - (-3) = 8 + 3 = 11$

37.  $8(-3)(-6) = -24(-6) = 144$

39.  $9 - 4(3 - 7) = 9 - 4(-4) = 9 + 16 = 25$

41.  $9 - 4 \cdot 3 - 7 = 9 - 12 - 7$
$= -3 - 7 = -10$

43.  $9 - (4 \cdot 3 - 7) = 9 - (12 - 7)$
$= 9 - 5 = 4$

45.  $(9 - 4)(3 - 7) = 5(-4) = -20$

47.  $|4 - 9| - |3 - 7| = |-5| - |-4| = 5 - 4 = 1$

49.  $\dfrac{-7 - 3}{-2(-5)} = \dfrac{-10}{+10} = -1$

51.  $\dfrac{-4(-2)(-8)}{-4(2) - 8} = \dfrac{8(-8)}{-8 - 8} = \dfrac{-64}{-16} = 4$

53.  $\dfrac{6 - 4(3 - 1)}{-2(-3) - 4} = \dfrac{6 - 4(2)}{6 - 4}$
$= \dfrac{6 - 8}{2} = \dfrac{-2}{2} = -1$

55.  $8 + 2[3 - 4(1 - 6)] = 8 + 2[3 - 4(-5)]$
$= 8 + 2[3 + 20]$
$= 8 + 2[23]$
$= 8 + 46 = 54$

57.  $-25.39$

59.  $0.97$

## Chapter 1 Practice Test

1.   (a)  True                        (b)  False

     (c)  False                     (d)  True (both have eight elements)

     (e)  $C = \emptyset$

3.   $|3 - 8| - |1 - 6| = |-5| - |-5| = 5 - 5 = 0$

5.   $-4(-3)(-2) = 12(-2) = -24$

7.   $-3(-5) - 2(-1) = 15 - (-2) = 15 + 2 = 17$

9.   $\dfrac{4(-8)}{4 - 8} - \dfrac{3 - 6}{-2 - 1} = \dfrac{-32}{-4} - \dfrac{-3}{-3} = 8 - 1 = 7$

11.   $\dfrac{(-2)(-3)(-4)}{(-2)(-3) - 4} = \dfrac{6(-4)}{6 - 4} = \dfrac{-24}{2} = -12$

13.   $8 - 5 \cdot 4 - 7 = 8 - 20 - 7 = -12 - 7 = -19$

15.   $\begin{aligned}[t] 8 - 3[8 - 3(8 - 3)] &= 8 - 3[8 - 3(5)] \\ &= 8 - 3[8 - 15] = 8 - 3(-7) \\ &= 8 - (-21) = 8 + 21 = 29 \end{aligned}$

# Chapter 2
# Algebraic Expressions

Exercises 2.1

1. $x \cdot x \cdot x \cdot x \cdot x \cdot x$

3. $(-x)(-x)(-x)(-x)$

5. $-(x \cdot x \cdot x \cdot x)$

7. $x \cdot x \cdot y \cdot y \cdot y$

9. $x \cdot x + y \cdot y \cdot y$

11. $x \cdot y \cdot y \cdot y$

13. $a^4$

15. $x^2 y^3$

17. $-r^2 s^3$

19. $-x^2(-y)^3$

21. $x^3 x^5 = x^{3+5} = x^8$

23. $3^5 = 3 \cdot 3 \cdot 3 \cdot 3 \cdot 3 = 9 \cdot 3 \cdot 3 \cdot 3$
$= 27 \cdot 3 \cdot 3 = 81 \cdot 3 = 243$

25. $-2^3 = -(2 \cdot 2 \cdot 2) = -(4 \cdot 2) = -8$

27. $(-2)^4 = (-2)(-2)(-2)(-2)$
$= 4(-2)(-2) = -8(-2) = 16$

29. $3^2 + 3^3 - 3^4 = 3 \cdot 3 + 3 \cdot 3 \cdot 3 - 3 \cdot 3 \cdot 3 \cdot 3$
$= 9 + 27 - 81 = -45$

31. $4 \cdot 3^2 - 2 \cdot 5^2 = 4 \cdot 3 \cdot 3 - 2 \cdot 5 \cdot 5$
$= 36 - 50 = -14$

33. $(3 - 7)^2 - (4 - 5)^3 = (-4)^2 - (-1)^3$
$= (-4)(-4) - (-1)(-1)(-1) = 16 - (-1) = 17$

35. $3^2 4^3 = 3 \cdot 3 \cdot 4 \cdot 4 \cdot 4 = 576$

37. $3^2 3^3 = 3^{2+3} = 3^5 = 3 \cdot 3 \cdot 3 \cdot 3 \cdot 3 = 243$

39. $x^3 x^5 = x^{3+5} = x^8$

41. $a a^2 a^4 = a^{1+2+4} = a^7$

43. $(3x)(5x)(4x) = (3 \cdot 5 \cdot 4)(x \cdot x \cdot x)$
$= 60x^3$

45. $(3r^2)(2r^3) = (3 \cdot 2)(r^2 r^3)$
$= 6r^{2+3} = 6r^5$

47. $(-3x^3)(5x^2) = (-3 \cdot 5)(x^3 x^2)$
$= -15x^{3+2} = -15x^5$

49. $(-c^4)(2^3)(-5c) = [-2^3(-5)](c^4 c)$
$= 40c^{4+1} = 40c^5$

51. $(8x^3 y^2)(4xy^5) = (8 \cdot 4)(x^3 x)(y^2 y^5)$
$= 32x^{3+1} y^{2+5} = 32x^4 y^7$

53. $(-2xy)(x^2 y^2)(-3xy) = [-2(-3)](xx^2 x)(yy^2 y)$
$= 6x^{1+2+1} y^{1+2+1} = 6x^4 y^4$

55. $(3a^4)^2 = (3a^4)(3a^4) = (3 \cdot 3)(a^4 a^4)$
$= 9a^{4+4} = 9a^8$

57. $(-4n^2)^3 = (-4n^2)(-4n^2)(-4n^2)$
$= [(-4)(-4)(-4)](n^2 n^2 n^2)$
$= -64n^{2+2+2} = -64n^6$

59. $(x^2)^3 (x^4)^2 = (x^2 x^2 x^2)(x^4 x^4) = x^{2+2+2+4+4} = x^{14}$

61. 0.073

63. 14.758

65. 107.916

67. 48.337

69. In $3 \cdot 2^4$, the exponent of 4 applies only to the 2, not to the 3; in $(3 \cdot 2)^4$, the exponent of 4 applies to both the 2 and the 3. Put another way, we compute $3 \cdot 2^4$ by first raising 2 to the fourth power and then multiplying the result by 3. We compute $(3 \cdot 2)^4$ by first multiplying 2 by 3 and then raising the result to the fourth power.

70. (a)  The exponent of 2 applies only to the 2, not to the 3.
    (b)  Only $x$ should be raised to the fourth power.
    (c)  $5^4$ is not the same as $5 \cdot 4$.
    (d)  When we square 3, we get 9, not 6. Addition is performed in the exponents.
    (e)  $-3^4$ is the opposite of $3^4$, so it is equal to $-81$.
    (f)  We add exponents when we multiply powers of $x$, not when we add powers of $x$.
    (g)  When multiplying powers of $x$, we add exponents. So $x^2 x^7 = x^{2+7} = x^9$.
    (h)  Here, we do not add exponents, since $(x^2)^4 = x^2 x^2 x^2 x^2 = x^{2+2+2+2} = x^8$.

71. $3 \cdot 5$ represents the product of 3 and 5. Its value is 15.
    $5^3$ means $5 \cdot 5 \cdot 5$, the product of three factors of 5. Its value is 125.

72. $3^5 = 3 \cdot 3 \cdot 3 \cdot 3 \cdot 3$, the product of five factors of 3. Its value is 243.
    $5^3 = 5 \cdot 5 \cdot 5$, the product of three factors of 5. Its value is 125.

73. $2^4 = 2 \cdot 2 \cdot 2 \cdot 2 = (2 \cdot 2) \cdot (2 \cdot 2) = 4 \cdot 4 = 4^2$. In general, if $a$ and $b$ are two different counting numbers, $a^b \neq b^a$ (see previous problem). The <u>only</u> distinct counting numbers for which $a^b = b^a$ are 2 and 4.

Exercises 2.2

1. $C = 22.75 + 0.12m$
   $= 22.75 + 0.12(86)$
   $= 22.75 + 10.32 = \$33.07$

3. $C = 18.95d + 0.14m$
   $= 18.95(5) + 0.14(264)$
   $= 94.75 + 36.96 = \$131.71$

5. $h = 80 + 40t - 16t^2$
   $= 80 + 40(3.6) - 16(3.6)^2$
   $= 80 + 144 - 207.36$
   $= 244 - 207.36 = 16.64$
   $= 16.6$ ft. (rounded to the nearest tenth)

7. $P = 12,600 + 6.35n - 0.002n^2$
   $= 12,600 + 6.35(4280) - 0.002(4280)^2$
   $= 12,600 + 27,178 - 36,636.8$
   $= 39,778 - 36,636.8 = \$3,141.20$

9. $I = 415 + 17.5h$
   $= 415 + 17.5(12)$
   $= 415 + 210 = \$625$

11. $C = 18.95 + 0.065n$
    $= 18.95 + 0.065(82)$
    $= 18.95 + 5.33 = \$24.28$

13. $A = s^2$
    $= (6.4)^2 = 40.96$
    $= 41.0$ sq. in. (rounded to the nearest tenth)

15. $V = LWH$
    $= (1.8)(2.3)(1.1) = (4.14)(1.1) = 4.554$
    $= 4.6$ cu. ft. (rounded to the nearest tenth)

17. $-y = -(-3) = 3$

19. $-|y| = -|-3| = -3$

21. $x + y = (2) + (-3) = -1$

23. $x - y = (2) - (-3) = 2 + 3 = 5$

25. $|x - y| = |(2) - (-3)| = |5| = 5$

27. $|x| - y = |(2)| - (-3) = 2 + 3 = 5$

29. $|x| - |y| = |(2)| - |(-3)| = 2 - 3 = -1$

31. $x + y + z = (2) + (-3) + (-4)$
$$= 2 - 3 - 4 = -5$$

33. $x + y - z = (2) + (-3) - (-4)$
$$= 2 - 3 + 4 = 3$$

35. $xy - z = (2)(-3) - (-4)$
$$= -6 + 4 = -2$$

37. $x(y - z) = (2)[(-3) - (-4)] = 2(1) = 2$

39. $x - (y - z) = (2) - [(-3) - (-4)]$
$$= 2 - 1 = 1$$

41. $xy^2 = (2)(-3)^2 = 2(9) = 18$

43. $x + y^2 = (2) + (-3)^2 = 2 + 9 = 11$

45. $x^2 + y^2 = (2)^2 + (-3)^2 = 4 + 9 = 13$

47. $(x + y + z)^2 = [(2) + (-3) + (-4)]^2$
$$= (-5)^2 = 25$$

49. $-y^2 = -(-3)^2 = -9$

51. $-x^2z = -(2)^2(-4) = -4(-4) = 16$

53. $xyz^2 = (2)(-3)(-4)^2 = -6(16) = -96$

55. $x(y - z)^2 = (2)[(-3) - (-4)]^2$
$$= 2(1)^2 = 2(1) = 2$$

57. $xy^2 - (xy)^2 = (2)(-3)^2 - [(2)(-3)]^2$
$$= 2(9) - (-6)^2$$
$$= 18 - 36 = -18$$

59. $(z - 3x)^2 = [(-4) - 3(2)]^2$
$$= (-4 - 6)^2 = (-10)^2 = 100$$

61. $(5x + y)(3x - y) = [5(2) + (-3)][3(2) - (-3)]$
$$= (10 - 3)(6 + 3) = (7)(9) = 63$$

63. $y^2 - 3y + 2 = (-3)^2 - 3(-3) + 2$
$$= 9 + 9 + 2 = 20$$

65. $3x^2 + 4x + 1 = 3(2)^2 + 4(2) + 1$
$$= 3(4) + 8 + 1$$
$$= 12 + 8 + 1 = 21$$

67. $x^2 + 3x^2y - 3xy^2 + y^3 = (2)^2 + 3(2)^2(-3) - 3(2)(-3)^2 + (-3)^3$
$$= 4 + 3(4)(-3) - 3(2)(9) + (-27)$$
$$= 4 - 36 - 54 - 27 = -113$$

69. $\dfrac{xy}{x + y} = \dfrac{(2)(-3)}{(2) + (-3)} = \dfrac{-6}{-1} = 6$

71. $\dfrac{2}{x} - \dfrac{y}{3} + \dfrac{z}{2} = \dfrac{2}{(2)} - \dfrac{(-3)}{3} + \dfrac{(-4)}{2}$
$$= 1 - (1) + (-2) = 0$$

73. $3x - 4y$ has two terms. The first term has a coefficient of 3 and a literal part of $x$; the second term has a coefficient of $-4$ and a literal part of $y$.

75. $3x(-4y) = -12xy$ has one term with a coefficient of $-12$ and a literal part of $xy$.

77. $3x(z - y)$ has one term with a coefficient of 3 and a literal part of $x(z - y)$.

79. $4x^2 - 3x + 2$ has three terms. The first term has a coefficient of 4 and a literal part of $x^2$; the second term has a coefficient of $-3$ and a literal part of $x$; the third term has no literal part—it is just the constant 2.

81. $-x^2 + y - 13$ has three terms. The first term has a coefficient of $-1$ and a literal part of $x^2$; the second term has a coefficient of 1 and a literal part of $y$; the third term has no literal part—it is just the constant $-13$.

83. $3x(2x) + 4y(5y) = 6x^2 + 20y^2$ has two terms. The first term has a coefficient of 6 and a literal part of $x^2$; the second term has a coefficient of 20 and a literal part of $y^2$.

85. 2.031

87. $-9.862$

89. 0.046

91. 25.667

93. 3.78

95. $-5.70$

97. A term is an algebraic expression that is connected by multiplication (and/or division). If a term is formed by multiplying two or more expressions, each is called a factor of that term.

98. $x + xy + xyz$ is one of many possible answers.

## Exercises 2.3

1. Essential

3. Nonessential

5. Essential

7. Nonessential

9. Essential

11. Essential (both)

13. The first is nonessential, the second is essential.

15. $x, 2x$: literal part is $x$, coefficients are 1 and 2.
    $y, 3y$: literal part is $y$, coefficients are 1 and 3.

17. $2x^2, -x^2$: literal part is $x^2$, coefficients are 2 and $-1$.
    $-3x, -x$ literal part is $x$, coefficients are $-3$ and $-1$.
    $4x^3$ : literal part is $x^3$, coefficient is 4.

19. 4, 5: constant terms.
    $4u, 5u$: literal part is $u$, coefficients are 4 and 5.
    $4u^2$: literal part is $u^2$, coefficient is 4.

21. $5x^2$: literal part is $x^2$, coefficient is 5.
    $5x^2y$: literal part is $x^2y$, coefficient is 5.
    $5y^2$: literal part is $y^2$, coefficient is 5.
    $5xy^2$: literal part is $xy^2$, coefficient is 5.

23. $-x^2y, -2x^2y$: literal part is $x^2y$, coefficients are $-1$ and $-2$.
    $2xy^2, 3xy^2$: literal part is $xy^2$, coefficients are 2 and 3.
    $x^2y^2$: literal part is $x^2y^2$, coefficient is 1.

25. $3(x + 4) = 3 \cdot x + 3 \cdot 4 = 3x + 12$

27. $5(y - 2) = 5 \cdot y - 5 \cdot 2 = 5y - 10$

29. $-2(x + 7) = -2 \cdot x + (-2) \cdot 7 = -2x - 14$

31. $3(5x + 2) = 3 \cdot 5x + 3 \cdot 2 = 15x + 6$

33. $-4(3x + 1) = -4 \cdot 3x + (-4) \cdot 1 = -12x - 4$

35. $x(x + 3) = x \cdot x + x \cdot 3 = x^2 + 3x$

37. $x(x^2 + 3x) = x \cdot x^2 + x \cdot 3x = x^3 + 3x^2$

39. $5x(2x - 4) = 5x \cdot 2x - 5x \cdot 4 = 10x^2 - 20x$

41. $0.42(40 - x) = 0.42 \cdot 40 - 0.42 \cdot x = 16.8 - 0.42x$

43. $1.28(3x + 60) = 1.28 \cdot 3x + 1.28 \cdot 60 = 3.84x + 76.8$

45. $2x + 10 = 2 \cdot x + 2 \cdot 5 = 2(x + 5)$

47. $5y - 20 = 5 \cdot y - 5 \cdot 4 = 5(y - 4)$

49. $9x + 3y - 6 = 3 \cdot 3x + 3 \cdot y - 3 \cdot 2 = 3(3x + y - 2)$

51. $x^2 + xy = x \cdot x + x \cdot y = x(x + y)$

53. $2x + 5x = (2 + 5)x = 7x$

55. $2x^2 + 5x^2 = (2 + 5)x^2 = 7x^2$

57. $3a - 8a + 2a = (3 - 8 + 2)a = -3a$

59. $-3y + y - 2y = (-3 + 1 - 2)y = -4y$

61. $-x - 2x - 3x = (-1 - 2 - 3)x = -6x$

63. $2x - 3y - 7x + 5y = 2x - 7x - 3y + 5y$
$$= (2 - 7)x + (-3 + 5)y$$
$$= -5x + 2y$$

65. $3x + 5y + 2z$ cannot be simplified

67. $3x^2 + 7x + x^2 + 3x = 3x^2 + x^2 + 7x + 3x$
$$= (3 + 1)x^2 + (7 + 3)x$$
$$= 4x^2 + 10x$$

69. $x^2 - 2x + x^2 - x = x^2 + x^2 - 2x - x$
$$= (1 + 1)x^2 + (-2 - 1)x = 2x^2 - 3x$$

71. $5x^2y - 3x^2 + x^2y - x^2 = 5x^2y + x^2y - 3x^2 - x^2$
$$= (5 + 1)x^2y + (-3 - 1)x^2$$
$$= 6x^2y - 4x^2$$

73. $2x + 5x - 3y + y - 7x = 2x + 5x - 7x - 3y + y$
$$= (2 + 5 - 7)x + (-3 + 1)y = 0x - 2y = -2y$$

75. $-5s^2 + 3st - s^2 + 6s^2 = -5s^2 - s^2 + 6s^2 + 3st$
$$= (-5 - 1 + 6)s^2 + 3st = 0s^2 + 3st = 3st$$

77. $3a^2b + ab^2 - ab^2 - 2a^2b - ab^2 = 3a^2b - 2a^2b + ab^2 - ab^2 - ab^2$
$$= (3 - 2)a^2b + (1 - 1 - 1)ab^2$$
$$= 1a^2b - 1ab^2 = a^2b - ab^2$$

79. $0.8x - 5 - x + 2.7 = 0.8x - x - 5 + 2.7$
$$= (0.8 - 1)x + (-5 + 2.7) = -0.2x - 2.3$$

81. $0.28x + 5.48 + 0.54x = 0.28x + 0.54x + 5.48$
$$= (0.28 + 0.54)x + 5.48 = 0.82x + 5.48$$

83. $3(x + y) + 4x - y = 3x + 3y + 4x - y$
$$= 3x + 4x + 3y - y$$
$$= (3 + 4)x + (3 - 1)y$$
$$= 7x + 2y$$

85. $3(x + y) + 4(x - y) = 3x + 3y + 4x - 4y$
$$= 3x + 4x + 3y - 4y$$
$$= (3 + 4)x + (3 - 4)y$$
$$= 7x - 1y = 7x - y$$

87. $3(x + y) + 4x(-y) = 3x + 3y - 4xy$

89. $5x(x^2 + 3) + 2x(3 + x^2) = 5x^3 + 15x + 6x + 2x^3 = 5x^3 + 2x^3 + 15x + 6x$
$$= (5 + 2)x^3 + (15 + 6)x = 7x^3 + 21x$$

91. $5x(x^2 + 3) + 2x(3x^2) = 5x^3 + 15x + 6x^3 = 5x^3 + 6x^3 + 15x$
$$= (5 + 6)x^3 + 15x = 11x^3 + 15x$$

93. $0.06x + 0.09(x + 5000) = 0.06x + 0.09x + 0.09(5000)$
$$= (0.06 + 0.09)x + 450 = 0.15x + 450$$

95. $0.35x + 0.42(2x) + 0.54(80 - 3x) = 0.35x + 0.84x + 43.2 - 1.62x$
$$= 0.35x + 0.84x - 1.62x + 43.2$$
$$= (0.35 + 0.84 - 1.62)x + 43.2 = -0.43x + 43.2$$

97. The distribution property allows us to combine like terms.

98. (a) Our order of operations requires that multiplication be done before addition. Therefore, we should not add 5 and 3 first.

    (b) The distribution property requires that we multiply each term of $x - 4$ by 3.

    (c) We cannot add 5 and $3x$ to get $8x$, since these are not like terms.

    (d) This is correct as written.

99. (a) We can write $45(116) - 45(16)$ as $45(116 - 16)$. This equals $45(100)$ or $4,500$.

    (b) We can write $60(278) - 60(28)$ as $60(278 - 28)$. This equals $60(250)$ or $15,000$.

## Exercises 2.4

1. $4x + y + 4(x + y) = 4x + y + 4x + 4y$
$$= 4x + 4x + y + 4y = 8x + 5y$$

3. $5(m + 2n) + 3(m - n) = 5m + 10n + 3m - 3n$
$$= 5m + 3m + 10n - 3n = 8m + 7n$$

5. $-2(x - 3y) + 5(y - x) = -2x + 6y + 5y - 5x$
$$= -2x - 5x + 6y + 5y = -7x + 11y$$

7. $5 + 3(x - 2) = 5 + 3x - 6 = 3x + 5 - 6 = 3x - 1$

9. $5 - 3(x - 2) = 5 - 3x + 6 = -3x + 5 + 6 = -3x + 11$

11. $(5 - 3)(x - 2) = 2(x - 2) = 2x - 4$

13. $4(m - 3n) + 2(5m + 6n) = 4m - 12n + 10m + 12n$
$$= 4m + 10m - 12n + 12n = 14m$$

15. $2(a - 2b) - 4(b - 2a) = 2a - 4b - 4b + 8a$
$$= 2a + 8a - 4b - 4b = 10a - 8b$$

17. $3(2x^2 - 4y) + 4(5y - 3x^2) = 6x^2 - 12y + 20y - 12x^2$
$$= 6x^2 - 12x^2 - 12y + 20y = -6x^2 + 8y$$

19. $8 - (3x - 4) = 8 - 3x + 4$
$$= -3x + 8 + 4 = -3x + 12$$

21. $5y - (1 - 2y) = 5y - 1 + 2y$
$$= 5y + 2y - 1 = 7y - 1$$

23. $5(x - 3y) - x - 3y = 5x - 15y - x - 3y$
$$= 5x - x - 15y - 3y$$
$$= 4x - 18y$$

25. $5(x - 3y) - (x - 3y) = 5x - 15y - x + 3y$
$$= 5x - x - 15y + 3y$$
$$= 4x - 12y$$

27. $5(x - 3y) - x(-3y) = 5x - 15y + 3xy$

29. $5x(-3y) - x(-3y) = -15xy + 3xy$
$$= -12xy$$

31. $5x(-3y)(-x)(-3y) = [(5)(-3)(-1)(-3)](x \cdot x)(y \cdot y) = -45x^2y^2$

33. $2x^2(x - 2) + x(3x^2 - 4x) = 2x^3 - 4x^2 + 3x^3 - 4x^2$
$$= 2x^3 + 3x^3 - 4x^2 - 4x^2 = 5x^3 - 8x^2$$

35. $3a(4a - 1) - a(4 - a) = 12a^2 - 3a - 4a + a^2$
$$= 12a^2 + a^2 - 3a - 4a = 13a^2 - 7a$$

37. $4(x^2 + 7x) - (x^2 + 7x) = 4x^2 + 28x - x^2 - 7x$
$$= 4x^2 - x^2 + 28x - 7x = 3x^2 + 21x$$

39. $3a(a^2 + 3b) + 4b^2(a^2 - b) = 3a^3 + 9ab + 4a^2b^2 - 4b^3$

41. $3x^2 - 7x + 4 - 8x^2 - 3 - x = 3x^2 - 8x^2 - 7x - x + 4 - 3$
$$= -5x^2 - 8x + 1$$

43. $x^2y(xy - x) - 5xy(x^2y - x^2) = x^3y^2 - x^3y - 5x^3y^2 + 5x^3y$
$$= x^3y^2 - 5x^3y^2 - x^3y + 5x^3y$$
$$= -4x^3y^2 + 4x^3y$$

45. $4u^2v(u - v) - (uv^3 + u^2v^2) = 4u^3v - 4u^2v^2 - uv^3 - u^2v^2$
$$= 4u^3v - 4u^2v^2 - u^2v^2 - uv^3$$
$$= 4u^3v - 5u^2v^2 - uv^3$$

47. $4(x + 3y) + (4x + 3y) + 4x(3y) = 4x + 12y + 4x + 3y + 12xy$
$$= 4x + 4x + 12y + 3y + 12xy$$
$$= 8x + 15y + 12xy$$

49. $6(m - 2n) + (6m - 2n) + 6m(-2n) = 6m - 12n + 6m - 2n - 12mn$
$$= 6m + 6m - 12n - 2n - 12mn$$
$$= 12m - 14n - 12mn$$

51. $3t^5(t^4 - 4) - (t^5 + t^4) - 2t^3(3t)(-t^5) = 3t^9 - 12t^5 - t^5 - t^4 + 6t^9$
$$= 3t^9 + 6t^9 - 12t^5 - t^5 - t^4$$
$$= 9t^9 - 13t^5 - t^4$$

53. $-3(-x + 2) + (8 - 5x) - (2 - 2x) = 3x - 6 + 8 - 5x - 2 + 2x$
$$= 3x - 5x + 2x - 6 + 8 - 2$$
$$= 0x + 0 = 0$$

55. $a - 2[a - 2(a - 2)] = a - 2[a - 2a + 4]$
$$= a - 2[-a + 4]$$
$$= a + 2a - 8 = 3a - 8$$

57. $x\{x - 4[x - (x - 4)]\} = x\{x - 4[x - x + 4]\}$
$$= x\{x - 4(4)\} = x\{x - 16\}$$
$$= x^2 - 16x$$

59. $3(x + 2) + 4[x - 3(2 - x)] = 3(x + 2) + 4[x - 6 + 3x]$
$$= 3(x + 2) + 4(x + 3x - 6) = 3(x + 2) + 4(4x - 6)$$
$$= 3x + 6 + 16x - 24$$
$$= 3x + 16x + 6 - 24 = 19x - 18$$

61. $4(y - 3) - 2[3y - 5(y - 1)] = 4(y - 3) - 2[3y - 5y + 5]$
$$= 4(y - 3) - 2(-2y + 5)$$
$$= 4y - 12 + 4y - 10$$
$$= 4y + 4y - 12 - 10 = 8y - 22$$

63. (a) The solution is correct up to $3 + 2[5x + 12]$. At this point, it is wrong to add $3 + 2 = 5$. Instead, use the distributive property:
$$3 + 2[5x + 12] = 3 + 10x + 24$$
$$= 10x + 3 + 24$$
$$= 10x + 27$$

  (b) Within the brackets, we must add $x$ and $x + 3$, not multiply them. As a result, we get:
$$3x + 5[x + (x + 3)] = 3x + 5[2x + 3]$$
$$= 3x + 10x + 15$$
$$= 13x + 15$$

64. (a) Do not subtract $5 - 3$; distribute $-3$ instead.

  (b) When $-3$ is distributed, it must multiply $-4$ as well as $x$.

  (c) When we multiply $-3$ by $-4$, we get 12, not $-12$.

  (d) This is done correctly.

## Exercises 2.5

1. $n + 4$      3. $n - 4$      5. $n - 4$      7. $5n + 6$

9. $2n - 9$      11. $n(n + 7)$      13. $(n + 2)(n - 6)$      15. $2n - 8 = 14$

17. $5n + 4 = n - 2$      19. $r + s = rs$      21. $2(r + s) = rs - 3$

23. $x + (x + 1) = 2x + 1$, where $x =$ smaller integer

25. $x + (x + 2) = 2x + 2$, where $x =$ smaller even integer

27. $x + (x + 2) = 2x + 2$, where $x =$ smallest of the odd integers

29. $8x = 7(x + 2) - 4$, where $x =$ smaller even integer

31. $x^2 + (x+1)^2 + (x+2)^2 = 5$, where $x = $ smallest of the integer

33. $A = 6.85h$. When $h = 14$, we get:
$A = 6.85(14) = \$95.90$

35. $C = 225 + 0.045p$. When $p = 4,800$, we get:
$C = 225 + 0.045(4,800)$
$= 225 + 216 = \$441$

37. $C = 35 + 8p$. When $p = 9$, we get:
$C = 35 + 8(9) = 35 + 72 = \$107$

39. $C = 0.65 + 0.42(m - 1)$. When $m = 12$,
we get:
$C = 0.65 + 0.42[(12) - 1]$
$= 0.65 + 0.42(11)$
$= 0.65 + 4.62 = \$5.27$

41. $T = 8d + 10k$. When $d = 12$ and $k = 15$,
we get:
$T = 8(12) + 10(15)$
$= 96 + 150 = 246$ km.

43. $T = 6L + 9D$. When $L = 5$ and $D = 8$,
we get:
$T = 6(5) + 9(8)$
$= 30 + 72 = 102$ lawns

45. $0.30d$ dollars

47. $d + 0.04d = 1.04d$ dollars

49. $p - 0.20p = 0.80p$ dollars

51. $285 + 0.52x$ dollars

53. $6h + 14(h - 7) = 6h + 14h - 98$
$= 20h - 98$ dollars

55. $48t + 54(t - 2) = 48t + 54t - 108$
$= 102t - 108$ miles

57. (a) 8 nickels
(b) 40¢ ($= 8 \times 5$¢)
(c) 12 dimes
(d) 120¢ ($= 12 \times 10$¢) or $1.20
(e) 9 quarters
(f) 225¢ ($= 9 \times 25$¢) or $2.25
(g) 29 coins
(h) 40¢ + 120¢ + 225¢ = 385¢ or $3.85

59. (a) A grades 200 exams per minute.
(b) A grades for 15 minutes.
(c) A grades 3,000 exams ($= 200 \times 15$).
(d) B grades 160 exams per minute.
(e) B grades for 20 minutes.
(f) B grades 3,200 exams ($= 160 \times 20$).
(g) A and B grade 6,200 exams altogether ($= 3,000 + 3,200$).

61. (a) She walks at the rate of 100 meters per minute.
(b) She walks for 25 minutes.
(c) She walks 2,500 meters ($= 100 \times 25$).
(d) She jogs at the rate of 220 meters per minute.
(e) She jogs for 35 minutes.
(f) She jogs 7,700 meters ($= 220 \times 35$).
(g) She covers a distance of 10,200 meters altogether ($= 2,500 + 7,700$).

## Chapter 2 Review Exercises

1. $x \cdot y \cdot y \cdot y$

3. $-(x \cdot x \cdot x \cdot x)$

5. $3 \cdot x \cdot x$

7. $x^2 y^3$

9. $a^2 - b^3$

11. $-z^4 = -(-2)^4 = -(-2)(-2)(-2)(-2)$
$= -16$

13. $xy^2 = (-3)(4)^2 = -3(16) = -48$

15. $xyz - (x + y + z) = (-3)(4)(-2) - [(-3) + (4) + (-2)]$
$= 24 - (-1) = 25$

17. $|xy| + z - |z| = |(-3)(4)| + (-2) - |(-2)|$
$= |-12| + (-2) - |-2|$
$= 12 - 2 - 2 = 8$

19. $2x^2 - (x + y)^2 = 2(-3)^2 - [(-3) + (4)]^2$
$= 2(9) - (1)^2$
$= 18 - 1 = 17$

21. $x^3 x^4 x = x^{3+4+1} = x^8$

23. $a^7 a^2 + a^3 a^6 = a^{7+2} + a^{3+6}$
$= a^9 + a^9 = 2a^9$

25. $4x^3 x^2 + 3x^2 x^4 = 4x^{3+2} + 3x^{2+4}$
$= 4x^5 + 3x^6$

27. $3x^2 - 7x + 7 - 5x^2 - x - 3 = 3x^2 - 5x^2 - 7x - x + 7 - 3$
$= -2x^2 - 8x + 4$

29. $2a^2 b(3ab^4) = (2 \cdot 3)(a^2 a)(bb^4)$
$= 6a^{2+1} b^{1+4} = 6a^3 b^5$

31. $2a^2 b(3a + b^4) = 2a^2 b(3a) + 2a^2 b(b^4)$
$= 6a^3 b + 2a^2 b^5$

33. $3x(2x + 4) + 5(x^2 - 3) = 6x^2 + 12x + 5x^2 - 15$
$= 6x^2 + 5x^2 + 12x - 15$
$= 11x^2 + 12x - 15$

35. $4y^2(y - 2) - y(y^2 - 5y) = 4y^3 - 8y^2 - y^3 + 5y^2$
$= 4y^3 - y^3 - 8y^2 + 5y^2 = 3y^3 - 3y^2$

37. $3xy(x^2 - 2y) + 4xy^2(y - x) = 3x^3 y - 6xy^2 + 4xy^3 - 4x^2 y^2$

39. $3(2x - 4y) - (x + 2y) - (x - 2y) = 6x - 12y - x - 2y - x + 2y$
$= 6x - x - x - 12y - 2y + 2y$
$= 4x - 12y$

41. $3x^4(x^3 - 2y^2) - 4x(x^2)(x^4) = 3x^7 - 6x^4 y^2 - 4x^7$
$= 3x^7 - 4x^7 - 6x^4 y^2$
$= -x^7 - 6x^4 y^2$

43. $3 - [x - 3 - (x - 3)] = 3 - [x - 3 - x + 3]$
$= 3 - [x - x - 3 + 3]$
$= 3 - 0 = 3$

45. $0.02x + 0.07(2x + 1250) = 0.02x + 0.07(2x) + 0.07(1250)$
$= 0.02x + 0.14x + 87.5 = 0.16x + 87.5$

47. $5x^2 + 10 = 5 \cdot x^2 + 5 \cdot 2 = 5(x^2 + 2)$

49. $3y - 6z + 9 = 3 \cdot y - 3 \cdot 2z + 3 \cdot 3$
$= 3(y - 2z + 3)$

51. $n + 7 = 3n - 4$

53. $n + (n + 2) = n - 5$

55. $d - 0.35d = 0.65d$ dollars

57. (a) He sells 12 newspaper subscriptions.
    (b) He earns \$2 for each newspaper subscription.
    (c) He sells 9 magazine subscriptions.
    (d) He earns \$5 for each magazine subscription.
    (e) He earns \$24 ( $= 12 \times 2$) for the newspaper subscriptions.
    (f) He earns \$45 ( $= 9 \times 5$) for the magazine subscriptions.
    (g) He earns \$69 ( $= 24 + 45$) altogether.

## Chapter 2 Practice Test

1. $-3^4 = -(3 \cdot 3 \cdot 3 \cdot 3) = -81$

3. $(-3 - 4 + 6)^5 = (-7 + 6)^5 = (-1)^5$
$$= (-1)(-1)(-1)(-1)(-1)$$
$$= -1$$

5. $x - y - z = (-2) - (3) - (-4)$
$$= -2 - 3 + 4 = -5 + 4 = -1$$

7. $x^3 - z^2 = (-2)^3 - (-4)^2$
$$= (-2)(-2)(-2) - (-4)(-4)$$
$$= -8 - (+16) = -8 - 16 = -24$$

9. $4x^2y - 5xy + y^2 - 3xy - 2y^2 - x^2y = 4x^2y - x^2y - 5xy - 3xy + y^2 - 2y^2$
$$= 3x^2y - 8xy - y^2$$

11. $2x(x^2 - y) - 3(x - xy) - (2x^3 - 3x) = 2x^3 - 2xy - 3x + 3xy - 2x^3 + 3x$
$$= 2x^3 - 2x^3 - 2xy + 3xy - 3x + 3x$$
$$= xy$$

13. $4 - [x - 4(x - 4)] = 4 - [x - 4x + 16]$
$$= 4 - [-3x + 16]$$
$$= 4 + 3x - 16 = 3x + 4 - 16$$
$$= 3x - 12$$

15. (a) $2n + 4$       (b) $5n - 20 = n$

17. $M = 99 + 0.15C$

# Chapter 3
# First Degree Equations and Inequalities

Exercises 3.1

1. $2(x-3) = 2x - 6$
   $2x - 6 = 2x - 6$
   identity

3. $5(x+2) = 5x + 2$
   $5x + 10 = 5x + 2$
   $10 = 2$
   contradiction

5. $2a + 4 + 3a = 6a + 4 - a$
   $5a + 4 = 5a + 4$
   identity

7. $2z^2 + 3z - 2z^2 - z = z + z + 1$
   $2z^2 - 2z^2 + 3z - z = z + z + 1$
   $2z = 2z + 1$
   $0 = 1$
   contradiction

9. $5u - 4(u-1) - u = u - 4 - (u-2)$
   $5u - 4u + 4 - u = u - 4 - u + 2$
   $5u - 4u - u + 4 = u - u - 4 + 2$
   $4 = -2$
   contradiction

11. $7 - 3(y-2) = y + 4(5-y) - 7$
    $7 - 3y + 6 = y + 20 - 4y - 7$
    $7 + 6 - 3y = 20 - 7 + y - 4y$
    $13 - 3y = 13 - 3y$
    identity

13. $w(w-2) - w^2 + 2w = 3(w+1) - (3w-1)$
    $w^2 - 2w - w^2 + 2w = 3w + 3 - 3w + 1$
    $w^2 - w^2 - 2w + 2w = 3w - 3w + 3 + 1$
    $0 = 4$
    contradiction

15. $2(x^2 - 3) - x(2x-1) + x = 2 - x - (x+8) + 4x$
    $2x^2 - 6 - 2x^2 + x + x = 2 - x - x - 8 + 4x$
    $2x^2 - 2x^2 + x + x - 6 = -x - x + 4x + 2 - 8$
    $2x - 6 = 2x - 6$
    identity

17. CHECK $x = -3$:
    $x + 5 = -2$
    $(-3) + 5 \overset{?}{=} -2$
    $2 \neq -2$
    Therefore $x = -3$ does not satisfy the equation.

    CHECK $x = -7$:
    $x + 5 = -2$
    $(-7) + 5 \overset{?}{=} -2$
    $-2 \overset{\checkmark}{=} -2$
    Therefore $x = -7$ does satisfy the equation

19. CHECK $a = -5$:
    $2 - a = 3$
    $2 - (-5) \overset{?}{=} 3$
    $7 \neq 3$
    Therefore $a = -5$ does not satisfy the equation.

    CHECK $a = 5$:
    $2 - a = 3$
    $2 - (5) \overset{?}{=} 3$
    $-3 \neq 3$
    Therefore $a = 5$ does not satisfy the equation.

CHECK $a = 1$:

$2 - a = 3$

$2 - (1) \overset{?}{=} 3$

$1 \neq 3$

Therefore $a = 1$ does not satisfy the equation.

21. CHECK $y = 2$:

$5y - 6 = y - 3$

$5(2) - 6 \overset{?}{=} (2) - 3$

$10 - 6 \overset{?}{=} 2 - 3$

$4 \neq -1$

Therefore $y = 2$ does not satisfy the equation.

CHECK $y = \dfrac{3}{4}$:

$5y - 6 = y - 3$

$5\left(\dfrac{3}{4}\right) - 6 \overset{?}{=} \left(\dfrac{3}{4}\right) - 3$

$\dfrac{15}{4} - 6 \overset{?}{=} \dfrac{3}{4} - 3$

$\dfrac{15}{4} - \dfrac{24}{4} \overset{?}{=} \dfrac{3}{4} - \dfrac{12}{4}$

$\dfrac{-9}{4} \overset{\checkmark}{=} \dfrac{-9}{4}$

Therefore $y = \dfrac{3}{4}$ does satisfy the equation.

23. CHECK $w = -4$:

$6 - 2w = 10 - 3w$

$6 - 2(-4) \overset{?}{=} 10 - 3(-4)$

$6 + 8 \overset{?}{=} 10 + 12$

$14 \neq 22$

Therefore w $= -4$ does not satisfy the equation.

CHECK $w = 1$:

$6 - 2w = 10 - 3w$

$6 - 2(1) \overset{?}{=} 10 - 3(1)$

$6 - 2 \overset{?}{=} 10 - 3$

$4 \neq 7$

Therefore $w = 1$ does not satisfy the equation

25. CHECK $x = \dfrac{2}{5}$:

$4(x - 7) - (x + 1) = 15 - x$

$4\left[\left(\dfrac{2}{5}\right) - 7\right] - \left[\left(\dfrac{2}{5}\right) + 1\right] \overset{?}{=} 15 - \left(\dfrac{2}{5}\right)$

$4\left(\dfrac{2}{5} - \dfrac{35}{5}\right) - \left(\dfrac{2}{5} + \dfrac{5}{5}\right) \overset{?}{=} \dfrac{75}{5} - \dfrac{2}{5}$

$4\left(\dfrac{-33}{5}\right) - \dfrac{7}{5} \overset{?}{=} \dfrac{73}{5}$

$\dfrac{-132}{5} - \dfrac{7}{5} \overset{?}{=} \dfrac{73}{5}$

$\dfrac{-139}{5} \neq \dfrac{73}{5}$

Therefore $x = \dfrac{2}{5}$ does not satisfy the equation.

CHECK $x = 7$:

$4(x - 7) - (x + 1) = 15 - x$

$4[(7) - 7] - [(7) + 1] \overset{?}{=} 15 - (7)$

$4(0) - 8 \overset{?}{=} 15 - 7$

$0 - 8 \overset{?}{=} 15 - 7$

$-8 \neq 8$

Therefore $x = 7$ does not satisfy the equation.

27. CHECK $z = -2$:

$3z + 2(z - 1) = 4(z + 2) - (z + 5)$

$3(-2) + 2[(-2) - 1] \overset{?}{=} 4[(-2) + 2] - [(-2) + 5]$

$-6 + 2(-3) \overset{?}{=} 4(0) - 3$

$-6 - 6 \overset{?}{=} 0 - 3$

$-12 \neq -3$

Therefore $z = -2$ does not satisfy the equation.

CHECK $z = 1$:

$3z + 2(z - 1) = 4(z + 2) - (z + 5)$

$3(1) + 2[(1) - 1] \overset{?}{=} 4[(1) + 2] - [(1) + 5]$

$3 + 2(0) \overset{?}{=} 4(3) - 6$

$3 + 0 \overset{?}{=} 12 - 6$

$3 \neq 6$

Therefore $z = 1$ does not satisfy the equation.

29. CHECK $x = -2$:
$$x^2 - 3x = 2x - 6$$
$$(-2)^2 - 3(-2) \stackrel{?}{=} 2(-2) - 6$$
$$4 + 6 \stackrel{?}{=} -4 - 6$$
$$10 \neq -10$$
Therefore $x = -2$ does not satisfy the equation.

CHECK $x = 2$:
$$x^2 - 3x = 2x - 6$$
$$(2)^2 - 3(2) \stackrel{?}{=} 2(2) - 6$$
$$4 - 6 \stackrel{?}{=} 4 - 6$$
$$-2 \stackrel{\checkmark}{=} -2$$
Therefore $x = 2$ does satisfy the equation.

31. CHECK $a = -1$:
$$a^2 - 4a = 4 - a$$
$$(-1)^2 - 4(-1) \stackrel{?}{=} 4 - (-1)$$
$$1 + 4 \stackrel{?}{=} 4 + 1$$
$$5 \stackrel{\checkmark}{=} 5$$
Therefore $a = -1$ does satisfy the equation.

CHECK $a = 4$:
$$a^2 - 4a = 4 - a$$
$$(4)^2 - 4(4) \stackrel{?}{=} 4 - (4)$$
$$16 - 16 \stackrel{?}{=} 4 - 4$$
$$0 \stackrel{\checkmark}{=} 0$$
Therefore $a = 4$ does satisfy the equation.

33. CHECK $y = -2$:
$$y(y + 6) = (y + 2)^2$$
$$(-2)[(-2) + 6] \stackrel{?}{=} [(-2) + 2]^2$$
$$-2(4) \stackrel{?}{=} 0^2$$
$$-8 \neq 0$$
Therefore $y = -2$ does not satisfy the equation.

CHECK $y = 2$:
$$y(y + 6) = (y + 2)^2$$
$$(2)[(2) + 6] \stackrel{?}{=} [(2) + 2]^2$$
$$2(8) \stackrel{?}{=} 4^2$$
$$16 \stackrel{\checkmark}{=} 16$$
Therefore $y = 2$ does satisfy the equation.

35. A value satisfies or is a solution to an equation if both sides of the equation are equal when the value is substituted for the variable.

36. An identity is an equation that is always true, no matter what value is chosen for the variable.
A contradiction is an equation that is never true, no mater what value is chosen for the variable.
A conditional equation is one that is true for some values of the variables and false for the others.

37. If two quantities are equal, then we will not disturb this equality if we change each quantity in exactly the same way.

38. Two equations are called equivalent if their solution sets are exactly the same.

39.
$$\begin{cases} x + 1 = 16 \\ x - 3 = 12 \\ x = 15 \\ 2x = 30 \end{cases}$$
$$\begin{cases} x = -1 \\ x + 7 = 6 \end{cases}$$
$$\begin{cases} 2x = 18 \\ x = 9 \\ x + 2 = 11 \\ x - 2 = 7 \end{cases}$$
$$\begin{cases} 2x = 10 \\ 5x = 25 \end{cases}$$

41. $-5$

43. $4$

45. $64$

47. $-11$

49. $10 + 4(-6) = 10 - 24 = -14$

51. $-5 - 2(-8) = -5 + 16 = 11$

## Exercises 3.2

1.
$$x + 3 = 8$$
$$\underline{-3 \quad -3}$$
$$x \quad = 5$$

CHECK $x = 5$:
$$(5) + 3 \overset{?}{=} 8$$
$$8 \overset{\checkmark}{=} 8$$

3.
$$y - 4 = 7$$
$$\underline{+4 \quad +4}$$
$$y \quad = 11$$

CHECK $y = 11$:
$$(11) - 4 \overset{?}{=} 7$$
$$7 \overset{\checkmark}{=} 7$$

5.
$$a + 3 = 1$$
$$\underline{-3 \quad -3}$$
$$a \quad = -2$$

CHECK $a = -2$:
$$(-2) + 3 \overset{?}{=} 1$$
$$1 \overset{\checkmark}{=} 1$$

7.
$$a - 5 = -8$$
$$\underline{+5 \quad +5}$$
$$a \quad = -3$$

CHECK $a = -3$:
$$(-3) - 5 \overset{?}{=} -8$$
$$-8 \overset{\checkmark}{=} -8$$

9.
$$3x = 21$$
$$\frac{\cancel{3}x}{\cancel{3}} = \frac{21}{3}$$
$$x = 7$$

CHECK $x = 7$:
$$3(7) \overset{?}{=} 21$$
$$21 \overset{\checkmark}{=} 21$$

11.
$$4x = 15$$
$$\frac{\cancel{4}x}{\cancel{4}} = \frac{15}{4}$$
$$x = \frac{15}{4}$$

CHECK $x = \frac{15}{4}$:
$$4\left(\frac{15}{4}\right) \overset{?}{=} 15$$
$$15 \overset{\checkmark}{=} 15$$

13.
$$-4x = -12$$
$$\frac{\cancel{-4}x}{\cancel{-4}} = \frac{-12}{-4}$$
$$x = 3$$

CHECK $x = 3$:
$$-4(3) \overset{?}{=} -12$$
$$-12 \overset{\checkmark}{=} -12$$

15.
$$3x + x = 8 - 12$$
$$4x = -4$$
$$\frac{\cancel{4}x}{\cancel{4}} = \frac{-4}{4}$$
$$x = -1$$

CHECK $x = -1$:
$$3(-1) + (-1) \overset{?}{=} 8 - 12$$
$$-3 - 1 \overset{?}{=} -4$$
$$-4 \overset{\checkmark}{=} -4$$

17.
$$4x - x = 2 - 8$$
$$3x = -6$$
$$\frac{\cancel{3}x}{\cancel{3}} = \frac{-6}{3}$$
$$x = -2$$

CHECK $x = -2$:
$$4(-2) - (-2) \overset{?}{=} 2 - 8$$
$$-8 + 2 \overset{?}{=} 2 - 8$$
$$-6 \overset{\checkmark}{=} -6$$

19.
$$2z - 3z - 11z = -4(6)$$
$$-12z = -24$$
$$\frac{\cancel{-12}z}{\cancel{-12}} = \frac{-24}{-12}$$
$$z = 2$$

CHECK $z = 2$:
$$2(2) - 3(2) - 11(2) \overset{?}{=} -4(6)$$
$$4 - 6 - 22 \overset{?}{=} -24$$
$$-24 \overset{\checkmark}{=} -24$$

21.
$$3w - 7w + 4w = 8 - 3$$
$$0w = 5$$
$$0 = 5 \quad \text{(contradiction)}$$
No solutions

23.
$$2x + 1 = 7$$
$$\underline{-1 \quad -1}$$
$$2x \quad = 6$$
$$\frac{\cancel{2}x}{\cancel{2}} = \frac{6}{2}$$
$$x = 3$$

CHECK $x = 3$:
$$2(3) + 1 \overset{?}{=} 7$$
$$6 + 1 \overset{?}{=} 7$$
$$7 \overset{\checkmark}{=} 7$$

25.
$$10x - 4 = 22$$
$$\underline{+4 \quad +4}$$
$$10x \quad = 26$$
$$\frac{\cancel{10}x}{\cancel{10}} = \frac{26}{10}$$
$$x = \frac{13}{5}$$

CHECK $x = \frac{13}{5}$:
$$10\left(\frac{13}{5}\right) - 4 \overset{?}{=} 22$$
$$26 - 4 \overset{?}{=} 22$$
$$22 \overset{\checkmark}{=} 22$$

27.
$$2t = 3t + 5$$
$$\underline{-3t \quad -3t}$$
$$-t = \quad 5$$
$$t = -5$$

CHECK $t = -5$:
$$2(-5) \overset{?}{=} 3(-5) + 5$$
$$-10 \overset{?}{=} -15 + 5$$
$$-10 \overset{\checkmark}{=} -10$$

29.
$$9 = 6 - 3a$$
$$\frac{-6 \quad -6}{3 = \quad -3a}$$
$$\frac{3}{-3} = \frac{\cancel{-3}a}{\cancel{-3}}$$
$$-1 = a$$

CHECK $a = -1$:
$$9 \overset{?}{=} 6 - 3(-1)$$
$$9 \overset{?}{=} 6 + 3$$
$$9 \overset{\checkmark}{=} 9$$

31.
$$20 = 3w - 1$$
$$\frac{+1 \qquad +1}{21 = 3w}$$
$$\frac{21}{3} = \frac{\cancel{3}w}{\cancel{3}}$$
$$7 = w$$

CHECK $w = 7$:
$$20 \overset{?}{=} 3(7) - 1$$
$$20 \overset{?}{=} 21 - 1$$
$$20 \overset{\checkmark}{=} 20$$

33. Let $w$ = width.
$$6w = 40$$
$$\frac{\cancel{6}w}{\cancel{6}} = \frac{40}{6}$$
$$w = \frac{20}{3} \text{ or } w = 6\frac{2}{3}$$

The width of the rectangle is $6\frac{2}{3}$ in.

35. Let $l$ = length.
$$3.5l = 73.5$$
$$\frac{\cancel{3.5}\,l}{\cancel{3.5}} = \frac{73.5}{3.5}$$
$$l = 21$$

The length of the rectangle is 21 yd.

37. Let $t$ = time needed.
$$52t = 234$$
$$\frac{\cancel{52}t}{\cancel{52}} = \frac{234}{52}$$
$$t = 4.5$$

It would take someone 4.5 hours.

39. Let $m$ = number of miles driven.
$$0.17m + 3(29.95) = 123.17$$
$$0.17m + 89.85 = 123.17$$
$$\frac{-89.85 \qquad -89.85}{0.17m \qquad = \quad 33.32}$$
$$\frac{0.\cancel{17}m}{0.\cancel{17}} = \frac{33.32}{0.17}$$
$$m = 196$$

The car was driven 196 miles.

41.
$$C = 0.003s + 0.21$$
$$0.35 = 0.003s + 0.21$$
$$\frac{-0.21 \qquad \qquad -0.21}{0.14 = 0.003s}$$
$$\frac{0.14}{0.003} = \frac{0.\cancel{003}s}{0.\cancel{003}}$$
$$\frac{140}{3} \text{ or } 46\frac{2}{3} = s$$

The speed of the truck is $46\frac{2}{3}$ mph.

43.
$$C = 4T - 160$$
$$150 = 4T - 160$$
$$\frac{+160 \qquad +160}{310 = 4T}$$
$$\frac{310}{4} = \frac{\cancel{4}T}{\cancel{4}}$$
$$\frac{155}{2} \text{ or } 77.5 = T$$

The temperature is 77.5°F.

45.
$$T = 46 - 0.0054A$$
$$32 = 46 - 0.0054A$$
$$\frac{-46 \quad -46}{-14 = \qquad -0.0054A}$$
$$\frac{-14}{-0.0054} = \frac{\cancel{-0.0054}A}{\cancel{-0.0054}}$$
$$2593.6 = A \text{ (rounded to the nearest tenth)}$$

The altitude is 2593.6 feet.

47.
$$N = 1.67s + 55$$
$$85 = 1.67s + 55$$
$$\frac{-55 \qquad -55}{30 = 1.67s}$$
$$\frac{30}{1.67} = \frac{1.\cancel{67}s}{1.\cancel{67}}$$
$$18.0 = s \text{ (rounded to the nearest tenth)}$$

The speed is 18.0 feet/sec.

49.  Let $w =$ width.
Then $2w + 7 =$ length.

$$2w + 2(2w + 7) = 50$$
$$2w + 4w + 14 = 50$$
$$6w + 14 = 50$$
$$\frac{-14 \quad -14}{6w \quad = \quad 36}$$
$$\frac{6w}{6} = \frac{36}{6}$$
$$w = 6$$

The width is 6 cm. and the
length is $2(6) + 7 = 19$ cm.

51.  Let $l =$ length.
Then $5l - 10 =$ width.

$$2l + 2(5l - 10) = 100$$
$$2l + 10l - 20 = 100$$
$$12l - 20 = 100$$
$$\frac{+20 \quad +20}{12l \quad = \quad 120}$$
$$\frac{12l}{12} = \frac{120}{12}$$
$$l = 10$$

The length is 10 yards and the
width is $5(10) - 10 = 40$ yards.

53.  To check your answer after you have solved an equation, write your answer in place of the variable in the original equation. If the result is a true statement, then your answer satisfies the equation.

54.  (a)  After subtracting 4 from both sides of the equation, we should get $2x = 4$, not $2x = 12$.

(b)  Subtracting 3 from $3x$ will not give $x$. (These are not like terms.)

(c)  When we divide both sides of the equation by 3, we must remember to divide both terms on the left. We would get $x - 2 = 4$, not $x - 6 = 4$.

(d)  We cannot divide both sides by $x$, since $x$ might be equal to 0. (In this case, it is!)

(e)  In the last step, we should divide both sides by $-3$. Then we get $\dfrac{-3x}{-3} = \dfrac{-6}{-3}$, which gives $x = 2$.

(f)  We need to divide both sides of the final equation by $-1$ to obtain $x = -2$.

(g)  This is correct as written.

55.  $|5 - 9| - |-3 - 8| = |-4| - |-11|$
$= 4 - 11 = -7$

57.  $\dfrac{-4(-2)(-6)}{-4(-2) - 6} = \dfrac{8(-6)}{8 - 6} = \dfrac{-48}{2} = -24$

59.  $72 = 2 \cdot 2 \cdot 2 \cdot 3 \cdot 3$

# Exercises 3.3

1.  $\begin{aligned} 8y + 4 &= 5y + 19 \\ \underline{-5y \qquad -5y} & \\ 3y + 4 &= \qquad 19 \\ \underline{-4 \qquad -4} & \\ 3y &= \qquad 15 \\ \frac{3y}{3} &= \frac{15}{3} \\ y &= 5 \end{aligned}$

CHECK $y = 5$:

$$8(5) + 4 \overset{?}{=} 5(5) + 19$$
$$40 + 4 \overset{?}{=} 25 + 19$$
$$44 \overset{\checkmark}{=} 44$$

3.
$$2a + 5 = 4a + 12$$
$$\underline{-2a \qquad -2a}$$
$$5 = 2a + 12$$
$$\underline{-12 \qquad -12}$$
$$-7 = 2a$$
$$\frac{-7}{2} = \frac{\cancel{2}a}{\cancel{2}}$$
$$-\frac{7}{2} = a$$

CHECK $a = -\dfrac{7}{2}$:
$$2\left(-\frac{7}{2}\right) + 5 \overset{?}{=} 4\left(-\frac{7}{2}\right) + 12$$
$$-7 + 5 \overset{?}{=} -14 + 12$$
$$-2 \overset{\checkmark}{=} -2$$

5.
$$5r - 8 = 3r - 20$$
$$\underline{-3r \qquad -3r}$$
$$2r - 8 = -20$$
$$\underline{+8 \qquad +8}$$
$$2r = -12$$
$$\frac{\cancel{2}r}{\cancel{2}} = \frac{-12}{2}$$
$$r = -6$$

CHECK $r = -6$:
$$5(-6) - 8 \overset{?}{=} 3(-6) - 20$$
$$-30 - 8 \overset{?}{=} -18 - 20$$
$$-38 \overset{\checkmark}{=} -38$$

7.
$$10 - x = 4 - 3x$$
$$\underline{+3x \qquad +3x}$$
$$10 + 2x = 4$$
$$\underline{-10 \qquad -10}$$
$$2x = -6$$
$$\frac{\cancel{2}x}{\cancel{2}} = \frac{-6}{2}$$
$$x = -3$$

CHECK $x = -3$:
$$10 - (-3) \overset{?}{=} 4 - 3(-3)$$
$$10 + 3 \overset{?}{=} 4 + 9$$
$$13 \overset{\checkmark}{=} 13$$

9.
$$-4 - 3u = -2 - u$$
$$\underline{+3u \qquad +3u}$$
$$-4 = -2 + 2u$$
$$\underline{+2 \qquad +2}$$
$$-2 = 2u$$
$$\frac{-2}{2} = \frac{\cancel{2}u}{\cancel{2}}$$
$$-1 = u$$

CHECK $u = -1$:
$$-4 - 3(-1) \overset{?}{=} -2 - (-1)$$
$$-4 + 3 \overset{?}{=} -2 + 1$$
$$-1 \overset{\checkmark}{=} -1$$

11.
$$x + 7 = 7 - x$$
$$\underline{+x \qquad +x}$$
$$2x + 7 = 7$$
$$\underline{-7 \qquad -7}$$
$$2x = 0$$
$$\frac{\cancel{2}x}{\cancel{2}} = \frac{0}{2}$$
$$x = 0$$

CHECK $x = 0$:
$$(0) + 7 \overset{?}{=} 7 - (0)$$
$$0 + 7 \overset{?}{=} 7 - 0$$
$$7 \overset{\checkmark}{=} 7$$

13.
$$x + 7 = 7 + x$$
$$\underline{-x \qquad -x}$$
$$7 = 7 \quad \text{(identity)}$$
Always true

15.
$$x - 7 = 7 + x$$
$$\underline{-x \qquad -x}$$
$$-7 = 7 \quad \text{(contradiction)}$$
No solutions

17.
$$x - 7 = 7 - x$$
$$\underline{+x \qquad\quad +x}$$
$$2x - 7 = 7$$
$$\underline{+7 \quad +7}$$
$$2x \quad = 14$$
$$\frac{\not{2}x}{\not{2}} = \frac{14}{2}$$
$$x = 7$$
CHECK $x = 7$:
$$(7) - 7 \overset{?}{=} 7 - (7)$$
$$7 - 7 \overset{?}{=} 7 - 7$$
$$0 \overset{\checkmark}{=} 0$$

19.
$$2(t + 1) + 4t - 7 = 29$$
$$2t + 2 + 4t = 29$$
$$6t + 2 = 29$$
$$\underline{-2 \quad -2}$$
$$6t \quad = 27$$
$$\frac{\not{6}t}{\not{6}} = \frac{27}{6}$$
$$t = \frac{9}{2}$$
CHECK $t = \frac{9}{2}$:
$$2\left[\left(\frac{9}{2}\right) + 1\right] + 4\left(\frac{9}{2}\right) \overset{?}{=} 29$$
$$2\left(\frac{11}{2}\right) + 4\left(\frac{9}{2}\right) \overset{?}{=} 29$$
$$11 + 18 \overset{?}{=} 29$$
$$29 \overset{\checkmark}{=} 29$$

21.
$$2(y + 3) + 4(y - 2) = 22$$
$$2y + 6 + 4y - 8 = 22$$
$$6y - 2 = 22$$
$$\underline{+2 \quad +2}$$
$$6y \quad = 24$$
$$\frac{\not{6}y}{\not{6}} = \frac{24}{6}$$
$$y = 4$$
CHECK $y = 4$:
$$2[(4) + 3] + 4[(4) - 2] \overset{?}{=} 22$$
$$2(7) + 4(2) \overset{?}{=} 22$$
$$14 + 8 \overset{?}{=} 22$$
$$22 \overset{\checkmark}{=} 22$$

23.
$$4 + 3(3y - 5) = 2y - 11 + y$$
$$4 + 9y - 15 = 2y - 11 + y$$
$$9y - 11 = 3y - 11$$
$$\underline{-3y \qquad\quad -3y}$$
$$6y - 11 = -11 \ .$$
$$\underline{+11 \qquad\quad +11}$$
$$6y \quad = 0$$
$$\frac{\not{6}y}{\not{6}} = \frac{0}{6}$$
$$y = 0$$
CHECK $y = 0$:
$$4 + 3[3(0) - 5] \overset{?}{=} 2(0) - 11 + (0)$$
$$4 + 3(0 - 5) \overset{?}{=} 0 - 11 + 0$$
$$4 + 3(-5) \overset{?}{=} -11$$
$$4 - 15 \overset{?}{=} -11$$
$$-11 \overset{\checkmark}{=} -11$$

25.
$$3(a - 2) + 4(2 - a) = a + 2(a + 1)$$
$$3a - 6 + 8 - 4a = a + 2a + 2$$
$$-a + 2 = 3a + 2$$
$$\underline{+a \qquad\quad +a}$$
$$2 = 4a + 2$$
$$\underline{-2 \qquad\quad -2}$$
$$0 = 4a$$
$$\frac{0}{4} = \frac{\not{4}a}{\not{4}}$$
$$0 = a$$
CHECK $a = 0$:
$$3[(0) - 2] + 4[2 - (0)] \overset{?}{=} (0) + 2[(0) + 1]$$
$$3(-2) + 4(2) \overset{?}{=} 0 + 2(1)$$
$$-6 + 8 \overset{?}{=} 0 + 2$$
$$2 \overset{\checkmark}{=} 2$$

27. 
$$
\begin{aligned}
8z - 3(z - 3) &= -9 \\
8z - 3z + 9 &= -9 \\
5z + 9 &= -9 \\
\underline{-9 \quad\; -9} & \\
5z &= -18 \\
\frac{\cancel{5}z}{\cancel{5}} &= \frac{-18}{5} \\
z &= -\frac{18}{5}
\end{aligned}
$$

CHECK $z = -\dfrac{18}{5}$:

$$
8\left(-\frac{18}{5}\right) - 3\left[\left(-\frac{18}{5}\right) - 3\right] \overset{?}{=} -9
$$

$$
8\left(-\frac{18}{5}\right) - 3\left(-\frac{33}{5}\right) \overset{?}{=} -9
$$

$$
-\frac{144}{5} + \frac{99}{5} \overset{?}{=} -9
$$

$$
-\frac{45}{5} \overset{?}{=} -9
$$

$$
-9 \overset{\checkmark}{=} -9
$$

29. 
$$
\begin{aligned}
20t - 5(t - 1) &= 0 \\
20t - 5t + 5 &= 0 \\
15t + 5 &= 0 \\
\underline{-5 \quad\; -5} & \\
15t &= -5 \\
\frac{\cancel{15}t}{\cancel{15}} &= \frac{-5}{15} \\
t &= -\frac{1}{3}
\end{aligned}
$$

CHECK $t = -\dfrac{1}{3}$:

$$
20\left(-\frac{1}{3}\right) - 5\left[\left(-\frac{1}{3}\right) - 1\right] \overset{?}{=} 0
$$

$$
20\left(-\frac{1}{3}\right) - 5\left(-\frac{4}{3}\right) \overset{?}{=} 0
$$

$$
-\frac{20}{3} + \frac{20}{3} \overset{?}{=} 0
$$

$$
0 \overset{\checkmark}{=} 0
$$

31. 
$$
\begin{aligned}
20 - 5(t - 1) &= 0 \\
20 - 5t + 5 &= 0 \\
-5t + 25 &= 0 \\
\underline{-25 \quad\; -25} & \\
-5t &= -25 \\
\frac{\cancel{-5}t}{\cancel{-5}} &= \frac{-25}{-5} \\
t &= 5
\end{aligned}
$$

CHECK $t = 5$:

$$
20 - 5[(5) - 1] \overset{?}{=} 0
$$

$$
20 - 5(4) \overset{?}{=} 0
$$

$$
20 - 20 \overset{?}{=} 0
$$

$$
0 \overset{\checkmark}{=} 0
$$

33. 
$$
\begin{aligned}
2(y - 3) - 3(y - 5) &= 5y - 5(y - 2) \\
2y - 6 - 3y + 15 &= 5y - 5y + 10 \\
-y + 9 &= 10 \\
\underline{-9 \quad\; -9} & \\
-y &= 1 \\
\frac{-y}{-1} &= \frac{1}{-1} \\
y &= -1
\end{aligned}
$$

CHECK $y = -1$:

$$
2[(-1) - 3] - 3[(-1) - 5] \overset{?}{=} 5(-1) - 5[(-1) - 2]
$$

$$
2(-4) - 3(-6) \overset{?}{=} 5(-1) - 5(-3)
$$

$$
-8 + 18 \overset{?}{=} -5 + 15
$$

$$
10 \overset{\checkmark}{=} 10
$$

35. 
$$
\begin{aligned}
4x - 3(x + 8) &= 5x - 2(x - 12) - 2x \\
4x - 3x - 24 &= 5x - 2x + 24 - 2x \\
x - 24 &= x + 24 \\
\underline{-x \qquad\; -x} & \\
-24 &= 24 \quad \text{(contradiction)}
\end{aligned}
$$

No solutions

**37.**
$$a - (5 - 3a) = 7a - (a - 3) - 8$$
$$a - 5 + 3a = 7a - a + 3 - 8$$
$$4a - 5 = 6a - 5$$
$$\underline{-4a \qquad\qquad -4a}$$
$$-5 = 2a - 5$$
$$\underline{+5 \qquad\qquad +5}$$
$$0 = 2a$$
$$\frac{0}{2} = \frac{\cancel{2}a}{\cancel{2}}$$
$$0 = a$$

CHECK $a = 0$:
$$(0) - [5 - 3(0)] \stackrel{?}{=} 7(0) - [(0) - 3] - 8$$
$$0 - (5 - 0) \stackrel{?}{=} 0 - (-3) - 8$$
$$0 - 5 \stackrel{?}{=} 0 + 3 - 8$$
$$-5 \stackrel{\checkmark}{=} -5$$

**39.**
$$x^2 + 3x - 7 = x^2 - 5x + 1$$
$$\underline{-x^2 \qquad\qquad -x^2}$$
$$3x - 7 = -5x + 1$$
$$\underline{+5x \qquad\qquad +5x}$$
$$8x - 7 = 1$$
$$\underline{+7 \qquad\qquad +7}$$
$$8x = 8$$
$$\frac{\cancel{8}x}{\cancel{8}} = \frac{8}{8}$$
$$x = 1$$

CHECK $x = 1$:
$$(1)^2 + 3(1) - 7 \stackrel{?}{=} (1)^2 - 5(1) + 1$$
$$1 + 3 - 7 \stackrel{?}{=} 1 - 5 + 1$$
$$-3 \stackrel{\checkmark}{=} -3$$

**41.**
$$x(x + 2) + 3x = x(x - 11) - 12$$
$$x^2 + 2x + 3x = x^2 - x - 12$$
$$x^2 + 5x = x^2 - x - 12$$
$$\underline{-x^2 \qquad\qquad -x^2}$$
$$5x = -x - 12$$
$$\underline{+x \qquad\qquad +x}$$
$$6x = -12$$
$$\frac{\cancel{6}x}{\cancel{6}} = \frac{-12}{6}$$
$$x = -2$$

CHECK $x = -2$:
$$(-2)[(-2) + 2] + 3(-2) \stackrel{?}{=} (-2)[(-2) - 1] - 12$$
$$-2(0) + 3(-2) \stackrel{?}{=} -2(-3) - 12$$
$$0 - 6 \stackrel{?}{=} 6 - 12$$
$$-6 \stackrel{\checkmark}{=} -6$$

**43.**
$$2z(z + 1) + 3(z + 2) = 3z(z + 2) - z^2$$
$$2z^2 + 2z + 3z + 6 = 3z^2 + 6z - z^2$$
$$2z^2 + 5z + 6 = 2z^2 + 6z$$
$$\underline{-2z^2 \qquad\qquad -2z^2}$$
$$5z + 6 = 6z$$
$$\underline{-5z \qquad\qquad -5z}$$
$$6 = z$$

CHECK $z = 6$:
$$2(6)[(6) + 1] + 3[(6) + 2] \stackrel{?}{=} 3(6)[(6) + 2] - (6)^2$$
$$12(7) + 3(8) \stackrel{?}{=} 18(8) - 36$$
$$84 + 24 \stackrel{?}{=} 144 - 36$$
$$108 \stackrel{\checkmark}{=} 108$$

**45.** $x = 6.5$      **47.** $t = 0.03$      **49.** $t = -8.19$

51.
$$2.8\,x = 1.60x + 2100$$
$$\underline{-1.60x \qquad -1.60x}$$
$$1.2\,x = \qquad\qquad 2100$$
$$\frac{\cancel{1.2}\,x}{\cancel{1.2}} = \frac{2100}{1.2}$$
$$x = 1750$$

The company must sell $1,750$ items in order to break even.

CHECK: When $1,750$ items are sold,
$R = 2.8(1750) = \$4900$.
When $1,750$ items are produced,
$C = 1.60(1750) + 2100$
$\quad = 2800 + 2100 = \$4900$.
So revenue $=$ cost.

53. Let $x =$ length of the shorter piece (in feet).
Then $x + 8 =$ length of the longer piece (in feet).
$$x + (x + 8) = 30$$
$$2x + 8 = 30$$
$$\underline{\quad -8 \quad -8}$$
$$2x \qquad = 22$$
$$\frac{\cancel{2}x}{\cancel{2}} = \frac{22}{2}$$
$$x = 11$$

Then $x + 8 = 11 + 8 = 19$. Thus, the two pieces are 11 ft. and 19 ft.
CHECK: 19 is 8 more than 11, and $11 + 19 = 30$.

55. Let $x =$ one of the numbers.
Then $3x + 4 =$ the other number.
$$x + (3x + 4) = 24$$
$$4x + 4 = 24$$
$$\underline{\quad -4 \quad -4}$$
$$4x \qquad = 20$$
$$\frac{\cancel{4}x}{\cancel{4}} = \frac{20}{4}$$
$$x = 5$$

Then $3x + 4 = 3(5) + 4 = 19$. Thus, the two numbers are 5 and 19.
CHECK: 19 is 4 more than 3 times 5, and the sum of 5 and 19 is 24.

57. Let $x =$ one of the numbers.
Then $4x - 5 =$ the other number.
$$x + (4x - 5) = 11$$
$$5x - 5 = 11$$
$$\underline{\quad +5 \quad +5}$$
$$5x \qquad = 16$$
$$\frac{\cancel{5}x}{\cancel{5}} = \frac{16}{5}$$
$$x = \frac{16}{5}$$

Then $4x - 5 = 4\left(\dfrac{16}{5}\right) - 5 = \dfrac{39}{5}$. Thus, the two numbers are $\dfrac{16}{5}$ and $\dfrac{39}{5}$.
CHECK: $\dfrac{39}{5}$ is 5 less than 4 times $\dfrac{16}{5}$ and the sum of $\dfrac{16}{5}$ and $\dfrac{39}{5}$ is $\dfrac{55}{5} = 11$.

59. Let $x =$ the number.
$$x + (5x - 4) = 27$$
$$6x - 4 = 27$$
$$\underline{\quad +4 \quad +4}$$
$$6x \qquad = 31$$
$$\frac{\cancel{6}x}{\cancel{6}} = \frac{31}{6}$$
$$x = \frac{31}{6}$$

CHECK: 4 less than 5 times $\dfrac{31}{6}$ is $\dfrac{131}{6}$, and $\dfrac{31}{6} + \dfrac{131}{6} = \dfrac{162}{6} = 27$.

61. Let $x =$ the number.
$$(2x + 4) - x = 12$$
$$x + 4 = 12$$
$$\underline{-4 \quad -4}$$
$$x \quad = 8$$

Thus, the number is 8.
CHECK: 4 more than twice 8 is 20, and.
$20 - 8 = 12.$

63. Let $x =$ the smallest number.
Then $2x - 5 =$ the middle number and
$2x + 10 =$ the largest number.
$$x + (2x - 5) + (2x + 10) = 80$$
$$5x + 5 = 80$$
$$\underline{-5 \quad -5}$$
$$5x = 75$$
$$\frac{\cancel{5}x}{\cancel{5}} = \frac{75}{5}$$
$$x = 15$$

Then $2x - 5 = 2(15) - 5 = 25$ and
$2x + 10 = 2(15) + 10 = 40.$ Thus the numbers
are 15, 25, and 40.
CHECK: 25 is 5 less than twice 15, and 40 is 10
more than twice 15. The sum of 15, 25, and 40 is 80.

65. Let $n =$ the first integer.
Then $n + 1 =$ the second integer and
$n + 2 =$ the third integer.
$$n + (n + 1) + (n + 2) = 45$$
$$3n + 3 = 45$$
$$\underline{-3 \quad -3}$$
$$3n = 42$$
$$\frac{\cancel{3}n}{\cancel{3}} = \frac{42}{3}$$
$$n = 14$$

Then $n + 1 = 14 + 1 = 15$ and $n + 2 = 14 + 2 = 16.$
Thus, the consecutive integers are 14, 15, and 16.
CHECK: 14, 15, and 16 are three consecutive
integers. The sum of 14, 15, and 16 is 45.

67. Let $n =$ the first odd integer.
Then $n + 2 =$ the second odd integer and
$n + 4 =$ the third odd integer.
$$n + (n + 2) + (n + 4) = 2(n + 4) + 29$$
$$3n + 6 = 2n + 8 + 29$$
$$3n + 6 = 2n + 37$$
$$\underline{-2n \qquad -2n}$$
$$n + 6 = 37$$
$$\underline{-6 \qquad -6}$$
$$n = 31$$

Then $n + 2 = 31 + 2 = 33$ and $n + 4 = 31 + 4 = 35.$
Thus, the consecutive odd integers are 31, 33, and 35.
CHECK: 31, 33, and 35 are three consecutive odd
integers. Their sum is 99, which is 29 more than
$2(35) = 70.$

69. Let $W =$ width of the rectangle.
$2W + 1 =$ length of the rectangle.
$$2W + 2(2W + 1) = 26$$
$$2W + 4W + 2 = 26$$
$$6W + 2 = 26$$
$$\underline{-2 \quad -2}$$
$$6W = 24$$
$$\frac{\cancel{6}W}{\cancel{6}} = \frac{24}{6}$$
$$W = 4$$

Then $2W + 1 = 2(4) + 1 = 9$ is the length of the
rectangle. Thus, the rectangle has a width of 4 cm
and a length of 9 cm.
CHECK: 9 is one more than twice 4, and
$2(4) + 2(9) = 8 + 18 = 26.$

71. Let $x$ = length of the shortest side.
Then $x + 1$ = length of the "middle" side and
$x + 2$ = length of the longest side.

$$x + (x + 1) + (x + 2) = 24$$
$$3x + 3 = 24$$
$$\underline{-3 = -3}$$
$$3x = 21$$
$$\frac{3x}{3} = \frac{21}{3}$$
$$x = 7$$

Then $x + 1 = 7 + 1 = 8$ and $x + 2 = 7 + 2 = 9$.
Thus, the sides of the triangle have lengths of
7 cm, 8 cm, and 9 cm.
CHECK: 7, 8, and 9 are three consecutive integers.
A triangle whose sides are 7 cm, 8 cm, and 9 cm
has a perimeter of $7 + 8 + 9 = 24$ cm.

73. Let $W$ = width of the rectangle.
Then $W + 6$ = length of the rectangle.

$$2(W + 10) + 2[3(W + 6)] = 2W + 2(W + 6) + 56$$
$$2W + 20 + 6W + 36 = 2W + 2W + 12 + 56$$
$$8W + 56 = 4W + 68$$
$$\underline{-4W \qquad\qquad -4W}$$
$$4W + 56 = +68$$
$$\underline{-56 \qquad -56}$$
$$4W = 12$$
$$\frac{4W}{4} = \frac{12}{4}$$
$$W = 3$$

Then $W + 6 = 3 + 6 = 9$, so that
the original rectangle has a width of
3 and a length of 9.
CHECK: 9 is 6 more than 3. The new
rectangle will have a width of
$3 + 10 = 13$ and a length of $3(9) = 27$,
so that its perimeter is
$2(13) + 2(27) = 80$. This is 56 more
than $2(3) + 2(9) = 24$, the original
perimeter.

75. Let $m$ = number of miles driven.

$$2(45) + 0.40m = 170$$
$$90 + 0.40m = 170$$
$$\underline{-90 \qquad\qquad -90}$$
$$0.40m = 80$$
$$\frac{0.40m}{0.40} = \frac{80}{0.40}$$
$$m = 200$$

The truck was driven 200 miles.
CHECK: To rent the truck for two days at $45 per
day costs $2(45) = \$90$. To drive 200 miles at $0.40
per mile costs $0.40(200) = \$80$. $\$90 + \$80 = \$170$.

77. Let $d$ = the daily rental charge.

$$125 + 5d = 275$$
$$\underline{-125 \qquad -125}$$
$$5d = 150$$
$$\frac{5d}{5} = \frac{150}{5}$$
$$d = 30$$

The daily rental charge is $30.
CHECK: To rent a computer for 5 days at $30 per
day costs $5(30) = \$150$. With an installation charge
of $125, the total five-day rental costs
$\$150 + \$125 = \$275$.

79.  Let $q$ = the number of quarters.
Then $20 - q$ = the number of dimes.

$25q + 10(20 - q) = 425$
$25q + 200 - 10q = 425$
$15q + 200 = 425$
$\underline{\phantom{15q} -200 \qquad -200}$
$15q \qquad\qquad 225$

$$\frac{\cancel{15}q}{\cancel{15}} = \frac{225}{15}$$
$$q = 15$$

Then $20 - q = 20 - 15 = 5$. Thus there are 15 quarters and 5 dimes in the collection.

CHECK: $15 + 5 = 20$. The values of 15 quarters is $15(0.25) = \$3.75$. The value of 5 dimes is $5(0.10) = \$0.50$.
$\$3.75 + \$0.50 = \$4.25$.

81.  Let $x$ = number of advanced-purchase tickets sold.
Then $150 - x$ = number of tickets sold at the door.

$10x + 12(150 - x) = 1580$
$10x + 1800 - 12x = 1580$
$-2x + 1800 = 1580$
$\underline{\phantom{-2x} -1800 \qquad -1800}$
$-2x \qquad = -220$

$$\frac{\cancel{-2}x}{\cancel{-2}} = \frac{-220}{-2}$$
$$x = 110$$

110 advanced-purchase tickets were sold.

CHECK: If 110 tickets were sold in advance, then 40 were sold at the door. $110(10) = \$1100$ was collected from the advance sale and $40(12) = \$480$ was collected at the door.
$\$1100 + \$480 = \$1580$.

83.  Let $x$ = number of half-dollars.
Then $x + 11$ = number of quarters.
So $45 - x - (x + 11)$ = number of dimes.

$50x + 25(x + 11) + 10[45 - x - (x + 11)] = 1110$
$50x + 25x + 275 + 10(34 - 2x) = 1110$
$75x + 275 + 340 - 20x = 1110$
$55x + 615 = 1110$
$\underline{\phantom{55x} -615 \qquad -615}$
$55x \qquad = 495$

$$\frac{55x}{55} = \frac{495}{55}$$
$$x = 9$$

Then $x + 11 = 9 + 11 = 20$ and
$45 - x - (x + 11) = 45 - 9 - (9 + 11)$
$= 16$. So there are 9 half-dollars, 20 quarters, and 16 dimes in the collection.

CHECK: $9 + 20 + 16 = 45$. The value of 9 half-dollars is $\$4.50$. The value of 20 quarters is $\$5.00$. The value of 16 dimes is $\$1.60$. Total value is $\$4.50 + \$5.00 + \$1.60 = \$11.10$.

85.  Let $t$ = number of hours the electrician worked.
Then $t + 3$ = number of hours her assistant worked.

$24t + 12(t + 3) = 228$
$24t + 12t + 36 = 228$
$36t + 36 = 228$
$\underline{\phantom{36t} -36 \qquad -36}$
$36t \qquad = 192$

$$\frac{36t}{36} = \frac{192}{36}$$
$$t = \frac{16}{3} \text{ or } 5\frac{1}{3}$$

The electrician worked for $5\frac{1}{3}$ hours.

CHECK: The electrician earns $\frac{16}{3}(\$24) = \$128$ for her time. Her assistant works $t + 3 = \frac{16}{3} + 3 = \frac{25}{3}$ hours and earns $\frac{25}{3}(\$12) = \$100$.
Together, they earn $\$128 + \$100 = \$228$.

87. Let $t$ = the time (in hours) that each train travels up to the point that they pass by each other

Using $d = rt$, $20t$ = distance traveled by the slower train and $40t$ = distance traveled by the faster train.

$$20t + 40t = 300$$
$$60t = 300$$
$$\frac{60t}{60} = \frac{300}{60}$$
$$t = 5$$

Thus the trains pass by each other 5 hours later than 10:00a.m., or at 3:00p.m.

CHECK: In 5 hours, the slower train travels $20(5) = 100$ miles and the faster train travels $40(5) = 200$ miles. Together they travel $100 + 200 = 300$ miles.

89. Let $t$ = the time (in hours) traveled by the person who left earlier.

Then $t - 1$ = the time (in hours) traveled by the person who left 1 hour later.

$$55t + 45(t - 1) = 280$$
$$55t + 45t - 45 = 280$$
$$100t - 45 = 280$$
$$\underline{+45 \quad +45}$$
$$100t \quad = 325$$
$$\frac{100t}{100} = \frac{325}{100}$$
$$t = \frac{13}{4} \text{ or } 3\frac{1}{4}$$

Thus, the people will be 280 km apart $3\frac{1}{4}$ hours later than 2:00p.m., or at 5:15 p.m.

CHECK: The person who left earlier travels $\frac{13}{4}(55) = \frac{715}{4}$ km. The person who left later travels $t - 1 = \frac{13}{4} - 1 = \frac{9}{4}$ hours and covers $\frac{9}{4}(45) = \frac{405}{4}$ km. The distance between them is $\frac{715}{4} + \frac{405}{4} = \frac{1120}{4} = 280$ km.

91. Let $t$ = number of hours needed to complete the running section.

Then $6 - t$ = number of hours needed to complete the bicycling section.

$$18t + 50(6 - t) = 172$$
$$18t + 300 - 50t = 172$$
$$-32t + 300 = 172$$
$$\underline{-300 \quad -300}$$
$$-32t \quad = -128$$
$$\frac{-32t}{-32} = \frac{-128}{-32}$$
$$t = 4$$

Thus, it takes 4 hours to complete the running section of the course. Since the running rate is 18 kph, the running section of the course is $18(4) = 72$ km.

CHECK: The first person runs a distance of $18(4) = 72$ km. The second person bicycles a distance of $50(6 - t) = 50(6 - 4) = 50(2) = 100$ km. The total distance covered is $72 + 100 = 172$ km.

93. Let $r$ = slower rate.

Then $r + 15$ = faster rate.

$$5r = 4(r + 15)$$
$$5r = 4r + 60$$
$$\underline{-4r = -4r}$$
$$r = 60$$

A car that travels for 5 hours at the rate of 60 kph covers $5(60) = 300$ km, which is the distance between town $A$ and town $B$.

CHECK: At the rate of 60 kph, the car travels $5(60) = 300$ km in 5 hours. At the rate of $60 + 15 = 75$ kph, the car travels $4(75) = 300$ km in 4 hours, which is the same distance.

95. Let $t =$ number of hours the trainee works. Then $t - 2 =$ number of hours that the secretary works.
$$7t + 15(t - 2) = 124$$
$$7t + 15t - 30 = 124$$
$$22t - 30 = 124$$
$$\underline{\phantom{22t}\;+30 \quad +30}$$
$$22t \phantom{-30} = 154$$
$$\frac{22t}{22} = \frac{154}{22}$$
$$t = 7$$

Thus, the pile of forms will be finished 7 hours after 9:00 a.m., or at 4:00 p.m.

CHECK: The trainee processes $7(7) = 49$ forms. The secretary processes $15(7 - 2) = 15(5) = 75$ forms. The total number of forms is $49 + 75 = 124$.

97. Let $x =$ number of shares Margaret buys at a lower price.
Then $200 - x =$ number of shares Margaret buys at a higher price.
$$8.125x + 9.375(200 - x) = 1725$$
$$8.125x + 1875 - 9.375x = 1725$$
$$-1.25x + 1875 = 1725$$
$$\underline{\phantom{-1.25x}\;-1875 = -1875}$$
$$-1.25x \phantom{+1875} = -150$$
$$\frac{-1.25x}{-1.25} = \frac{-150}{-1.25}$$
$$x = 120$$

Then $200 - x = 200 - 120 = 80$. Thus Margaret buys 120 shares at the lower price and 80 shares at the higher price.

CHECK: 120 shares at $8.125 per share cost $120(\$8.125) = \$975$.
80 shares at $9.375 per share cost $80(\$9.375) = \$750$.
Total cost of all shares is $\$975 + \$750 = \$1725$.

99. Let $x =$ number of lighter boxes.
Then $x + 89 =$ number of heavier boxes.
$$6.58x + 9.32(x + 89) = 1974.28$$
$$6.58x + 9.32x + 829.48 = 1974.28$$
$$15.9x + 829.48 = 1974.28$$
$$\underline{\phantom{15.9x}\;-829.48 \quad -829.48}$$
$$15.9x \phantom{+829.48} = 1144.80$$
$$\frac{15.9x}{15.9} = \frac{1144.80}{15.9}$$
$$x = 72$$

Then $x + 89 = 72 + 89 = 161$. Since there are 72 lighter boxes and 161 heavier boxes on the truck, there are $72 + 161 = 233$ boxes altogether.

CHECK: 72 boxes that weigh 6.58 kg each weigh 473.76 kg altogether; 161 boxes that weigh 9.32 kg each weigh 1500.52 kg altogether. The total weight of the boxes is $473.76 + 1500.52 = 1974.28$ kg.

## Exercises 3.4

1. $x + 4 < 3$
$-2 + 4 \overset{?}{<} 3$
$2 \overset{\checkmark}{<} 3$
Therefore, $-2$ satisfies the inequality.

3. $a - 2 > -1$
$-3 - 2 \overset{?}{>} -1$
$-5 \not> -1$
Therefore, $-3$ does not satisfy the inequality.

5. 
$$-y + 3 \le 5$$
$$-(-2) + 3 \overset{?}{\le} 5$$
$$2 + 3 \overset{?}{\le} 5$$
$$5 \overset{\checkmark}{\le} 5$$
Therefore, $-2$ satisfies the inequality.

7. 
$$-8 \le 2 - x$$
$$-8 \overset{?}{\le} 2 - 6$$
$$-8 \overset{\checkmark}{\le} -4$$
Therefore, 6 satisfies the inequality.

9. 
$$2z - 5 < -3$$
$$2(1) - 5 \overset{?}{<} -3$$
$$2 - 5 \overset{?}{<} -3$$
$$-3 \not< -3$$
Therefore, 1 does not satisfy the inequality.

11. 
$$12 < 5 + 2u$$
$$12 \overset{?}{<} 5 + 2(3)$$
$$12 \overset{?}{<} 5 + 6$$
$$12 \not< 11$$
Therefore, 3 does not satisfy the inequality.

13. 
$$7 - 4x < 8$$
$$7 - 4(-4) \overset{?}{<} 8$$
$$7 + 16 \overset{?}{<} 8$$
$$23 \not< 8$$
Therefore, $-4$ does not satisfy the inequality.

15. 
$$-2 < 8 - x < 3$$
$$-2 \overset{?}{<} 8 - 6 \overset{?}{<} 3$$
$$-2 \overset{\checkmark}{<} 2 \overset{\checkmark}{<} 3$$
Therefore, 6 satisfies the inequality.

17. 
$$6 + 2(a - 3) < 1$$
$$6 + 2[(-2) - 3] \overset{?}{<} 1$$
$$6 + 2(-5) \overset{?}{<} 1$$
$$6 - 10 \overset{?}{<} 1$$
$$-4 \overset{\checkmark}{<} 1$$
Therefore, $-2$ satisfies the inequality.

19. 
$$-12 < 9 - 5(x + 1) < -5$$
$$-12 \overset{?}{<} 9 - 5[(3) + 1] \overset{?}{<} -5$$
$$-12 \overset{?}{<} 9 - 5(4) \overset{?}{<} -5$$
$$-12 \overset{?}{<} 9 - 20 \overset{?}{<} -5$$
$$-12 \overset{\checkmark}{<} -11 \overset{\checkmark}{<} -5$$
Therefore, 3 satisfies the inequality.

21. 
$$\begin{array}{rcl} x - 3 &<& 2 \\ +3 & & +3 \\ \hline x &<& 5 \end{array}$$

23. 
$$\begin{array}{rcl} a + 7 &>& 4 \\ -7 & & -7 \\ \hline a &>& -3 \end{array}$$

25. 
$$\begin{array}{rcl} -4 &>& w + 2 \\ -2 & & -2 \\ \hline -6 &>& w \end{array}$$

27. 
$$\begin{array}{rcl} z + 3 &>& 0 \\ -3 & & -3 \\ \hline z &>& -3 \end{array}$$

29. 
$$3x \le 12$$
$$\frac{\cancel{3}x}{\cancel{3}} \le \frac{12}{3}$$
$$x \le 4$$

31. 
$$4y > -8$$
$$\frac{\cancel{4}y}{\cancel{4}} > \frac{-8}{4}$$
$$y > -2$$

33. $-3x < 6$

$$\dfrac{\cancel{-3}x}{\cancel{-3}} > \dfrac{6}{-3}$$

$$x > -2$$

35. $-3x < -6$

$$\dfrac{\cancel{-3}x}{\cancel{-3}} > \dfrac{-6}{-3}$$

$$x > 2$$

37. $7a > 0$

$$\dfrac{\cancel{7}a}{\cancel{7}} > \dfrac{0}{7}$$

$$a > 0$$

39. $-7a \geq 0$

$$\dfrac{\cancel{-7}a}{\cancel{-7}} \leq \dfrac{0}{-7}$$

$$a \leq 0$$

41. $\qquad -x < 3$

$$(-1)(-x) > (-1)3$$

$$x > -3$$

43. $\qquad -x < -3$

$$(-1)(-x) > (-1)(-3)$$

$$x > 3$$

45. $-5 < a - 4 \leq \ \ 2$

$$+4 \qquad +4 \quad \ +4$$

$$\overline{-1 < a \qquad \ \ \leq \ \ 6}$$

47. $1 \ \ \leq x + 3 \leq \ \ 5$

$$-3 \qquad \ \ -3 \quad \ -3$$

$$\overline{-2 \leq x \qquad \ \leq \ \ 2}$$

49. $-6 < 3y < \ 3$

$$\dfrac{-6}{3} < \dfrac{\cancel{3}y}{\cancel{3}} < \dfrac{3}{3}$$

$$-2 < \ \ y < 1$$

51. $0 \ \leq -2x < \ \ 2$

$$\dfrac{0}{-2} \geq \dfrac{\cancel{-2}x}{\cancel{-2}} > \dfrac{2}{-2}$$

$$0 \ \geq \ \ x \ \ > -1 \quad \text{or} \quad -1 < x \leq 0$$

53. $\qquad\qquad -5 < -x < -1$

$$(-1)(-5) > (-1)(-x) > (-1)(-1)$$

$$5 > x > 1 \quad \text{or} \quad 1 < x < 5$$

55. One number is less than a second number if the first number is located to the left of the second on the number line. (Equivalently, we could say that the second number is located to the right of the first on the number line.)

56. If we begin with an inequality, we may add the same quantity to both sides, subtract the same quantity from both sides, or multiply or divide both sides by the same positive quantity and get another inequality with the same symbol. If we multiply or divide both sides by the same negative quantity, we get another inequality with the reversed symbol.

57. When we multiply or divide both sides of an equality by the same nonzero quantity, it does not matter whether the quantity is positive or negative. That is not the case for inequalities (see Exercise 56).

58. $2 < x \leq 5$ means that $x$ is between 2 and 5 on the number line, possibly equal to 5, but not equal to 2. That is, $x$ is greater than 2 <u>and</u> less than or equal to 5.

59. $(2x^2)(3x^3) = (2 \cdot 3)(x^2 x^3) = 6x^5$

61. $4(x - 3y) + 2(3x - y) = 4x - 12y + 6x - 2y$
$$= 4x + 6x - 12y - 2y$$
$$= 10x - 14y$$

63. $x^2(x^3 - 4x) - (2x^5 - 4x^3) = x^5 - 4x^3 - 2x^5 + 4x^3$
$$= x^5 - 2x^5 - 4x^3 + 4x^3$$
$$= -x^5$$

## Exercises 3.5

1. $x + 5 < 3$
   $\phantom{x}\ \ -5 \quad -5$
   $\overline{x \qquad < -2}$

3. $a - 2 > -3$
   $\phantom{a}\ \ +2 \quad +2$
   $\overline{a \qquad > -1}$

5. $2y < 7$
   $\dfrac{\not{2}y}{\not{2}} < \dfrac{7}{2}$
   $y < \dfrac{7}{2}$

7. $2y > -7$
   $\dfrac{\not{2}y}{\not{2}} > \dfrac{-7}{2}$
   $y > -\dfrac{7}{2}$

9. $-2y < 7$
   $\dfrac{\not{-2}y}{\not{-2}} > \dfrac{7}{-2}$
   $y > -\dfrac{7}{2}$

11. $-2y > -7$
   $\dfrac{\not{-2}y}{\not{-2}} < \dfrac{-7}{-2}$
   $y < \dfrac{7}{2}$

13. $-x < 4$
   $\dfrac{-x}{-1} > \dfrac{4}{-1}$
   $x > -4$
   or
   $-x < 4$
   $(-1)(-x) > (-1)(4)$
   $x > -4$

15. $-1 > -y$
   $\dfrac{-1}{-1} < \dfrac{-y}{-1}$
   $1 < x$
   or
   $-1 > -y$
   $(-1)(-1) < (-1)(-y)$
   $1 < y$

17. $5x + 3 \le 8$
   $\phantom{5x}\ \ -3 \quad -3$
   $\overline{5x \qquad \le 5}$
   $\dfrac{\not{5}x}{\not{5}} \le \dfrac{5}{5}$
   $x \le 1$

19. $2x - 9 \ge 16$
   $\phantom{2x}\ \ +9 \quad +9$
   $\overline{2x \qquad \ge 25}$
   $\dfrac{\not{2}x}{\not{2}} \ge \dfrac{25}{2}$
   $x \ge \dfrac{25}{2}$

21. $2(z-3)+4 \geq -6$
$$2z - 6 + 4 \geq -6$$
$$2z - 2 \geq -6$$
$$\underline{\phantom{2z}+2 \quad +2}$$
$$2z \geq -4$$
$$\frac{\cancel{2}z}{\cancel{2}} \geq \frac{-4}{2}$$
$$z \geq -2$$

23. $3(x+4)+2(x-1) < \phantom{0}20$
$$3x + 12 + 2x - 2 < \phantom{0}20$$
$$5x + 10 < \phantom{0}20$$
$$\underline{\phantom{5x}-10 \quad -10}$$
$$5x < \phantom{0}10$$
$$\frac{\cancel{5}x}{\cancel{5}} < \frac{10}{5}$$
$$x < 2$$

25. $5(w+3)-7w \leq \phantom{0}7$
$$5w + 15 - 7w \leq \phantom{0}7$$
$$-2w + 15 \leq \phantom{0}7$$
$$\underline{\phantom{-2w}-15 \quad -15}$$
$$-2w \phantom{+15} \leq -8$$
$$\frac{\cancel{-2}w}{\cancel{-2}} \geq \frac{-8}{-2}$$
$$w \geq 4$$

27. $3(a+4)-4(a-1) < \phantom{0}10$
$$3a + 12 - 4a + 4 < \phantom{0}10$$
$$-a + 16 < \phantom{0}10$$
$$\underline{\phantom{-a}-16 \quad -16}$$
$$-a \phantom{+16} < -6$$
$$\frac{-a}{-1} > \frac{-6}{-1}$$
$$a > 6$$

29. $4(y-3)-(3y-12) \geq 2$
$$4y - 12 - 3y + 12 \geq 2$$
$$y \geq 2$$

31. $2(u+2)-2(u-1) < 5$
$$2u + 4 - 2u + 2 < 5$$
$$6 < 5$$

Contradiction

33. $4(x-2)-(4x-3) < 6$
$$4x - 8 - 4x + 3 < 6$$
$$-5 < 6$$

Identity

35. $x + 3 < 2x + 7$
$$\underline{-x \phantom{+3} \quad -x \phantom{+7}}$$
$$3 < \phantom{0}x + 7$$
$$\underline{-7 \phantom{<} \quad -7}$$
$$-4 < \phantom{0}x$$

37. $5t - 3 \geq \phantom{0}3t + 10$
$$\underline{-3t \phantom{-3} \quad -3t \phantom{+10}}$$
$$2t - 3 \geq \phantom{3t+} 10$$
$$\underline{+3 \phantom{-3} \quad +3 \phantom{10}}$$
$$2t \phantom{-3} \geq \phantom{3t+} 13$$
$$\frac{\cancel{2}t}{\cancel{2}} \geq \frac{13}{2}$$
$$t \geq \frac{13}{2}$$

39. $2(a-5)+3a > \phantom{0}6a - 6$
$$2a - 10 + 3a > \phantom{0}6a - 6$$
$$5a - 10 > \phantom{0}6a - 6$$
$$\underline{-5a \phantom{-10} \quad -5a \phantom{-6}}$$
$$-10 > \phantom{00}a - 6$$
$$\underline{+6 \phantom{-10} \quad +6 \phantom{a}}$$
$$-4 > \phantom{00}a$$

41. $4(w+2)-3(w-1) > 5(w-1)-5w$
$$4w + 8 - 3w + 3 > 5w - 5 - 5w$$
$$w + 11 > \phantom{0}-5$$
$$\underline{\phantom{w+}-11 \quad -11}$$
$$w \phantom{+11} > -16$$

43. $2y - 4(4y + 1 \leq 8 - (y + 2)$
$$2y - 4y - 4 \leq 6 - y - 2$$
$$-2y - 4 \leq \phantom{0}6 - y$$
$$\underline{+2y \phantom{-4} \quad +2y}$$
$$-4 \leq \phantom{0}6 + y$$
$$\underline{-6 \phantom{\leq} \quad -6 \phantom{+y}}$$
$$-10 \leq \phantom{06+} y$$

**45.**
$$2 < x + 7 < 10$$
$$\underline{-7 \qquad -7 \quad -7}$$
$$-5 < x \qquad < 3$$

**47.**
$$3 < 2a + 5 < 7$$
$$\underline{-5 \qquad -5 \quad -5}$$
$$-2 < 2a \qquad < 2$$
$$\dfrac{-2}{2} < \dfrac{2a}{2} \quad < \dfrac{2}{2}$$
$$-1 < a \qquad < 1$$

**49.**
$$-5 \le -4y + 3 < 7$$
$$\underline{-3 \qquad\quad -3 \quad -3}$$
$$-8 \le -4y \qquad < 4$$
$$\dfrac{-8}{-4} \ge \dfrac{-4y}{-4} \qquad > \dfrac{4}{-4}$$
$$2 \ge y \qquad > -1$$
$$\text{or}$$
$$-1 < y \qquad \le 2$$

**51.**
$$1 \le 6 - x < 3$$
$$\underline{-6 \quad -6 \qquad -6}$$
$$-5 \le \qquad -x < -3$$
$$\dfrac{-5}{-1} \ge \dfrac{-x}{-1} > \dfrac{-3}{-1}$$
$$5 \ge \qquad x > 3$$
$$\text{or}$$
$$3 < \qquad x \le 5$$

**53.**
$$x + 4 < 2x - 1$$
$$\underline{-x \qquad -x}$$
$$4 < x - 1$$
$$\underline{+1 \qquad +1}$$
$$5 < x$$

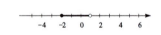

**55.**
$$3(a + 2) \qquad - 5a \ge 2 - a$$
$$3a + 6 - 5a \ge 2 - a$$
$$-2a + 6 \ge 2 - a$$
$$\underline{+2a \qquad\qquad +2a}$$
$$6 \ge 2 + a$$
$$\underline{-2 \quad -2}$$
$$4 \ge a$$

**57.**
$$-1 < x + 3 < 2$$
$$\underline{-3 \qquad -3 \quad -3}$$
$$-4 < x \qquad < -1$$

**59.**
$$-1 \le y - 3 \le 2$$
$$\underline{+3 \qquad +3 \quad +3}$$
$$2 \le y \qquad \le 5$$

**61.**
$$-3 \le 4t + 5 < 9$$
$$\underline{-5 \qquad -5 \quad -5}$$
$$-8 \le 4t \qquad < 4$$
$$\dfrac{-8}{4} \le \dfrac{4t}{4} \qquad < \dfrac{4}{4}$$
$$-2 \le t \qquad < 1$$

**63.**
$$-5 < 3 - 2x \le 9$$
$$\underline{-3 \quad -3 \qquad -3}$$
$$-8 < \qquad -2x \le 6$$
$$\dfrac{-8}{-2} > \dfrac{-2x}{-2} \ge \dfrac{6}{-2}$$
$$4 > \qquad x \ge -3$$
$$\text{or}$$
$$-3 \le \qquad x < 4$$

**65.** $x \ge 3.2$

**67.** $6.32 < t < 11.51$

**69.** $7 > x$

**71.** $t - 10$

**73.** $x \ge 5$

**75.** $y \le 2$

77. Let $n =$ the number.

$$3n - 4 < 17$$
$$\underline{\phantom{3n} +4 \quad +4}$$
$$3n \quad\;\; < 21$$
$$\frac{\cancel{3}n}{\cancel{3}} < \frac{21}{3}$$
$$n < \;\; 7$$

All numbers less than 7 satisfy the condition.

79. Let $n =$ the number.

$$6n + 12 > \quad 3n$$
$$\underline{-6n \qquad\qquad -6n}$$
$$12 > -3n$$
$$\frac{12}{-3} < \frac{\cancel{-3}n}{\cancel{-3}}$$
$$-4 < n$$

The number must be larger than $-4$.

81. Let $l =$ the length.

$$2l + 2(8) \geq \quad 80$$
$$2l + 16 \geq \quad 80$$
$$\underline{\phantom{2l} -16 \quad -16}$$
$$2l \quad\;\; \geq \quad 64$$
$$\frac{\cancel{2}l}{\cancel{2}} \geq \frac{64}{2}$$
$$l \geq 32$$

The length must be at least 32 cm.

83. Let $w =$ the width.

$$50 \leq 2w + 2(18) \leq \quad 70$$
$$50 \leq \quad 2w + 36 \leq \quad 70$$
$$\underline{-36 \qquad\qquad\;\; -36 \quad -36}$$
$$14 \leq \quad 2w \qquad \leq \quad 34$$
$$\frac{14}{2} \leq \frac{\cancel{2}w}{\cancel{2}} \leq \frac{34}{2}$$
$$7 \leq \qquad w \qquad \leq \quad 17$$

The width is at least 7 in. and at most 17 in.

85. Let $d =$ the price of a ticket at the door.
Then $d + 2 =$ price of a reserved seat ticket.

$$150d + 300(d + 2) \geq 3750$$
$$150d + 300d + 600 \geq 3750$$
$$450d + 600 \geq 3750$$
$$\underline{\phantom{450d} -600 \quad -600}$$
$$450d \qquad \geq 3150$$
$$\frac{\cancel{450}d}{\cancel{450}} \geq \frac{3150}{450}$$
$$d \geq 7$$

So $d + 2 \geq 9$

The minimum price of a reserved seat is \$9.

87. Let $s =$ annual sales (in dollars).

$$0.016(s - 82,000) \geq \quad 1800$$
$$0.016s - 0.016(82,000) \geq \quad 1800$$
$$0.016s - 1312 \geq \quad 1800$$
$$\underline{\phantom{0.016s} +1312 \quad +1312}$$
$$0.016s \qquad \geq \quad 3112$$
$$\frac{\cancel{0.016}s}{\cancel{0.016}} \geq \frac{3112}{0.016}$$
$$s \geq 194,500$$

Marc's annual sales must be at least \$194,500.

89. (a) Starting with $2 \leq -2x$, we must divide both sides of this inequality by $-2$, obtaining
$$\frac{2}{-2} \geq \frac{\cancel{-2}x}{\cancel{-2}} \text{ or } -1 \geq x.$$

   (b) When we subtract 4 from both sides of the original inequality, the resulting inequality should read $2x < -2$, not $2x > -2$. (We only reverse the inequality sign when we multiply or divide by a negative quantity.)

   (c) We must divide both sides of the inequality $-9 > 3x$ by 3 giving $\dfrac{-9}{3} > \dfrac{\cancel{3}x}{\cancel{3}}$ or $-3 > x$.

90. Only part (a) makes sense, since it requires that $x$ be a number between $-3$ and 2. Parts (b), (c), (e) and (f) all lead to contradictions. For instance, the inequality $-5 < x < -8$ implies that $-5 < -8$, which is false. (the others are similar.) Part (d) makes no sense, since we cannot write double inequalities in which the inequality symbols point in opposite directions.

## Chapter 3 Review Exercises

1. 
$$5(x-4) - 3(x-3) = 3 - (14-2x)$$
$$5x - 20 - 3x + 9 = 3 - 14 + 2x$$
$$2x - 11 = 2x - 11$$
$$\underline{-2x \qquad\qquad -2x}$$
$$-11 = -11$$

Identity

3. 
$$3a(a+3) - a(2a+4) = a^2 + 10$$
$$3a^2 + 9a - 2a^2 - 4a = a^2 + 10$$
$$a^2 + 5a = a^2 + 10$$
$$\underline{-a^2 \qquad\qquad -a^2}$$
$$5a = 10$$
$$\frac{\cancel{5}a}{\cancel{5}} = \frac{10}{5}$$
$$a = 2$$

Conditional equation

5. $2x - 5 = -7$
CHECK $x = -6$:
$$2(-6) - 5 \overset{?}{=} -7$$
$$-12 - 5 \overset{?}{=} -7$$
$$-17 \neq -7$$
So $-6$ does not satisfy the equation.

CHECK $x = -1$:
$$2(-1) - 5 \overset{?}{=} -7$$
$$-2 - 5 \overset{?}{=} -7$$
$$-7 \overset{\checkmark}{=} -7$$
So $-1$ satisfies the equation.

7. $4y + 3 \leq 10 + 2y$
CHECK $y = 4$:
$$4(4) + 3 \overset{?}{\leq} 10 + 2(4)$$
$$16 + 3 \overset{?}{\leq} 10 + 8$$
$$19 \not\leq 18$$
So 4 does not satisfy the inequality.

CHECK $y = \dfrac{5}{2}$:
$$4\left(\frac{5}{2}\right) + 3 \overset{?}{\leq} 10 + 2\left(\frac{5}{2}\right)$$
$$10 + 3 \overset{?}{\leq} 10 + 5$$
$$13 \overset{\checkmark}{\leq} 15$$
So $\dfrac{5}{2}$ satisfies the inequality.

9. $3t + 2(t + 1) = 3t + 3$
CHECK $t = -2$:
$$3(-2) + 2[(-2) + 1] \overset{?}{=} 3(-2) + 3$$
$$3(-2) + 2(-1) \overset{?}{=} 3(-2) + 3$$
$$-6 - 2 \overset{?}{=} -6 + 3$$
$$-8 \neq -3$$
So $-2$ does not satisfy the equation.

CHECK $t = \dfrac{1}{2}$:
$$3\left(\frac{1}{2}\right) + 2\left[\left(\frac{1}{2}\right) + 1\right] \overset{?}{=} 3\left(\frac{1}{2}\right) + 3$$
$$3\left(\frac{1}{2}\right) + 2\left(\frac{3}{2}\right) \overset{?}{=} 3\left(\frac{1}{2}\right) + 3$$
$$\frac{3}{2} + 3 \overset{?}{=} \frac{3}{2} + 3$$
$$\frac{9}{2} \overset{\checkmark}{=} \frac{9}{2}$$
So $\dfrac{1}{2}$ satisfies the equation.

11. $8 - 3(x - 2) > x - 4$

CHECK $x = -5$:

$$8 - 3[(-5) - 2] \overset{?}{>} (-5) - 4$$
$$8 - 3(-7) \overset{?}{>} -5 - 4$$
$$8 + 21 \overset{?}{>} -5 - 4$$
$$29 \overset{\checkmark}{>} -9$$

So $-5$ satisfies the inequality.

CHECK $x = 5$:

$$8 - 3[(5) - 2] \overset{?}{>} (5) - 4$$
$$8 - 3(3) \overset{?}{>} 5 - 4$$
$$8 - 9 \overset{?}{>} 5 - 4$$
$$-1 \not> 1$$

So 5 does not satisfy the inequality.

13. $a^2 + (a - 2)^2 = 20$

CHECK $a = -2$:

$$(-2)^2 + [(-2) - 2]^2 \overset{?}{=} 20$$
$$(-2)^2 + (-4)^2 \overset{?}{=} 20$$
$$4 + 16 \overset{?}{=} 20$$
$$20 \overset{\checkmark}{=} 20$$

So $-2$ satisfies the equation.

CHECK $a = 2$:

$$(2)^2 + [(2) - 2]^2 \overset{?}{=} 20$$
$$(2)^2 + (0)^2 \overset{?}{=} 20$$
$$4 + 0 \overset{?}{=} 20$$
$$4 \neq 20$$

So 2 does not satisfy the equation.

15.

$$
\begin{aligned}
5x + 8 &= 2x - 7 \\
-2x &\quad -2x \\
\hline
3x + 8 &= -7 \\
-8 &\quad -8 \\
\hline
3x &= -15 \\
\frac{\cancel{3}x}{\cancel{3}} &= \frac{-15}{3} \\
x &= -5
\end{aligned}
$$

17.

$$
\begin{aligned}
2(y + 4) - 2y &= 8 \\
2y + 8 - 2y &= 8 \\
8 &= 8
\end{aligned}
$$

Identity

19.

$$
\begin{aligned}
2(3a + 4) + 8 &= 5(3a - 1) \\
6a + 8 + 8 &= 15a - 5 \\
6a + 16 &= 15a - 5 \\
-6a &\quad -6a \\
\hline
16 &= 9a - 5 \\
+5 &\quad +5 \\
\hline
21 &= 9a \\
\frac{21}{9} &= \frac{\cancel{9}a}{\cancel{9}} \\
\frac{7}{3} &= a
\end{aligned}
$$

21.

$$
\begin{aligned}
8x - 3(x - 4) &= 4(x + 3) + 28 \\
8x - 3x + 12 &= 4x + 12 + 28 \\
5x + 12 &= 4x + 40 \\
-4x &\quad -4x \\
\hline
x + 12 &= 40 \\
-12 &\quad -12 \\
\hline
x &= 28
\end{aligned}
$$

**23.**
$$a(a+3) - 2(a-1) = a(a-1) + 7$$
$$a^2 + 3a - 2a + 2 = a^2 - a + 7$$
$$a^2 + a + 2 = a^2 - a + 7$$
$$\underline{-a^2 \qquad\qquad -a^2}$$
$$a + 2 = -a + 7$$
$$\underline{+a \qquad\qquad +a}$$
$$2a + 2 = 7$$
$$\underline{-2 \qquad\qquad -2}$$
$$2a = 5$$
$$\frac{\cancel{2}a}{\cancel{2}} = \frac{5}{2}$$
$$a = \frac{5}{2}$$

**25.**
$$8 - 3(x-1) < 2$$
$$8 - 3x + 3 < 2$$
$$-3x + 11 < 2$$
$$\underline{-11 \quad -11}$$
$$-3x < -9$$
$$\frac{\cancel{-3}x}{\cancel{-3}} > \frac{-9}{-3}$$
$$x > 3$$

**27.**
$$2(x-3) - 4(x-1) \geq 7 - x$$
$$2x - 6 - 4x + 4 \geq 7 - x$$
$$-2x - 2 \geq 7 - x$$
$$\underline{+2x \qquad\qquad +2x}$$
$$-2 \geq 7 + x$$
$$\underline{-7 \quad -7}$$
$$-9 \geq x$$

**29.**
$$2 \leq 3a + 8 < 20$$
$$\underline{-8 \qquad -8 \quad -8}$$
$$-6 \leq 3a < 12$$
$$\frac{-6}{3} \leq \frac{\cancel{3}a}{\cancel{3}} < \frac{12}{3}$$
$$-2 \leq a < 4$$

**31.** Let $x =$ one of the numbers.
Then $2x - 3 =$ the other number.
$$x + (2x - 3) = 18$$
$$3x - 3 = 18$$
$$\underline{+3 \quad +3}$$
$$3x = 21$$
$$\frac{\cancel{3}x}{\cancel{3}} = \frac{21}{3}$$
$$x = 7$$

Then $2x - 3 = 2(7) - 3 = 14 - 3 = 11$.
Thus, the numbers are 7 and 11.

CHECK: 11 is 3 less than twice 7.
The sum of 7 and 11 is 18.

**33.** Let $W =$ width of the rectangle.
Then $5W + 4 =$ length of the rectangle.
$$2W + 2(5W + 4) = 80$$
$$2W + 10W + 8 = 80$$
$$12W + 8 = 80$$
$$\underline{-8 \quad -8}$$
$$12W = 72$$
$$\frac{\cancel{12}W}{\cancel{12}} = \frac{72}{12}$$
$$W = 6$$

Then $5W + 4 = 5(6) + 4 = 30 + 4 = 34$.
Thus, the dimensions of the rectangle are
6 cm. and 34 cm.

CHECK: 34 is 4 more than 5 times 6.
$2(6) + 2(34) = 80$.

35. Let $x$ = number of first quality skirts bought.
Then $150 - x$ = number of irregular skirts bought.

$$12x + 7(150 - x) = 1500$$
$$12x + 1050 - 7x = 1500$$
$$5x + 1050 = 1500$$
$$\underline{-1050 \qquad -1050}$$
$$5x \quad = \quad 450$$
$$\frac{5x}{5} = \frac{450}{5}$$
$$x = 90$$

Then $150 - x = 150 - (90) = 60$. Thus, the wholesaler bought 90 first quality skirts and 60 irregular skirts.

CHECK: 90 first quality skirts cost $12(90) = \$1080$. 60 irregular skirts cost $7(60) = \$420$. The total cost of the skirts is $\$1080 + \$420 = \$1500$.

37. Let $x$ = number of overtime hours the laborer must work.

$$6(40) + 9x \geq 348$$
$$240 + 9x \geq 348$$
$$\underline{-240 \qquad\quad -240}$$
$$9x \geq 108$$
$$\frac{9x}{9} \geq \frac{108}{9}$$
$$x \geq 12$$

Thus, the minimum number of overtime hours that the laborer must work is 12.

## Chapter 3 Practice Test

1. (a) $3x - 5(x - 2) = -2x + 8$
$$3x - 5x + 10 = -2x + 8$$
$$-2x + 10 = -2x + 8$$
$$\underline{+2x \qquad\qquad +2x}$$
$$10 = \qquad 8$$

Contradiction

(b) $3x - 5(x - 2) = -2x + 10$
$$3x - 5x + 10 = -2x + 10$$
$$-2x + 10 = -2x + 10$$
$$\underline{+2x \qquad\qquad +2x}$$
$$10 = \qquad\quad 10$$

Identity

(c) $3x - 5(x - 2) = -2x - 10$
$$3x - 5x + 10 = -2x - 10$$
$$-2x + 10 = -2x - 10$$
$$\underline{+2x \qquad\qquad +2x}$$
$$10 = \quad 4x - 10$$
$$\underline{+10 \qquad\qquad +10}$$
$$20 = \quad 4x$$
$$\frac{20}{4} = \frac{4x}{4}$$
$$5 = x$$

Conditional

3. (a)
$$6 - 3x = 3x - 10$$
$$\underline{+3x \quad\quad +3x}$$
$$6 = 6x - 10$$
$$\underline{+10 \quad\quad\quad\quad +10}$$
$$16 = 6x$$
$$\frac{16}{6} = \frac{\cancel{6}x}{\cancel{6}}$$
$$\frac{8}{3} = x$$

(b)
$$2(3y - 5) - 4y = 2 - (y + 12)$$
$$6y - 10 - 4y = 2 - y - 12$$
$$2y - 10 = -y - 10$$
$$\underline{+y \quad\quad\quad +y}$$
$$3y - 10 = -10$$
$$\underline{+10 \quad\quad\quad +10}$$
$$3y = 0$$
$$\frac{\cancel{3}y}{\cancel{3}} = \frac{0}{3}$$
$$y = 0$$

(c)
$$2a^2 - 3(a - 4 = 2a(a - 6) + 3a$$
$$2a^2 - 3a + 12 = 2a^2 - 12a + 3a$$
$$2a^2 - 3a + 12 = 2a^2 - 9a$$
$$\underline{-2a^2 \quad\quad\quad -2a^2}$$
$$-3a + 12 = -9a$$
$$\underline{+3a \quad\quad\quad +3a}$$
$$12 = -6a$$
$$\frac{12}{-6} = \frac{\cancel{-6}a}{\cancel{-6}}$$
$$-2 = a$$

(d)
$$9 - 5(x - 2) \geq 4$$
$$9 - 5x + 10 \geq 4$$
$$-5x + 19 \geq 4$$
$$\underline{-19 \quad\quad -19}$$
$$-5x \geq -15$$
$$\frac{\cancel{-5}x}{\cancel{-5}} \leq \frac{-15}{-5}$$
$$x \leq 3$$

(e)
$$1 < 3 - x \leq 5$$
$$\underline{-3 \quad -3 \quad\quad -3}$$
$$-2 < -x \leq 2$$
$$\frac{-2}{-1} > \frac{-x}{-1} \geq \frac{2}{-1}$$
$$2 > x \geq -2$$
or
$$-2 \leq x < 2$$

5. Let $x$ = number of used cassettes.
Then $20 - x$ = number of new cassettes.
$$1x + 3(20 - x) = 46$$
$$x + 60 - 3x = 46$$
$$-2x + 60 = 46$$
$$\underline{-60 \quad -60}$$
$$-2x = -14$$
$$\frac{\cancel{-2}x}{\cancel{-2}} = \frac{-14}{-2}$$
$$x = 7$$

Then $20 - x = 20 - 7 = 13$. Thus the person bought 7 used cassettes and 13 new cassettes.

CHECK: $7 + 13 \overset{\checkmark}{=} 20$.
7 cassettes at $1 each cost $7.
13 cassettes at $3 each cost $13(3) = \$39$.
$\$7 + \$39 \overset{\checkmark}{=} \$46$

7.  Let $x$ = sales (in dollars).

$$0.0125(x - 10,000) \geq 1500$$
$$0.0125x - 125 \geq 1500$$
$$\underline{+125 \qquad +125}$$
$$0.0125x \qquad \geq 1625$$
$$\frac{0.0125x}{0.0125} \geq \frac{1625}{0.0125}$$
$$x \geq 130,000$$

Jan's sales must be at least $130,000.

# Chapters 1 - 3
# Cumulative Review

1. $-8 - 5 - 7 = -13 - 7 = -20$

3. $12 - 4(3 - 5) = 12 - 4(-2)$
$$= 12 + 8$$
$$= 20$$

5. $-5^2 = -(5 \cdot 5) = -25$

7. $xx^2x^3 = x^{1+2+3} = x^6$

9. $x^2y - 2xy^2 - xy^2 - 3x^2y = x^2y - 3x^2y - 2xy^2 - xy^2$
$$= (1 - 3)x^2y + (-2 - 1)xy^2$$
$$= -2x^2y - 3xy^2$$

11. $2x(3x^2 - 4y) = 2x(3x^2) - 2x(4y)$
$$= 6x^3 - 8xy$$

13. $-3u^2(u^3)(-5v) = [(-3)(-5)](u^2u^3)v$
$$= 15u^5v$$

15. $4(m - 3n) + 3(2m - n) = 4m - 12n + 6m - 3n$
$$= 4m + 6m - 12n - 3n$$
$$= 10m - 15n$$

17. $2ab(a^2 - ab) - 4a^2(ab - b^2) = 2a^3b - 2a^2b^2 - 4a^3b + 4a^2b^2$
$$= 2a^3b - 4a^3b - 2a^2b^2 + 4a^2b^2 = (2 - 4)a^3b + (-2 + 4)a^2b^2$$
$$= -2a^3b + 2a^2b^2$$

19. $x^2y - xy^2 - (xy^2 - x^2y) = x^2y - xy^2 - xy^2 + x^2y$
$$= x^2y + x^2y - xy^2 - xy^2 = (1 + 1)x^2y + (-1 - 1)xy^2$$
$$= 2x^2y - 2xy^2$$

21. $x - 3[x - 4(x - 5)] = x - 3[x - 4x + 20]$
$$= x - 3[-3x + 20] = x + 9x - 60$$
$$= 10x - 60$$

23. $3xy(4x^3y - 2y) - 2x(3y^2)(2x^3) = 12x^4y^2 - 6xy^2 - 12x^4y^2$
$$= 12x^4y^2 - 12x^4y^2 - 6xy^2 = (12 - 12)x^4y^2 - 6xy^2$$
$$= -6xy^2$$

25. $x^2 = (-2)^2 = 4$

27. $xy^2 - (xy)^2 = (-2)(-3)^2 - [(-2)(-3)]^2$
$$= -2(9) - (6)^2 = -18 - 36$$
$$= -54$$

29. $|x - y - z| = |(-2) - (-3) - (5)|$
$$= |-2 + 3 - 5| = |-4|$$
$$= 4$$

31. $2x - 4y^2 = 2(-2) - 4(-3)^2$
$$= 2(-2) - 4(9) = -4 - 36$$
$$= -40$$

33.
$$2x + 11 = 5x + 10$$
$$\underline{-2x \qquad\quad -2x}$$
$$11 = 3x + 10$$
$$\underline{-10 \qquad\quad -10}$$
$$1 = 3x$$
$$\frac{1}{3} = \frac{\cancel{3}x}{\cancel{3}}$$
$$\frac{1}{3} = x$$

35.
$$9(a+1) - 3(2a-2) = 12$$
$$9a + 9 - 6a + 6 = 12$$
$$3a + 15 = 12$$
$$\underline{-15 \qquad -15}$$
$$3a \quad = -3$$
$$\frac{\cancel{3}a}{\cancel{3}} = \frac{-3}{3}$$
$$a = -1$$

37.
$$4(5-x) - 2(6-2x) = 8$$
$$20 - 4x - 12 + 4x = 8$$
$$8 = 8$$

Identity

39.
$$2(s+4) + 3(2s+2) > 6s$$
$$2s + 8 + 6s + 6 > 6s$$
$$8s + 14 > 6s$$
$$\underline{-8s \qquad\qquad -8s}$$
$$14 > -2s$$
$$\frac{14}{-2} < \frac{\cancel{-}2s}{\cancel{-}2}$$
$$-7 < s$$

41.
$$1 < 2y - 5 \le 3$$
$$\underline{+5 \qquad\quad +5 \quad +5}$$
$$6 < 2y \qquad \le 8$$
$$\frac{6}{2} \le \frac{\cancel{2}y}{\cancel{2}} \qquad \le \frac{8}{2}$$
$$3 < y \qquad \le 4$$

43.
$$4(3d-2) + 6(8-d) = 9d - 4(d-10)$$
$$12d - 8 + 48 - 6d = 9d - 4d + 40$$
$$6d + 40 = 5d + 40$$
$$\underline{-5d \qquad\qquad -5d}$$
$$d + 40 = 40$$
$$\underline{-40 \qquad\qquad -40}$$
$$d = 0$$

45. Let $x$ = the number
$$2x - 8 \ge x + 4$$
$$\underline{-x \qquad\quad -x}$$
$$x - 8 \ge 4$$
$$\underline{+8 \qquad\quad +8}$$
$$x \ge 12$$
The number is at least 12.

47. Let $x$ = number of danishes that Louise bought.
$18 - x$ = number of pastries that Louise bought.
$$40x + 55(18 - x) = 825$$
$$40x + 990 - 55x = 825$$
$$-15x + 990 = 825$$
$$\underline{-990 \qquad -990}$$
$$-15x = -165$$
$$\frac{\cancel{-15}x}{\cancel{-15}} = \frac{-165}{-15}$$
$$x = 11$$

Then $18 - x = 18 - 11 = 7$.
Thus, Louise bought 11 danishes and 7 pastries.

CHECK: 11 danishes cost $11(\$0.40) = \$4.40$.
7 pastries cost $7(\$0.55) = \$3.85$. The total cost of the assortment is $\$4.40 + \$3.85 = \$8.25$

# Chapters 1 - 3
# Cumulative Practice Test

1.   (a)   $-3^4 + 3(-2)^3 = -(3 \cdot 3 \cdot 3 \cdot 3) + 3((-2)(-2)(-2))$
$$= -81 + 3(-8) = -81 - 24$$
$$= -105$$

    (b)   $(3 - 7 - 2)^2 = (-4 - 2)^2 = (-6)^2$
$$= (-6)(-6) = 36$$

3.   (a)   $5x^2 - 4x - 8 - 7x^2 - x + 11 = 5x^2 - 7x^2 - 4x - x - 8 + 11$
$$= -2x^2 - 5x + 3$$

    (b)   $2(x - 3y) + 5(y - x) = 2x - 6y + 5y - 5x = 2x - 5x - 6y + 5y$
$$= -3x - y$$

    (c)   $3a(a^2 - 2b) - 5(a^3 - ab) = 3a^3 - 6ab - 5a^3 + 5ab = 3a^3 - 5a^3 - 6ab + 5ab$
$$= -2a^3 - ab$$

    (d)   $2(u - 4v) - 3(v - 2u) - (8u - 11v) = 2u - 8v - 3v + 6u - 8u + 11v$
$$= 2u + 6u - 8u - 8v - 3v + 11v$$
$$= 0u + 0v = 0$$

    (e)   $4x^2y^3(x - 5y) - 2xy(5y^2)(-2xy) = 4x^3y^3 - 20x^2y^4 + 20x^2y^4$
$$= 4x^3y^3$$

    (f)   $6 - a[6 - a(6 - a)] = 6 - a[6 - 6a + a^2]$
$$= 6 - 6a + 6a^2 - a^3$$

5.   (a)

$$
\begin{array}{rcl}
7 - 3a & \geq & 13 \\
-7 & & -7 \\
\hline
-3a & \geq & 6 \\
\dfrac{-3a}{-3} & \leq & \dfrac{6}{-3} \\
a & \leq & -2
\end{array}
$$

    (b)

$$
\begin{array}{ccccc}
1 & \leq & 4x - 3 & < & 17 \\
+3 & & +3 & & +3 \\
\hline
4 & \leq & 4x & < & 20 \\
\dfrac{4}{4} & \leq & \dfrac{4x}{4} & < & \dfrac{20}{4} \\
1 & \leq & x & < & 5
\end{array}
$$

# Chapter 4
# Rational Expressions

Exercises 4.1

1. $\dfrac{18}{30} = \dfrac{3 \cdot \cancel{6}}{5 \cdot \cancel{6}} = \dfrac{3}{5}$

3. $\dfrac{-9}{21} = \dfrac{-3 \cdot \cancel{3}}{7 \cdot \cancel{3}} = \dfrac{-3}{7} = -\dfrac{3}{7}$

5. $\dfrac{-15}{-6} = \dfrac{5(\cancel{-3})}{2(\cancel{-3})} = \dfrac{5}{2}$

7. $\dfrac{-5+8}{10-4} = \dfrac{3}{6} = \dfrac{1 \cdot \cancel{3}}{2 \cdot \cancel{3}} = \dfrac{1}{2}$

9. $\dfrac{8-5(2)}{6-4(3)} = \dfrac{8-10}{6-12} = \dfrac{-2}{-6}$
$= \dfrac{1(\cancel{-2})}{3(\cancel{-2})} = \dfrac{1}{3}$

11. $\dfrac{x^3}{x} = \dfrac{x^2 \cdot \cancel{x}}{1 \cdot \cancel{x}} = \dfrac{x^2}{1} = x^2$

13. $\dfrac{x}{x^3} = \dfrac{1 \cdot \cancel{x}}{x^2 \cdot \cancel{x}} = \dfrac{1}{x^2}$

15. $\dfrac{10x}{4x^2} = \dfrac{5 \cdot \cancel{2x}}{2x \cdot \cancel{2x}} = \dfrac{5}{2x}$

17. $\dfrac{-3z^6}{5z^2} = \dfrac{-3z^4 \cdot \cancel{z^2}}{5 \cdot \cancel{z^2}} = \dfrac{-3z^4}{5} = -\dfrac{3z^4}{5}$

19. $\dfrac{12t^5}{30t^{10}} = \dfrac{2 \cdot \cancel{6t^5}}{5t^5 \cdot \cancel{6t^5}} = \dfrac{2}{5t^5}$

21. $\dfrac{6ab^5}{-2a^3b^2} = \dfrac{3b^3 \cdot \cancel{2ab^2}}{-a^2 \cdot \cancel{2ab^2}} = \dfrac{3b^3}{-a^2} = -\dfrac{3b^3}{a^2}$

23. $\dfrac{(2x^3)(6x^2)}{(4x)(3x^4)} = \dfrac{12x^5}{12x^5} = \dfrac{1 \cdot \cancel{12x^5}}{1 \cdot \cancel{12x^5}} = \dfrac{1}{1} = 1$

25. $\dfrac{(r^3t^2)(-rt^3)}{2r^2t^7} = \dfrac{-r^4t^5}{2r^2t^7} = \dfrac{-r^2 \cdot \cancel{r^2t^5}}{2t^2 \cdot \cancel{r^2t^5}}$
$= \dfrac{-r^2}{2t^2} = -\dfrac{r^2}{2t^2}$

27. $\dfrac{3a(5b)(-4ab^3)}{6ab(2a^2b^2)} = \dfrac{-60a^2b^4}{12a^3b^3}$
$= \dfrac{-5b \cdot \cancel{12a^2b^3}}{a \cdot \cancel{12a^2b^3}}$
$= \dfrac{-5b}{a} = -\dfrac{5b}{a}$

29. $\dfrac{(2x)^5}{(4x)^3} = \dfrac{32x^5}{64x^3} = \dfrac{x^2 \cdot \cancel{32x^3}}{2 \cdot \cancel{32x^3}} = \dfrac{x^2}{2}$

31. $\dfrac{(-4x^3)^2}{(-2x^4)^3} = \dfrac{16x^6}{-8x^{12}} = \dfrac{2 \cdot \cancel{8x^6}}{-x^6 \cdot \cancel{8x^6}}$
$= \dfrac{2}{-x^6} = -\dfrac{2}{x^6}$

33. $\dfrac{(2xy^2)^3}{(4x^2y^3)^3} = \dfrac{8x^3y^6}{64x^6y^9} = \dfrac{1 \cdot \cancel{8x^3y^6}}{8x^3y^3 \cdot \cancel{8x^3y^6}}$
$= \dfrac{1}{8x^3y^3}$

35. $\dfrac{5x-2x}{10x-4x} = \dfrac{3x}{6x} = \dfrac{1 \cdot \cancel{3x}}{2 \cdot \cancel{3x}} = \dfrac{1}{2}$

37. $\dfrac{5a(2x)}{15a(8x)} = \dfrac{10ax}{120ax} = \dfrac{1 \cdot \cancel{10ax}}{12 \cdot \cancel{10ax}} = \dfrac{1}{12}$

39. $\dfrac{4s-3t}{8s-9t}$ cannot be reduced.

41. $\dfrac{7a^2 - 5a^2 - 6a^2}{4a - 8a} = \dfrac{-4a^2}{-4a} = \dfrac{a(\cancel{-4a})}{1(\cancel{-4a})}$

$\qquad\qquad = \dfrac{a}{1} = a$

43. $\dfrac{5x^2 - 3x - x^2 + 3x}{6x^2 - 5x - 2x^2 + 5x} = \dfrac{4x^2}{4x^2}$

$\qquad\qquad = \dfrac{1 \cdot \cancel{4x^2}}{1 \cdot \cancel{4x^2}} = \dfrac{1}{1} = 1$

45. The value of a fraction is unchanged when its numerator and denominator are both either multiplied or divided by the same nonzero quantity.

46. We would conclude that $2 = 4$. It was incorrect to write

$$\dfrac{4 + \cancel{2}}{1 + \cancel{2}} = \dfrac{4}{1}.$$

The 2's cannot be crossed out, since 2 is a common term, not a common factor.

47. (a) The factor of $x$ in the reduced fraction should be in its numerator.

(b) When both factors in the numerator are crossed out, a factor of 1 remains.

(c) As in part (b), a factor of 1 remains in the numerator, so that the reduced fraction is $\dfrac{1}{3x^2y}$.

(d) Crossing out $x$'s amounts to crossing out terms rather than factors.

48. The fraction in (a), (c), (d), and (e) are equivalent. So are the fractions in (b), (f), and (g).

49. Let $w = $ width of the rectangle. Then $2w + 5 = $ length of the rectangle.

$$2w + 2(2w + 5) = 34$$
$$2w + 4w + 10 = 34$$
$$6w + 10 = 34$$
$$\underline{\quad -10 \quad -10\quad}$$
$$6w = 24$$
$$\dfrac{\cancel{6}w}{\cancel{6}} = \dfrac{24}{6}$$
$$w = 4$$

Then $2w + 5 = 2(4) + 5 = 13$. Thus, the rectangle has a width of 4 cm and a length of 13 cm.

## Exercises 4.2

1. $\dfrac{-4}{9} \cdot \dfrac{-2}{3} = \dfrac{(-4)(-2)}{9 \cdot 3} = \dfrac{8}{27}$

3. $\dfrac{\overset{-2}{\cancel{6}}}{\underset{2}{10}} \cdot \dfrac{\overset{3}{\cancel{15}}}{\underset{3}{\cancel{9}}} = \dfrac{\overset{-1}{\cancel{2}}}{\underset{1}{\cancel{2}}} \cdot \dfrac{\overset{1}{\cancel{3}}}{\underset{1}{\cancel{3}}} = \dfrac{-1}{1} = -1$

5. $\dfrac{2}{3y} \cdot \dfrac{x}{5} = \dfrac{2x}{3y \cdot 5} = \dfrac{2x}{15y}$

7. $\dfrac{x^2}{4y} \cdot \dfrac{5x}{3y} = \dfrac{x^2 \cdot 5x}{4y \cdot 3y} = \dfrac{5x^3}{12y^2}$

9. $\dfrac{4}{5} \div \dfrac{5}{4} = \dfrac{4}{5} \cdot \dfrac{4}{5} = \dfrac{4 \cdot 4}{5 \cdot 5} = \dfrac{16}{25}$

11. $\dfrac{6x}{y} \div \dfrac{y^2}{2x^2} = \dfrac{6x}{y} \cdot \dfrac{2x^2}{y^2} = \dfrac{6x \cdot 2x^2}{y \cdot y^2} = \dfrac{12x^3}{y^3}$

13. $\dfrac{\overset{1}{\cancel{3}}}{2\cancel{t}} \cdot \dfrac{\cancel{t}w}{\underset{2}{\cancel{6}}} = \dfrac{1 \cdot w}{2 \cdot 2} = \dfrac{w}{4}$

15. $4 \cdot \dfrac{x}{12} = \dfrac{\overset{1}{\cancel{4}}}{1} \cdot \dfrac{x}{\underset{3}{\cancel{12}}} = \dfrac{1 \cdot x}{1 \cdot 3} = \dfrac{x}{3}$

**17.** $4 \div \dfrac{x}{12} = \dfrac{4}{1} \div \dfrac{x}{12} = \dfrac{4}{1} \cdot \dfrac{12}{x}$

$\qquad = \dfrac{4 \cdot 12}{1 \cdot x} = \dfrac{48}{x}$

**19.** $\dfrac{x}{12} \div 4 = \dfrac{x}{12} \div \dfrac{4}{1} = \dfrac{x}{12} \cdot \dfrac{1}{4}$

$\qquad = \dfrac{x \cdot 1}{12 \cdot 4} = \dfrac{x}{48}$

**21.** $\dfrac{\overset{-1}{\cancel{2x}}}{\underset{y}{3y^2}} \cdot \dfrac{\overset{-3}{\cancel{9y}}}{\underset{2}{4\cancel{x}}} = \dfrac{(-1)(-3)}{2y} = \dfrac{3}{2y}$

**23.** $\dfrac{\overset{m^2}{\cancel{m^3}}n^2}{\underset{1}{2\cancel{m}}} \cdot \dfrac{\overset{3}{\cancel{6}}}{\underset{n}{n^3}} = \dfrac{m^2 \cdot 3}{1 \cdot n} = \dfrac{3m^2}{n}$

**25.** $\dfrac{3uv^2}{5w} \div \dfrac{6u^2v}{15w} = \dfrac{\overset{1v}{\cancel{3uv^2}}}{\underset{1}{5\cancel{w}}} \cdot \dfrac{\overset{3}{\cancel{15w}}}{\underset{2u}{\cancel{6u^2v}}} :$

$\qquad = \dfrac{v \cdot 3}{2 \cdot u} = \dfrac{3v}{2u}$

**27.** $6xy \cdot \dfrac{2x}{3y} = \dfrac{\overset{2}{\cancel{6xy}}}{1} \cdot \dfrac{2x}{\underset{1}{\cancel{3y}}} = \dfrac{2x \cdot 2x}{1 \cdot 1}$

$\qquad = \dfrac{4x^2}{1} = 4x^2$

**29.** $6xy \div \dfrac{2x}{3y} = \dfrac{\overset{3}{\cancel{6xy}}}{1} \cdot \dfrac{3y}{\underset{1}{\cancel{2x}}} = \dfrac{3y \cdot 3y}{1 \cdot 1}$

$\qquad = \dfrac{9y^2}{1} = 9y^2$

**31.** $\dfrac{2x}{3y} \div (6xy) = \dfrac{2x}{3y} \div \dfrac{6xy}{1} = \dfrac{\overset{1}{\cancel{2x}}}{3y} \cdot \dfrac{1}{\underset{3}{\cancel{6xy}}}$

$\qquad = \dfrac{1 \cdot 1}{3y \cdot 3y} = \dfrac{1}{9y^2}$

**33.** $\dfrac{\overset{-2}{\cancel{4x}}}{\underset{3}{9y}} \cdot \dfrac{x^2}{y^2} \cdot \dfrac{\overset{1}{\cancel{3y}}}{\underset{1}{2\cancel{x}}} = \dfrac{-2 \cdot x^2}{3 \cdot y^2}$

$\qquad = -\dfrac{2x^2}{3y^2}$

**35.** $\dfrac{9}{a^2} \cdot \left(\dfrac{a}{3} \div \dfrac{3}{a}\right) = \dfrac{9}{a^2} \cdot \left(\dfrac{a}{3} \cdot \dfrac{a}{3}\right)$

$\qquad = \dfrac{\overset{1}{\cancel{9}}}{\underset{1}{a^2}} \cdot \dfrac{\overset{1}{a^2}}{\underset{1}{\cancel{9}}} = \dfrac{1 \cdot 1}{1 \cdot 1} = \dfrac{1}{1} = 1$

**37.** $\dfrac{9}{a^2} \div \left(\dfrac{\overset{1}{\cancel{a}}}{\cancel{3}} \cdot \dfrac{\overset{1}{\cancel{3}}}{\cancel{a}}\right) = \dfrac{9}{a^2} \div \dfrac{1}{1} = \dfrac{9}{a^2} \cdot \dfrac{1}{1}$

$\qquad = \dfrac{9 \cdot 1}{a^2 \cdot 1} = \dfrac{9}{a^2}$

**39.** $\dfrac{9}{a^2} \div \left(\dfrac{a}{3} \div \dfrac{3}{a}\right) = \dfrac{9}{a^2} \div \left(\dfrac{a}{3} \cdot \dfrac{a}{3}\right)$

$\qquad = \dfrac{9}{a^2} \div \dfrac{a^2}{9} = \dfrac{9}{a^2} \cdot \dfrac{9}{a^2}$

$\qquad = \dfrac{9 \cdot 9}{a^2 \cdot a^2} = \dfrac{81}{a^4}$

**41.** $\left(\dfrac{9}{a^2} \div \dfrac{a}{3}\right) \div \dfrac{a}{3} = \left(\dfrac{9}{a^2} \cdot \dfrac{3}{a}\right) \div \dfrac{a}{3}$

$\qquad = \dfrac{27}{a^3} \div \dfrac{a}{3} = \dfrac{27}{a^3} \cdot \dfrac{3}{a}$

$\qquad = \dfrac{27 \cdot 3}{a^3 \cdot a} = \dfrac{81}{a^4}$

**43.** $\dfrac{\dfrac{2x}{3}}{\dfrac{10x}{9}} = \dfrac{2x}{3} \div \dfrac{10x}{9} = \dfrac{\overset{1}{\cancel{2x}}}{\underset{1}{\cancel{3}}} \cdot \dfrac{\overset{3}{\cancel{9}}}{\underset{5}{\cancel{10x}}}$

$\qquad = \dfrac{1 \cdot 3}{1 \cdot 5} = \dfrac{3}{5}$

**45.** $\dfrac{\dfrac{x^2}{3}}{\dfrac{x}{6}} = \dfrac{x^2}{3} \div \dfrac{x}{6} = \dfrac{\overset{x}{\cancel{x^2}}}{\underset{1}{\cancel{3}}} \cdot \dfrac{\overset{2}{\cancel{6}}}{\underset{1}{\cancel{x}}}$

$\qquad = \dfrac{x \cdot 2}{1 \cdot 1} = \dfrac{2x}{1} = 2x$

**47.** $\dfrac{\dfrac{x}{y^2}}{\dfrac{y}{x^2}} = \dfrac{x}{y^2} \div \dfrac{y}{x^2} = \dfrac{x}{y^2} \cdot \dfrac{x^2}{y}$

$\qquad = \dfrac{x \cdot x^2}{y^2 \cdot y} = \dfrac{x^3}{y^3}$

**49.** $\dfrac{\dfrac{2u}{z^2}}{\dfrac{4z}{u}} = \dfrac{2u}{z^2} \div \dfrac{4z}{u} = \dfrac{\overset{1}{\cancel{2}}u}{z^2} \cdot \dfrac{u}{\underset{2}{\cancel{4}}z}$

$\qquad = \dfrac{u \cdot u}{z^2 \cdot 2z} = \dfrac{u^2}{2z^3}$

**51.** $\dfrac{3x^2 - x^2}{4y^2 - y^2} \cdot \dfrac{2y + y}{x^2 + x^2} = \dfrac{\overset{1}{\cancel{2x^2}}}{\underset{y}{\cancel{3y^2}}} \cdot \dfrac{\overset{1}{\cancel{3y}}}{\underset{1}{\cancel{2x^2}}}$

$\qquad = \dfrac{1 \cdot 1}{y \cdot 1} = \dfrac{1}{y}$

**53.** $\dfrac{3x^2 \cdot x^2}{4y^2 \cdot y^2} \cdot \dfrac{2y \cdot y}{x^2 \cdot x^2} = \dfrac{3x^4}{\underset{2y^2}{\cancel{4y^4}}} \cdot \dfrac{\overset{1}{\cancel{2y^2}}}{\underset{1}{\cancel{x^4}}}$

$\qquad = \dfrac{3}{2y^2}$

**55.** Let $x = $ number of miles.

$x = 100 \div \dfrac{8}{5} = 100 \cdot \dfrac{5}{8}$

$\quad = \dfrac{\overset{25}{\cancel{100}}}{1} \cdot \dfrac{5}{\underset{2}{\cancel{8}}} = \dfrac{25 \cdot 5}{1 \cdot 2}$

$\quad = \dfrac{125}{2}$ or $62.5$

So 100 kilometers is equivalent to 62.5 miles.

**57.** Let $x = $ part of the cake each person receives.

$x = \dfrac{3}{8} \div 4 = \dfrac{3}{8} \cdot \dfrac{1}{4}$

$\quad = \dfrac{3 \cdot 1}{8 \cdot 4} = \dfrac{3}{32}$

Each person will receive $\dfrac{3}{32}$ of the cake.

**59.** 0.48      **61.** 0.45      **63.** 0.556

**65.** To divide by a fraction, multiply by its reciprocal. This rule works because division is defined to be the inverse of multiplication.

**66.** (a) The fraction $\dfrac{2y}{3x}$ must be inverted, leading to $\dfrac{3x}{2y} \cdot \dfrac{3x}{2y} = \dfrac{9x^2}{4y^2}$.

(b) The factor 5 is $\dfrac{5}{1}$, not $\dfrac{5}{5}$. Thus, $5 \cdot \dfrac{3x}{2} = \dfrac{5}{1} \cdot \dfrac{3x}{2} = \dfrac{5 \cdot 3x}{1 \cdot 2} = \dfrac{15x}{2}$.

**67.** A collection of dimes and quarters has a total value of $2.75. If there are three more dimes than quarters, how many of each type of coin are in the collection?

Let $x = $ the number of quarters.

69. Let $x$ = smallest angle. Then $2x$ = largest angle and $x + 20$ = third angle.

$$x + 2x + (x + 20) = 180$$
$$4x + 20 = 180$$
$$\underline{\phantom{4x}-20 \quad -20}$$
$$4x = 160$$
$$\frac{\cancel{4}x}{\cancel{4}} = \frac{160}{4}$$
$$x = 40$$

Then $2x = 2(40) = 80$ and $x + 20 = 40 + 20 = 60$. So the three angles are 40°, 60°, and 80°.

## Exercises 4.3

1. $\dfrac{5}{3} + \dfrac{4}{3} = \dfrac{5+4}{3} = \dfrac{\overset{3}{\cancel{9}}}{\underset{1}{\cancel{3}}} = \dfrac{3}{1} = 3$

3. $\dfrac{3}{5} - \dfrac{7}{5} = \dfrac{3-7}{5} = \dfrac{-4}{5} = -\dfrac{4}{5}$

5. $\dfrac{7}{9} - \dfrac{5}{9} - \dfrac{8}{9} = \dfrac{7-5-8}{9} = \dfrac{\overset{-2}{\cancel{-6}}}{\underset{3}{\cancel{9}}} = -\dfrac{2}{3}$

7. $\dfrac{2}{3} + \dfrac{4}{5} = \dfrac{2(5)}{3(5)} + \dfrac{4(3)}{5(3)} = \dfrac{10}{15} + \dfrac{12}{15}$

   $\qquad = \dfrac{10+12}{15} = \dfrac{22}{15}$

9. $\dfrac{2}{3} \cdot \dfrac{4}{5} = \dfrac{2 \cdot 4}{3 \cdot 5} = \dfrac{8}{15}$

11. $\dfrac{2}{3} - \dfrac{5}{6} = \dfrac{2(2)}{3(2)} - \dfrac{5}{6} = \dfrac{4}{6} - \dfrac{5}{6}$

    $\qquad = \dfrac{4-5}{6} = \dfrac{-1}{6} = -\dfrac{1}{6}$

13. $3 + \dfrac{3}{4} - \dfrac{3}{8} = \dfrac{3}{1} + \dfrac{3}{4} - \dfrac{3}{8}$

    $\qquad = \dfrac{3(8)}{1(8)} + \dfrac{3(2)}{4(2)} - \dfrac{3}{8}$

    $\qquad = \dfrac{24}{8} + \dfrac{6}{8} - \dfrac{3}{8}$

    $\qquad = \dfrac{24+6-3}{8} = \dfrac{27}{8}$

15. $\dfrac{3}{2} - \dfrac{4}{5} + \dfrac{7}{10} = \dfrac{3(5)}{2(5)} - \dfrac{4(2)}{5(2)} + \dfrac{7}{10}$

    $\qquad = \dfrac{15}{10} - \dfrac{8}{10} + \dfrac{7}{10}$

    $\qquad = \dfrac{15-8+7}{10} = \dfrac{\overset{7}{\cancel{14}}}{\underset{5}{\cancel{10}}} = \dfrac{7}{5}$

17. $\dfrac{5}{6} - \dfrac{3}{8} + \dfrac{1}{4} = \dfrac{5(4)}{6(4)} - \dfrac{3(3)}{8(3)} + \dfrac{1(6)}{4(6)}$

    $\qquad = \dfrac{20}{24} - \dfrac{9}{24} + \dfrac{6}{24}$

    $\qquad = \dfrac{20-9+6}{24} = \dfrac{17}{24}$

19. $2 + \dfrac{1}{3} - \dfrac{1}{2} = \dfrac{2}{1} + \dfrac{1}{3} - \dfrac{1}{2}$

    $\qquad = \dfrac{2(6)}{1(6)} + \dfrac{1(2)}{3(2)} - \dfrac{1(3)}{2(3)}$

    $\qquad = \dfrac{12}{6} + \dfrac{2}{6} - \dfrac{3}{6}$

    $\qquad = \dfrac{12+2-3}{6} = \dfrac{11}{6}$

21. $\dfrac{8}{3x} + \dfrac{4}{3x} = \dfrac{8+4}{3x} = \dfrac{\overset{4}{\cancel{12}}}{\underset{1}{\cancel{3x}}} = \dfrac{4}{x}$

23. $\dfrac{8}{3x} \cdot \dfrac{4}{3x} = \dfrac{8 \cdot 4}{3x \cdot 3x} = \dfrac{32}{9x^2}$

25. $\dfrac{7}{6x} + \dfrac{13}{6x} - \dfrac{11}{6x} = \dfrac{7+13-11}{6x}$

$= \dfrac{\overset{3}{\cancel{9}}}{\underset{2}{\cancel{6}x}} = \dfrac{3}{2x}$

27. $\dfrac{3y}{7x} - \dfrac{5y}{7x} + \dfrac{4y}{7x} = \dfrac{3y - 5y + 4y}{7x} = \dfrac{2y}{7x}$

29. $\dfrac{w}{9z} - \dfrac{5w}{9z} + \dfrac{4w}{9z} = \dfrac{w - 5w + 4w}{9z}$

$= \dfrac{0}{9z} = 0$

31. $\dfrac{-5a}{7b} + \dfrac{3a}{7b} - \dfrac{12a}{7b} = \dfrac{-5a + 3a - 12a}{7b}$

$= \dfrac{\overset{-2}{\cancel{-14}a}}{\underset{1}{\cancel{7}b}}$

$= \dfrac{-2a}{b} = -\dfrac{2a}{b}$

33. $\dfrac{x+3}{3x} + \dfrac{x-6}{3x} = \dfrac{x+3+x-6}{3x}$

$= \dfrac{2x-3}{3x}$

35. $\dfrac{3y^2 - 5}{4y} + \dfrac{5 - 4y^2}{4y} = \dfrac{3y^2 - 5 + 5 - 4y^2}{4y}$

$= \dfrac{\overset{-y}{\cancel{-y^2}}}{\cancel{4y}} = \dfrac{-y}{4} = -\dfrac{y}{4}$

37. $\dfrac{5x+2}{10x} - \dfrac{x+2}{10x} = \dfrac{5x+2-(x+2)}{10x}$

$= \dfrac{5x+2-x-2}{10x}$

$= \dfrac{\overset{2}{\cancel{4x}}}{\underset{5}{\cancel{10x}}} = \dfrac{2}{5}$

39. $\dfrac{2a-1}{3a} - \dfrac{5a-1}{3a} = \dfrac{2a-1-(5a-1)}{3a}$

$= \dfrac{2a-1-5a+1}{3a}$

$= \dfrac{\overset{-1}{\cancel{-3a}}}{\underset{1}{\cancel{3a}}} = \dfrac{-1}{1} = -1$

41. $\dfrac{w-4}{6w} - \dfrac{w-3}{6w} + \dfrac{5}{6w} = \dfrac{w-4-(w-3)+5}{6w} = \dfrac{w-4-w+3+5}{6w}$

$= \dfrac{\overset{2}{\cancel{4}}}{\underset{3}{\cancel{6}w}} = \dfrac{2}{3w}$

43. $\dfrac{t^2 - 3t + 2}{5t^2} - \dfrac{7t + t^2}{5t^2} = \dfrac{t^2 - 3t + 2 - (7t + t^2)}{5t^2}$

$= \dfrac{t^2 - 3t + 2 - 7t - t^2}{5t^2} = \dfrac{-10t + 2}{5t^2}$

45. $\dfrac{3}{x} + \dfrac{2}{y} = \dfrac{3(y)}{x(y)} + \dfrac{2(x)}{y(x)} = \dfrac{3y}{xy} + \dfrac{2x}{xy}$

$= \dfrac{3y + 2x}{xy}$

47. $\dfrac{3}{x} \cdot \dfrac{2}{y} = \dfrac{3 \cdot 2}{x \cdot y} = \dfrac{6}{xy}$

**49.** $\dfrac{5}{3x} - \dfrac{7}{2} = \dfrac{5(2)}{3x(2)} - \dfrac{7(3x)}{2(3x)} = \dfrac{10}{6x} - \dfrac{21x}{6x}$

$$= \dfrac{10 - 21x}{6x}$$

**51.** $\dfrac{5}{4x} + \dfrac{3}{2y} = \dfrac{5(y)}{4x(y)} + \dfrac{3(2x)}{2y(2x)}$

$$= \dfrac{5y}{4xy} + \dfrac{6x}{4xy} = \dfrac{5y + 6x}{4xy}$$

**53.** $\dfrac{4}{x^2} - \dfrac{3}{2x} = \dfrac{4(2)}{x^2(2)} - \dfrac{3(x)}{2x(x)}$

$$= \dfrac{8}{2x^2} - \dfrac{3x}{2x^2} = \dfrac{8 - 3x}{2x^2}$$

**55.** $\dfrac{\overset{2}{\cancel{4}}}{x^2} \cdot \dfrac{3}{\cancel{2}x} = \dfrac{2 \cdot 3}{x^2 \cdot x} = \dfrac{6}{x^3}$

**57.** $\dfrac{2}{3x^2} + \dfrac{3}{2x^2} = \dfrac{2(2)}{3x^2(2)} + \dfrac{3(3)}{2x^2(3)}$

$$= \dfrac{4}{6x^2} + \dfrac{9}{6x^2} = \dfrac{4 + 9}{6x^2} = \dfrac{13}{6x^2}$$

**59.** $\dfrac{7}{4a^2} - \dfrac{9}{20a} = \dfrac{7(5)}{4a^2(5)} - \dfrac{9(a)}{20a(a)}$

$$= \dfrac{35}{20a^2} - \dfrac{9a}{20a^2} = \dfrac{35 - 9a}{20a^2}$$

**61.** $\dfrac{1}{x} + 2 = \dfrac{1}{x} + \dfrac{2}{1} = \dfrac{1}{x} + \dfrac{2(x)}{1(x)}$

$$= \dfrac{1}{x} + \dfrac{2x}{x} = \dfrac{1 + 2x}{x}$$

**63.** $\dfrac{5}{3xy} + \dfrac{1}{6y^2} = \dfrac{5(2y)}{3xy(2y)} + \dfrac{1(x)}{6y^2(x)}$

$$= \dfrac{10y}{6xy^2} + \dfrac{x}{6xy^2} = \dfrac{10y + x}{6xy^2}$$

**65.** $\dfrac{7}{6a^2b} + \dfrac{3}{4ab^3} = \dfrac{7(2b^2)}{6a^2b(2b^2)} + \dfrac{3(3a)}{4ab^3(3a)}$

$$= \dfrac{14b^2}{12a^2b^3} + \dfrac{9a}{12a^2b^3}$$

$$= \dfrac{14b^2 + 9a}{12a^2b^3}$$

**67.** $\dfrac{7}{\underset{2}{\cancel{6}}a^2b} \cdot \dfrac{\overset{1}{\cancel{3}}}{4ab^3} = \dfrac{7 \cdot 1}{2a^2b \cdot 4ab^3} = \dfrac{7}{8a^3b^4}$

**69.** $\dfrac{4y}{3x^2} - \dfrac{3}{2x} + \dfrac{y}{x^2} = \dfrac{4y(2)}{3x^2(2)} - \dfrac{3(3x)}{2x(3x)} + \dfrac{y(6)}{x^2(6)}$

$$= \dfrac{8y}{6x^2} - \dfrac{9x}{6x^2} + \dfrac{6y}{6x^2} = \dfrac{14y - 9x}{6x^2}$$

**71.** $\dfrac{3}{4m^2n} - \dfrac{5}{6mn^3} + \dfrac{1}{8n^2} = \dfrac{3(6n^2)}{4m^2n(6n^2)} - \dfrac{5(4m)}{6mn^3(4m)} + \dfrac{1(3m^2n)}{8n^2(3m^2n)}$

$$= \dfrac{18n^2}{24m^2n^3} - \dfrac{20m}{24m^2n^3} + \dfrac{3m^2n}{24m^2n^3} = \dfrac{18n^2 - 20m + 3m^2n}{24m^2n^3}$$

**73.** $\dfrac{x}{y} + \dfrac{y}{x} + \dfrac{3x}{2y} = \dfrac{x(2x)}{y(2x)} + \dfrac{y(2y)}{x(2y)} + \dfrac{3x(x)}{2y(x)} = \dfrac{2x^2}{2xy} + \dfrac{2y^2}{2xy} + \dfrac{3x^2}{2xy}$

$$= \dfrac{2x^2 + 2y^2 + 3x^2}{2xy} = \dfrac{5x^2 + 2y^2}{2xy}$$

75. $\quad t - \dfrac{3}{t} = \dfrac{t}{1} - \dfrac{3}{t} = \dfrac{t(t)}{1(t)} - \dfrac{3}{t} = \dfrac{t^2}{t} - \dfrac{3}{t} = \dfrac{t^2 - 3}{t}$

77. $\quad 3x^2 + \dfrac{1}{x} - \dfrac{2}{x^2} = \dfrac{3x^2}{1} + \dfrac{1}{x} - \dfrac{2}{x^2} = \dfrac{3x^2(x^2)}{1(x^2)} + \dfrac{1(x)}{x(x)} - \dfrac{2}{x^2}$

$$= \dfrac{3x^4}{x^2} + \dfrac{x}{x^2} - \dfrac{2}{x^2} = \dfrac{3x^4 + x - 2}{x^2}$$

79. $\quad \dfrac{2x+3}{x} + \dfrac{x}{2} = \dfrac{(2x+3)(2)}{x(2)} + \dfrac{x(x)}{2(x)}$

$$= \dfrac{4x+6}{2x} + \dfrac{x^2}{2x}$$

$$= \dfrac{4x+6+x^2}{2x}$$

81. $\quad \dfrac{a-5}{2} + \dfrac{3}{a} = \dfrac{(a-5)(a)}{2(a)} + \dfrac{3(2)}{a(2)}$

$$= \dfrac{a^2 - 5a}{2a} + \dfrac{6}{2a}$$

$$= \dfrac{a^2 - 5a + 6}{2a}$$

83. A least common denominator (LCD) is the "smallest" expression that is exactly divisible by each of the denomiantors in a problem. We need the LCD in order to add or subtract two or more fractions. (Actually, a common denominator is really all that is needed, but the LCD is the most efficient one to use.)

84. Each of the original denominators is a factor of the LCD, so this is a common denominator. Since each factor of the LCD was chosen the maximum number of times that it appears as a factor in any one of the denominators, there are no extra factors. Thus, this must be the smallest common denominators possible.

85. The LCD for $\dfrac{3}{10}$ and $\dfrac{7}{9}$ is 90. The LCD for $\dfrac{5}{6}$ and $\dfrac{3}{4}$ is 12. The LCD of two fractions will simply be the product of the two denominators when these denominators have no prime factor in common.

86. (a) We must subtract the entire expression $(5 - x)$ from $(x + 3)$ in the numerator. This gives
$$\dfrac{x + 3 - (5 - x)}{x} = \dfrac{x + 3 - 5 + x}{x} = \dfrac{2x - 2}{x}.$$

(b) We cannot cancel when performing addition.

(c) When building fractions, we must multiply both numerator and denominator by the same quantity. So $\dfrac{5x}{2y} = \dfrac{5x(3x)}{2y(3x)} = \dfrac{15x^2}{6xy}$ and $\dfrac{7y}{6x} = \dfrac{7y(y)}{6x(y)} = \dfrac{7y^2}{6xy}$. Then
$$\dfrac{5x}{2y} + \dfrac{7y}{6x} = \dfrac{15x^2}{6xy} + \dfrac{7y^2}{6xy} = \dfrac{15x^2 + 7y^2}{6xy}.$$

(d) The cancellation step undoes the building step and brings the problem back to its original form. The final step is incoreect, since we cannot add two fractions with unlike denominators.

(e) The cancellation is not allowed, since there are no common factors to be crossed out.

87. Reducing these fractions would reverse the building process and bring us back to the original problem.

89. Let $p =$ weight of package in pounds.

$$10 + 4p = \quad 28$$
$$\underline{-10 \qquad\quad -10}$$
$$4p = \quad 18$$
$$\frac{\cancel{4}p}{\cancel{4}} = \frac{18}{4}$$
$$p = 4\frac{1}{2}$$

The package weighed $4\frac{1}{2}$ pounds.

## Exercises 4.4

1. $\dfrac{x}{3} = 9 \qquad \text{LCD} = 3$

$$3\left(\frac{x}{3}\right) = 3 \cdot 9$$

$$\frac{\overset{1}{\cancel{3}}}{1} \cdot \frac{x}{\underset{1}{\cancel{3}}} = 27$$

$$x = 27$$

3. $\dfrac{a}{4} = -6 \qquad \text{LCD} = 4$

$$4\left(\frac{a}{4}\right) = 4 \cdot (-6)$$

$$\frac{\overset{1}{\cancel{4}}}{1} \cdot \frac{a}{\underset{1}{\cancel{4}}} = -24$$

$$a = -24$$

5. $\dfrac{y}{6} = \dfrac{5}{4} \qquad \text{LCD} = 12$

$$12\left(\frac{y}{6}\right) = 12\left(\frac{5}{4}\right)$$

$$\frac{\overset{2}{\cancel{12}}}{1} \cdot \frac{y}{\underset{1}{\cancel{6}}} = \frac{\overset{3}{\cancel{12}}}{1} \cdot \frac{5}{\underset{1}{\cancel{4}}}$$

$$2y = 15$$

$$y = \frac{15}{2}$$

7. $\dfrac{w}{8} = \dfrac{-7}{6} \qquad \text{LCD} = 24$

$$24\left(\frac{w}{8}\right) = 24\left(\frac{-7}{6}\right)$$

$$\frac{\overset{3}{\cancel{24}}}{1} \cdot \frac{w}{\underset{1}{\cancel{8}}} = \frac{\overset{4}{\cancel{24}}}{1} \cdot \frac{-7}{\underset{1}{\cancel{6}}}$$

$$3w = -28$$

$$w = \frac{-28}{3} = -\frac{28}{3}$$

9. $\dfrac{3x}{2} = -18 \qquad \text{LCD} = 2$

$$2\left(\frac{3x}{2}\right) = 2 \cdot (-18)$$

$$\frac{\overset{1}{\cancel{2}}}{1} \cdot \frac{3x}{\underset{1}{\cancel{2}}} = -36$$

$$3x = -36$$

$$x = -12$$

11. $16 = \dfrac{4}{5}x \qquad \text{LCD} = 5$

$$5 \cdot 16 = 5\left(\frac{4}{5}x\right)$$

$$80 = \frac{\overset{1}{\cancel{5}}}{1} \cdot \frac{4}{\underset{1}{\cancel{5}}}x$$

$$80 = 4x$$

$$20 = x$$

13. $\dfrac{x}{3} - 2 = \dfrac{2}{3}$   LCD = 3

$$3\left(\dfrac{x}{3} - 2\right) = 3\left(\dfrac{2}{3}\right)$$

$$\dfrac{\cancel{3}^1}{1} \cdot \dfrac{x}{\cancel{3}_1} - 3 \cdot 2 = \dfrac{\cancel{3}^1}{1} \cdot \dfrac{2}{\cancel{3}_1}$$

$$x - 6 = 2$$

$$x = 8$$

15. $\dfrac{3a}{4} + 2 = \dfrac{5}{4}$   LCD = 4

$$4\left(\dfrac{3a}{4} + 2\right) = 4\left(\dfrac{5}{4}\right)$$

$$\dfrac{\cancel{4}^1}{1} \cdot \dfrac{3a}{\cancel{4}_1} + 4 \cdot 2 = \dfrac{\cancel{4}^1}{1} \cdot \dfrac{5}{\cancel{4}_1}$$

$$3a + 8 = 5$$

$$3a = -3$$

$$a = -1$$

17. $\dfrac{u}{2} - \dfrac{u}{4} = 2$   LCD = 4

$$4\left(\dfrac{u}{2} - \dfrac{u}{4}\right) = 4 \cdot 2$$

$$\dfrac{\cancel{4}^2}{1} \cdot \dfrac{u}{\cancel{2}_1} - \dfrac{\cancel{4}^1}{1} \cdot \dfrac{u}{\cancel{4}_1} = 8$$

$$2u - u = 8$$

$$u = 8$$

19. $\dfrac{y}{3} + \dfrac{y}{5} < \dfrac{6}{5}$   LCD = 15

$$15\left(\dfrac{y}{3} + \dfrac{y}{5}\right) < 15 \cdot \dfrac{6}{5}$$

$$\dfrac{\cancel{15}^5}{1} \cdot \dfrac{y}{\cancel{3}_1} + \dfrac{\cancel{15}^3}{1} \cdot \dfrac{y}{\cancel{5}_1} < \dfrac{\cancel{15}^3}{1} \cdot \dfrac{6}{\cancel{5}_1}$$

$$5y + 3y < 18$$

$$8y < 18$$

$$y < \dfrac{9}{4}$$

21. $3x - \dfrac{2}{3}x = \dfrac{4}{3}$   LCD = 3

$$3\left(3x - \dfrac{2}{3}x\right) = 3 \cdot \dfrac{4}{3}$$

$$3 \cdot 3x - \dfrac{\cancel{3}^1}{1} \cdot \dfrac{2x}{\cancel{3}_1} = \dfrac{\cancel{3}^1}{1} \cdot \dfrac{4}{\cancel{3}_1}$$

$$9x - 2x = 4$$

$$7x = 4$$

$$x = \dfrac{4}{7}$$

23. $\dfrac{a}{4} - \dfrac{a}{3} \geq \dfrac{5}{2}$   LCD = 12

$$12\left(\dfrac{a}{4} - \dfrac{a}{3}\right) \geq 12 \cdot \dfrac{5}{2}$$

$$\dfrac{\cancel{12}^3}{1} \cdot \dfrac{a}{\cancel{4}_1} - \dfrac{\cancel{12}^4}{1} \cdot \dfrac{a}{\cancel{3}_1} \geq \dfrac{\cancel{12}^6}{1} \cdot \dfrac{5}{\cancel{2}_1}$$

$$3a - 4a \geq 30$$

$$-a \geq 30$$

$$a \leq -30$$

(Remember to reverse the inequality symbol when multiplying or dividing by a negative quantity.)

**25.**
$$0.7x + 0.4x = 5.5 \qquad \text{LCD} = 10$$
$$10(0.7x + 0.4x) = 10(5.5)$$
$$10(0.7x) + 10(0.4x) = 10(5.5)$$
$$7x + 4x = 55$$
$$11x = 55$$
$$x = 5$$

**27.**
$$0.3x - 0.25x = 2 \qquad \text{LCD} = 100$$
$$100(0.3x - 0.25x) = 100(2)$$
$$100(0.3x) - 100(0.25x) = 100(2)$$
$$30x - 25x = 200$$
$$5x = 200$$
$$x = 40$$

**29.**
$$0.8m + 0.05m = 0.34 \qquad \text{LCD} = 100$$
$$100(0.8m + 0.05m) = 100(0.34)$$
$$100(0.8m) + 100(0.05m) = 100(0.34)$$
$$80m + 5m = 34$$
$$85m = 34$$
$$m = \frac{34}{85} = \frac{2 \cdot \cancel{17}}{5 \cdot \cancel{17}} = \frac{2}{5}$$

**31.**
$$\frac{w+3}{4} = \frac{w+4}{3} \qquad \text{LCD} = 12$$
$$12\left(\frac{w+3}{4}\right) = 12\left(\frac{w+4}{3}\right)$$
$$\frac{\overset{3}{\cancel{12}}}{1} \cdot \frac{w+3}{\underset{1}{\cancel{4}}} = \frac{\overset{4}{\cancel{12}}}{1} \cdot \frac{w+4}{\underset{1}{\cancel{3}}}$$
$$3(w+3) = 4(w+4)$$
$$3w + 9 = 4w + 16$$
$$9 = w + 16$$
$$-7 = w$$

**33.**
$$\frac{w+3}{4} + 1 = \frac{w+4}{3} \qquad \text{LCD} = 12$$
$$12\left(\frac{w+3}{4} + 1\right) = 12\left(\frac{w+4}{3}\right)$$
$$\frac{\overset{3}{\cancel{12}}}{1} \cdot \frac{w+3}{\underset{1}{\cancel{4}}} + 12 \cdot 1 = \frac{\overset{4}{\cancel{12}}}{1} \cdot \frac{w+4}{\underset{1}{\cancel{3}}}$$
$$3(w+3) + 12 = 4(w+4)$$
$$3w + 9 + 12 = 4w + 16$$
$$3w + 21 = 4w + 16$$
$$21 = w + 16$$
$$5 = w$$

**35.**
$$\frac{x+1}{2} + x = 6 \qquad \text{LCD} = 2$$
$$2\left(\frac{x+1}{2} + x\right) = 2 \cdot 6$$
$$\frac{\overset{1}{\cancel{2}}}{1} \cdot \frac{x+1}{\underset{1}{\cancel{2}}} + 2 \cdot x = 12$$
$$x + 1 + 2x = 12$$
$$3x + 1 = 12$$
$$3x = 11$$
$$x = \frac{11}{3}$$

**37.**
$$\frac{y}{6} - \frac{y-2}{4} > 1 \qquad \text{LCD} = 12$$
$$12\left(\frac{y}{6} - \frac{y-2}{4}\right) > 12 \cdot 1$$
$$\frac{\overset{2}{\cancel{12}}}{1} \cdot \frac{y}{\underset{1}{\cancel{6}}} + \frac{\overset{3}{\cancel{12}}}{1} \cdot \frac{y-2}{\underset{1}{\cancel{4}}} > 12 \cdot 1$$
$$2y - 3(y-2) > 12$$
$$2y - 3y + 6 > 12$$
$$-y + 6 > 12$$
$$-y > 6$$
$$y < -6$$

**39.** 
$$3 - \frac{a+1}{4} = \frac{a+4}{2} \qquad \text{LCD} = 4$$

$$4\left(3 - \frac{a+1}{4}\right) = 4\left(\frac{a+4}{2}\right)$$

$$4 \cdot 3 - \frac{\overset{1}{\cancel{4}}}{1} \cdot \frac{a+1}{\underset{1}{\cancel{4}}} = \frac{\overset{2}{\cancel{4}}}{1} \cdot \frac{a+4}{\underset{1}{\cancel{2}}}$$

$$12 - (a+1) = 2(a+4)$$

$$12 - a - 1 = 2a + 8$$

$$11 - a = 2a + 8$$

$$11 = 3a + 8$$

$$3 = 3a$$

$$1 = a$$

**41.** 
$$\frac{2y-3}{2} - \frac{y-5}{3} = \frac{1}{6} \qquad \text{LCD} = 6$$

$$6\left(\frac{2y-3}{2} - \frac{y-5}{3}\right) = 6 \cdot \frac{1}{6}$$

$$\frac{\overset{3}{\cancel{6}}}{1} \cdot \frac{2y-3}{\underset{1}{\cancel{2}}} - \frac{\overset{2}{\cancel{6}}}{1} \cdot \frac{y-5}{\underset{1}{\cancel{3}}} = \frac{\overset{1}{\cancel{6}}}{1} \cdot \frac{1}{\underset{1}{\cancel{6}}}$$

$$3(2y-3) - 2(y-5) = 1$$

$$6y - 9 - 2y + 10 = 1$$

$$4y + 1 = 1$$

$$4y = 0$$

$$y = 0$$

**43.** 
$$\frac{x+2}{3} - \frac{2x+3}{4} = \frac{x+4}{8} \qquad \text{LCD} = 24$$

$$24\left(\frac{x+2}{3} - \frac{2x+3}{4}\right) = 24\left(\frac{x+4}{8}\right)$$

$$\frac{\overset{8}{\cancel{24}}}{1} \cdot \frac{x+2}{\underset{1}{\cancel{3}}} - \frac{\overset{6}{\cancel{24}}}{1} \cdot \frac{2x+3}{\underset{1}{\cancel{4}}} = \frac{\overset{3}{\cancel{24}}}{1} \cdot \frac{x+4}{\underset{1}{\cancel{8}}}$$

$$8(x+2) - 6(2x+3) = 3(x+4)$$

$$8x + 16 - 12x - 18 = 3x + 12$$

$$-4x - 2 = 3x + 12$$

$$-2 = 7x + 12$$

$$-14 = 7x$$

$$-2 = x$$

**45.** 
$$\frac{t}{2} + \frac{t-1}{3} + \frac{t-6}{4} = t - 2 \qquad \text{LCD} = 12$$

$$12\left(\frac{t}{2} + \frac{t-1}{3} + \frac{t-6}{4}\right) = 12(t-2)$$

$$\frac{\overset{6}{\cancel{12}}}{1} \cdot \frac{t}{\underset{1}{\cancel{2}}} + \frac{\overset{4}{\cancel{12}}}{1} \cdot \frac{t-1}{\underset{1}{\cancel{3}}} + \frac{\overset{3}{\cancel{12}}}{1} \cdot \frac{t-6}{\underset{1}{\cancel{4}}} = 12(t-2)$$

$$6t + 4(t-1) + 3(t-6) = 12(t-2)$$

$$6t + 4t - 4 + 3t - 18 = 12t - 24$$

$$13t - 22 = 12t - 24$$

$$t - 22 = -24$$

$$t = -2$$

**47.**
$$0.5(x+2) - 0.3(x-4) = 3 \qquad \text{LCD} = 10$$
$$10[0.5(x+2) - 0.3(x-4)] = 10 \cdot 3$$
$$10[0.5(x+2)] - 10[0.3(x-4)] = 10 \cdot 3$$
$$5(x+2) - 3(x-4) = 30$$
$$5x + 10 - 3x + 12 = 30$$
$$2x + 22 = 30$$
$$2x = 8$$
$$x = 4$$

**49.**
$$3(y+2) + \frac{y+3}{5} = \frac{9y+8}{2} \qquad \text{LCD} = 10$$
$$10\left[3(y+2) + \frac{y+3}{5}\right] = 10\left(\frac{9y+8}{2}\right)$$
$$10[3(y+2)] + \frac{\overset{2}{\cancel{10}}}{1} \cdot \frac{y+3}{\underset{1}{\cancel{5}}} = \frac{\overset{5}{\cancel{10}}}{1} \cdot \frac{9y+8}{\underset{1}{\cancel{2}}}$$
$$30(y+2) + 2(y+3) = 5(9y+8)$$
$$30y + 60 + 2y + 6 = 45y + 40$$
$$32y + 66 = 45y + 40$$
$$66 = 13y + 40$$
$$26 = 13y$$
$$2 = y$$

**51.**
$$z + \frac{z+5}{3} - \frac{z-2}{6} = \frac{z+4}{4} + 1 \qquad \text{LCD} = 12$$
$$12\left(z + \frac{z+5}{3} - \frac{z-2}{6}\right) = 12\left(\frac{z+4}{4} + 1\right)$$
$$12 \cdot z + \frac{\overset{4}{\cancel{12}}}{1} \cdot \frac{z+5}{\underset{1}{\cancel{3}}} - \frac{\overset{2}{\cancel{12}}}{1} \cdot \frac{z-2}{\underset{1}{\cancel{6}}} = \frac{\overset{3}{\cancel{12}}}{1} \cdot \frac{z+4}{\underset{1}{\cancel{4}}} + 12 \cdot 1$$
$$12z + 4(z+5) - 2(z-2) = 3(z+4) + 12$$
$$12z + 4z + 20 - 2z + 4 = 3z + 12 + 12$$
$$14z + 24 = 3z + 24$$
$$11z + 24 = 24$$
$$11z = 0$$
$$z = 0$$

**53.**
$$3 \leq \frac{x}{3} - \frac{x+1}{2} \leq 6 \qquad \text{LCD} = 6$$

$$6 \cdot 3 \leq 6\left(\frac{x}{3} - \frac{x+1}{2}\right) \leq 6 \cdot 6$$

$$18 \leq \frac{\overset{2}{\cancel{6}}}{1} \cdot \frac{x}{\underset{1}{\cancel{3}}} - \frac{\overset{3}{\cancel{6}}}{1} \cdot \frac{x+1}{\underset{1}{\cancel{2}}} \leq 36$$

$$18 \leq 2x - 3(x+1) \leq 36$$

$$18 \leq 2x - 3x - 3 \leq 36$$

$$18 \leq -x - 3 \leq 36$$

$$21 \leq -x \leq 39$$

$$-21 \geq x \geq -39 \quad \text{or} \quad -39 \leq x \leq -21$$

**55.** The LCD of 3, 2, and 5 is 30. So

$$\frac{x}{3} + \frac{x}{2} + \frac{x}{5} = \frac{x(10)}{3(10)} + \frac{x(15)}{2(15)} + \frac{x(6)}{5(6)}$$

$$= \frac{10x}{30} + \frac{15x}{30} + \frac{6x}{30} = \frac{31x}{30}$$

**57.**
$$\frac{x}{3} + \frac{x}{2} + \frac{x}{5} = 62 \qquad \text{LCD} = 30$$

$$30\left(\frac{x}{3} + \frac{x}{2} + \frac{x}{5}\right) = 30 \cdot 62$$

$$\frac{\overset{10}{\cancel{30}}}{1} \cdot \frac{x}{\underset{1}{\cancel{3}}} + \frac{\overset{15}{\cancel{30}}}{1} \cdot \frac{x}{\underset{1}{\cancel{2}}} + \frac{\overset{6}{\cancel{30}}}{1} \cdot \frac{x}{\underset{1}{\cancel{5}}} = 1860$$

$$10x + 15x + 6x = 1860$$

$$31x = 1860$$

$$x = 60$$

**59.**
$$\frac{x+5}{2} - \frac{x-1}{4} = 2 \qquad \text{LCD} = 4$$

$$4\left(\frac{x+5}{2} - \frac{x-1}{4}\right) = 4 \cdot 2$$

$$\frac{\overset{2}{\cancel{4}}}{1} \cdot \frac{x+5}{\underset{1}{\cancel{2}}} - \frac{\overset{1}{\cancel{4}}}{1} \cdot \frac{x-1}{\underset{1}{\cancel{4}}} = 4 \cdot 2$$

$$2(x+5) - (x-1) = 8$$

$$2x + 10 - x + 1 = 8$$

$$x + 11 = 8$$

$$x = -3$$

**61.** The LCD of 2 and 4 is 4. So

$$\frac{x+5}{2} - \frac{x-1}{4} = \frac{(x+5)(2)}{2(2)} - \frac{x-1}{4}$$

$$= \frac{2(x+5)}{4} - \frac{x-1}{4}$$

$$= \frac{2(x+5) - (x-1)}{4}$$

$$= \frac{2x + 10 - x + 1}{4}$$

$$= \frac{x+11}{4}$$

63. The shipment of wooden boxes contains 80 pieces altogether. If oak boxes sell for $30 apiece and mahogany boxes sell for $50 apiece, how many of each type are in the shipment if its total value is $3600?

Let $x$ = number of oak boxes.

65. Let $x$ = weight of the package.

$$7 + 4(x - 1) = 43$$
$$7 + 4x - 4 = 43$$
$$4x + 3 = 43$$
$$4x = 40$$
$$x = 10$$

The package weighs 10 lbs.

## Exercises 4.5

1. $\dfrac{\#\text{ red}}{\#\text{ black}} = \dfrac{7}{5}$

3. $\dfrac{\#\text{ black}}{\#\text{ red}} = \dfrac{5}{7}$

5. $\dfrac{\#\text{ with}}{\#\text{ without}} = \dfrac{11}{5}$

7. $\dfrac{\overset{1}{\cancel{t}}}{3\cancel{t}} = \dfrac{1}{3}$

9. $\dfrac{a}{b + c}$

11. $\dfrac{x}{5} = \dfrac{12}{3}$

$$15 \cdot \dfrac{x}{5} = 15 \cdot \dfrac{12}{3}$$

$$\dfrac{\overset{3}{\cancel{15}}}{1} \cdot \dfrac{x}{\cancel{5}_{1}} = \dfrac{\overset{5}{\cancel{15}}}{1} \cdot \dfrac{12}{\cancel{3}_{1}}$$

$$3x = 60$$
$$x = 20$$

13. $\dfrac{a}{6} = \dfrac{5}{3}$

$$6 \cdot \dfrac{a}{6} = 6 \cdot \dfrac{5}{3}$$

$$\dfrac{\overset{1}{\cancel{6}}}{1} \cdot \dfrac{a}{\cancel{6}_{1}} = \dfrac{\overset{2}{\cancel{6}}}{1} \cdot \dfrac{5}{\cancel{3}_{1}}$$

$$a = 10$$

15. $\dfrac{y}{15} = \dfrac{20}{6}$

$$30 \cdot \dfrac{y}{15} = 30 \cdot \dfrac{20}{6}$$

$$\dfrac{\overset{2}{\cancel{30}}}{1} \cdot \dfrac{y}{\cancel{15}_{1}} = \dfrac{\overset{5}{\cancel{30}}}{1} \cdot \dfrac{20}{\cancel{6}_{1}}$$

$$2y = 100$$
$$y = 50$$

17. $\dfrac{y}{9} = \dfrac{4}{3}$

$$9 \cdot \dfrac{y}{9} = 9 \cdot \dfrac{4}{3}$$

$$\dfrac{\overset{1}{\cancel{9}}}{1} \cdot \dfrac{y}{\cancel{9}_{1}} = \dfrac{\overset{3}{\cancel{9}}}{1} \cdot \dfrac{4}{\cancel{3}_{1}}$$

$$y = 12$$

19. $x = 0.12$

21. $t = 11.59$

23. $y = -2.05$

25. Let $x$ = number of red marbles in the jar.

$$\frac{x}{210} = \frac{7}{5}$$

$$210 \cdot \frac{x}{210} = 210 \cdot \frac{7}{5}$$

$$\frac{\overset{1}{\cancel{210}}}{1} \cdot \frac{x}{\underset{1}{\cancel{210}}} = \frac{\overset{42}{\cancel{210}}}{1} \cdot \frac{7}{\underset{1}{\cancel{5}}}$$

$$x = 294$$

There are 294 red marbles in the jar.

27. Let $x$ = number of students who got 90 or above.

$$\frac{x}{24} = \frac{3}{8}$$

$$24 \cdot \frac{x}{24} = 24 \cdot \frac{3}{8}$$

$$\frac{\overset{1}{\cancel{24}}}{1} \cdot \frac{x}{\underset{1}{\cancel{24}}} = \frac{\overset{3}{\cancel{24}}}{1} \cdot \frac{3}{\underset{1}{\cancel{8}}}$$

$$x = 9$$

There were 9 students who got 90 or above.

29. Let $W$ = width of the rectangle.

$$\frac{18}{W} = \frac{9}{4}$$

$$4W \cdot \frac{18}{W} = 4W \cdot \frac{9}{4}$$

$$\frac{4\cancel{W}}{1} \cdot \frac{18}{\underset{1}{\cancel{W}}} = \frac{\cancel{4}W}{1} \cdot \frac{9}{\underset{1}{\cancel{4}}}$$

$$72 = 9W$$

$$8 = W$$

The width of the rectangle is 8 cm.

31. $$\frac{\text{shorter side}}{\text{longer side}} = \frac{\overset{2}{\cancel{6x}}}{\underset{5}{\cancel{15x}}} = \frac{2}{5}$$

33. Let $x$ = length of the scale drawing of the 20-meter wall (in cm).

$$\frac{x}{20} = \frac{5}{12}$$

$$60 \cdot \frac{x}{20} = 60 \cdot \frac{5}{12}$$

$$\frac{\overset{3}{\cancel{60}}}{1} \cdot \frac{x}{\underset{1}{\cancel{20}}} = \frac{\overset{5}{\cancel{60}}}{1} \cdot \frac{5}{\underset{1}{\cancel{12}}}$$

$$3x = 25$$

$$x = \frac{25}{3} \approx 8.33$$

The scale drawing of the 20-meter wall is 8.33 cm.

35. Let $x$ = number of kilograms in 10 lbs.

$$\frac{10}{x} = \frac{2.2}{1}$$

$$x \cdot \frac{10}{x} = x \cdot \frac{2.2}{1}$$

$$\frac{\cancel{x}}{1} \cdot \frac{10}{\cancel{x}} = 2.2x$$

$$10 = 2.2x$$

$$\frac{10}{2.2} = x$$

$$4.55 \approx x$$

There are about 4.55 kilograms in 10 lbs.

37. Let $x$ = number of yards in 100 meters.

$$\frac{100}{x} = \frac{0.92}{1}$$

$$x \cdot \frac{100}{x} = x \cdot \frac{0.92}{1}$$

$$\frac{\cancel{x}}{1} \cdot \frac{100}{\cancel{x}} = 0.92x$$

$$100 = 0.92x$$

$$\frac{100}{0.92} = x$$

$$108.70 \approx x$$

There are about 108.70 yards in 100 meters.

39. Let $x$ = distance between the two cities (in miles).

$$\frac{12.6}{x} = \frac{5}{10}$$

$$10x \cdot \frac{12.6}{x} = 10x \cdot \frac{5}{10}$$

$$\frac{10\cancel{x}}{1} \cdot \frac{12.6}{\cancel{x}} = \frac{\cancel{10}x}{1} \cdot \frac{5}{\underset{1}{\cancel{10}}}$$

$$126 = 5x$$

$$25.2 = x$$

The distance between the two cities is 25.2 miles.

41. Let $x$ = Rolfe's share of the profits (in dollars).

$$\frac{18,500}{2,800} = \frac{12,800}{x}$$

$$\frac{185}{28} = \frac{12,800}{x}$$

$$28x \cdot \frac{185}{28} = 28\cancel{x} \cdot \frac{12,800}{\cancel{x}}$$

$$185x = 358,400$$

$$x = 1,937.30$$

Rolfe's share of the profits was $1,937.30.

43. Let $x$ = number of teaspoons of garlic needed.

$$\frac{2}{6} = \frac{x}{20}$$

$$\overset{10}{\cancel{60}} \cdot \frac{2}{\cancel{6}} = \overset{3}{\cancel{60}} \cdot \frac{x}{\underset{1}{\cancel{20}}}$$

$$20 = 3x$$

$$\frac{20}{3} = x$$

$$6.67 \approx x$$

Let $y$ = number of tablespoons of olive oil needed.

$$\frac{5}{6} = \frac{y}{20}$$

$$\overset{10}{\cancel{60}} \cdot \frac{5}{\cancel{6}} = \overset{3}{\cancel{60}} \cdot \frac{y}{\underset{1}{\cancel{20}}}$$

$$50 = 3y$$

$$\frac{50}{3} = y$$

$$16.67 \approx y$$

So 6.67 teaspoons of garlic and 16.67 tablespoons of olive oil are needed for 20 servings.

45. Let $x$ = tax due (in dollars).

$$\frac{788}{126,400} = \frac{x}{98,200}$$

$$98,200 \cdot \frac{788}{126,400} = \cancel{98,200} \cdot \frac{x}{\cancel{98,200}}$$

$$612.20 \approx x$$

So the tax due on the property is $612.20.

47. Let $x$ = ounces of copper in the alloy.
Then $52 - x$ = ounces of tin in the alloy.

$$\frac{x}{52 - x} = \frac{11}{2}$$

$$2(52 - x) \cdot \frac{x}{52 - x} = \cancel{2}(52 - x) \cdot \frac{11}{\cancel{2}}$$

$$2x = 572 - 11x$$

$$13x = 572$$

$$x = \frac{572}{13} = 44$$

Then $52 - x = 52 - 44 = 8$. So the alloy contains 44 ounces of copper and 8 ounces of tin.

49. Let $x$ = Samantha's share of the profits.
Then $128,650 - x$ = Greg's share of the profits.

$$\frac{x}{128,650 - x} = \frac{7}{5}$$

$$5(128,650 - x) \cdot \frac{x}{128,650 - x} = \cancel{5}(128,650 - x) \cdot \frac{7}{\cancel{5}}$$

$$5x = 7(128,650 - x)$$

$$5x = 900,550 - 7x$$

$$12x = \frac{900,550}{12} \approx 75,045.83$$

Then $128,650 - x = 128,650 - 75,045.83 = 53,604.17$. So Samantha's share of the profits is $75,045.83 and Greg's share of the profits is $53,604.17.

51. Let $x$ = number of losers.

$$\frac{8}{3} = \frac{976}{x}$$

$$\cancel{3}x \cdot \frac{8}{\cancel{3}} = 3\cancel{x} \cdot \frac{976}{\cancel{x}}$$

$$8x = 2928$$

$$x = \frac{2928}{8} = 366$$

There were 366 losers.

53. Let $x$ = total value of the estate.

$$\frac{7}{11} = \frac{56,300}{x}$$

$$\cancel{11}x \cdot \frac{7}{\cancel{11}} = 11\cancel{x} \cdot \frac{56,300}{\cancel{x}}$$

$$7x = 619,300$$

$$x = \frac{619,300}{7} \approx 88,471.43$$

The total value of the estate was $88,471.43.

**55.** Let $x$ = estimated number of deer in the entire area.

$$\frac{48}{x} = \frac{18}{60}$$

$$60\not{x} \cdot \frac{48}{\not{x}} = \not{60}x \cdot \frac{18}{\not{60}x}$$

$$2880 = 18x$$

$$160 = x$$

The estimated number of deer in the entire area is 160.

**57.**

$$\frac{x}{6} = \frac{18}{9} \qquad\qquad \frac{y}{9} = \frac{8}{6}$$

$$\overset{3}{\not{18}} \cdot \frac{x}{\not{6}} = \overset{2}{\not{18}} \cdot \frac{18}{\not{9}} \qquad \overset{2}{\not{18}} \cdot \frac{y}{\not{9}} = \overset{3}{\not{18}} \cdot \frac{8}{\not{6}}$$

$$3x = 36 \qquad\qquad 2y = 24$$

$$x = 12 \qquad\qquad y = 12$$

**59.**

$$\frac{x}{12.5} = \frac{5.2}{8.4} \qquad\qquad \frac{y}{8.4} = \frac{9.6}{12.5}$$

$$(8.4)(\not{12.5}) \cdot \frac{x}{\not{12.5}} = (\not{8.4})(12.5) \cdot \frac{5.2}{\not{8.4}} \qquad (12.5)(\not{8.4}) \cdot \frac{y}{\not{8.4}} = (\not{12.5})(8.4) \cdot \frac{9.6}{\not{12.5}}$$

$$8.4x = 65 \qquad\qquad 12.5y = 80.64$$

$$x = \frac{65}{8.4} \approx 7.74 \qquad\qquad y = \frac{80.64}{12.5} \approx 6.45$$

**61.** Let $x$ = height of the street light (in feet).

$$\frac{6}{8} = \frac{x}{23}$$

$$(23)(\not{8}) \cdot \frac{6}{\not{8}} = (\not{23})(8) \cdot \frac{x}{\not{23}}$$

$$138 = 8x$$

$$17.25 = x$$

The height of the street light is 17.25 feet.

**63.** Let $w$ = width of the river (in meters).

$$\frac{w}{1800} = \frac{500}{750}$$

$$(\not{1800}) \cdot \frac{w}{\not{1800}} = \frac{\overset{2}{\not{500}}}{\underset{\not{3}}{\not{750}}} \cdot \overset{600}{\underset{1}{\not{1800}}}$$

$$w = 1200$$

The width of the river is 1200 meters.

**65.** Let $x$ = Cherise's share (in dollars).
Then $85,600 - x$ = Jerome's share (in dollars).

$$\frac{x}{85,600 - x} = \frac{3}{2}$$

$$2(\not{85,600 - x}) \cdot \frac{x}{\not{85,600 - x}} = \not{2}(85,600 - x) \cdot \frac{3}{\not{2}}$$

$$2x = 3(85,600 - x)$$

$$2x = 256,800 - 3x$$

$$5x = 256,800$$

$$x = 51,360$$

Then $85,600 - x = 85,600 - 51,360 = 34,240$.
So Cherise's share is $51,360 and Jerome's share is $34,240.

## Exercises 4.6

1. Let $x =$ the number.

$$\frac{2}{3}x + 5 = 9$$

$$3\left(\frac{2}{3}x + 5\right) = 3 \cdot 9$$

$$\overset{1}{\cancel{3}} \cdot \frac{2}{\cancel{3}}x + 3 \cdot 5 = 3 \cdot 9$$

$$2x + 15 = 27$$

$$2x = 12$$

$$x = 6$$

Thus, the number is 6.

CHECK: $\frac{2}{3}$ of 6 is 4, and 5 more than 4 is 9.

3. Let $x =$ the number.

$$\frac{3}{4}x - 2 = \frac{1}{8}x - 7$$

$$8\left(\frac{3}{4}x - 2\right) = 8\left(\frac{1}{8}x - 7\right)$$

$$\overset{2}{\cancel{8}} \cdot \frac{3}{\cancel{4}}x - 8 \cdot 2 = \overset{1}{\cancel{8}} \cdot \frac{1}{\cancel{8}}x - 8 \cdot 7$$

$$6x - 16 = x - 56$$

$$5x - 16 = -56$$

$$5x = -40$$

$$x = -8$$

Thus, the number is $-8$.

CHECK: 2 less than $\frac{3}{4}$ of $-8$ is 2 less than $-6$ or $-8$. $\frac{1}{8}$ of $-8$ is $-1$, and $-8$ is 7 less than $-1$.

5. Let $L =$ length of the rectangle (in meters).

Then $\frac{1}{2}L =$ width of the rectangle (in meters).

$$2(L) + 2\left(\frac{1}{2}L\right) = 36$$

$$2L + \overset{1}{\cancel{2}} \cdot \frac{1}{\cancel{2}}L = 36$$

$$2L + L = 36$$

$$3L = 36$$

$$L = 12$$

Then $\frac{1}{2}L = \frac{1}{2}(12) = 6$.

Thus, the dimensions of the rectangle are 12 meters and 6 meters.

CHECK: The width of the 6 m $\times$ 12 m rectangle is half its length. The perimeter of this rectangle is $2(6) + 2(12) = 12 + 24 = 36$ meters.

7. Let $x =$ length of the longest side of the triangle (in inches).

Then $\frac{3}{4}x =$ length of the medium side of the triangle (in inches),

and $\frac{1}{2}\left(\frac{3}{4}x\right) =$ length of the medium side of the triangle (in inches).

$$x + \frac{3}{4}x + \frac{1}{2}\left(\frac{3}{4}x\right) = 17$$

$$x + \frac{3}{4}x + \frac{3}{8}x = 17$$

$$8\left(x + \frac{3}{4}x + \frac{3}{8}x\right) = 8 \cdot 17$$

$$8 \cdot x + \overset{2}{\cancel{8}} \cdot \frac{3}{\cancel{4}}x + \overset{1}{\cancel{8}} \cdot \frac{3}{\cancel{8}}x = 8 \cdot 17$$

$$8x + 6x + 3x = 136$$

$$17x = 136$$

$$x = 8$$

Then $\frac{3}{4}x = \frac{3}{4}(8) = 6$, and

$\frac{1}{2}\left(\frac{3}{4}x\right) = \frac{1}{2}\left(\frac{3}{4}(8)\right) = 3$.

Thus, the sides of the triangle are 8 inches, 6 inches, and 3 inches.

CHECK: 6 is $\frac{3}{4}$ of 8,

and 3 is $\frac{1}{2}$ of 6.

The sum of 8, 6, and 3 is 17.

9.  Let $x$ = number of regular tickets sold.
    Then $350 - x$ = number of combination tickets sold.
    $$15x + 22(350 - x) = 6895$$
    $$15x + 7700 - 22x = 6895$$
    $$-7x + 7700 = 6895$$
    $$-7x = -805$$
    $$x = 115$$

    Then $350 - x = 350 - (115) = 235$.
    Thus, there were 115 regular tickets and 235 combination tickets sold.

    CHECK: 115 regular tickets at $15 each gives $1725. 235 combination tickets at $22 each gives $5170. The total amount collected is $1725 + $5170 = $6985.

11. Let $x$ = number of quarter in the collection.
    Then $2x + 3$ = number of dimes in the collection.
    $$25x + 10(2x + 3) = 255$$
    $$25x + 20x + 30 = 255$$
    $$45x + 30 = 255$$
    $$45x = 225$$
    $$x = 5$$

    Then $2x + 3 = 2(5) + 3 = 13$.
    Thus, there are 5 quarters and 13 dimes in the collection.

    CHECK: 13 is 3 more than twice 5. 5 quarters are worth $5(\$0.25) = \$1.25$, while 13 dimes are worth $13(\$0.10) = \$1.30$.
    Total value $= \$1.25 + \$1.30 = \$2.55$.

13. Let $t$ = number of minutes that the older machine works.
    Then $t - 15$ = number of minutes that the newer machine works.
    $$175t + 250(t - 15) = 13675$$
    $$175t + 250t - 3750 = 13675$$
    $$425t - 3750 = 13675$$
    $$425t = 17425$$
    $$t = 41$$

    Since the older machine began the sorting process at 10:00 a.m. and worked for 41 minutes, the sorting is completed at 10:41 a.m.

    CHECK: The older machine sorts $175(41) = 7175$ screws, while the newer machine sorts $250(41 - 15) = 250(26) = 6500$ screws. In all, $7175 + 6500 = 13,675$ screws are sorted.

15. Let $x$ = amount of money invested 6.35%.
    Then $x + 4000$ = amount of money invested at 7.28%.
    $$0.0635x + 0.0728(x + 4000) = 972.70$$
    $$10000[0.0635x + 0.0728(x + 4000)] = 10000(972.70)$$
    $$10000(0.0635x) + 10000[0.0728(x + 4000)] = 10000(972.70)$$
    $$635x + 728(x + 4000) = 9727000$$
    $$635x + 728x + 2912000 = 9727000$$
    $$1363x + 2912000 = 9727000$$
    $$1363x = 6815000$$
    $$x = 5000$$

    Thus, $5000 was invested at 6.35%.
    CHECK: $5000 at 6.35% yields $317.50 in interest. $5000 + $4000 = $9000 at 7.28% yields $655.20 in interest. The total interest earned is $317.50 + $655.20 = $972.70.

17. Let $x$ = amount of money invested 9%.
Then $800 - x$ = amount of money invested at 6%.
$$0.09x + 0.06(800 - x) = 67.50$$
$$100[0.09x + 0.06(800 - x)] = 100(67.50)$$
$$100(0.09x) + 100[0.06(800 - x)] = 100(67.50)$$
$$9x + 6(800 - x) = 6750$$
$$9x + 4800 - 6x = 6750$$
$$3x + 4800 = 6750$$
$$3x = 1950$$
$$x = 650$$

Then $800 - x = 800 - 650 = 150$.
Thus, \$650 was invested at 9% and \$150 was invested at 6%.

CHECK: The interest on the 9% investment is $0.09(\$650) = \$58.50$. The interest on the 6% investment is $0.06(\$150) = \$9.00$. Total interest $= \$58.50 + \$9.00 = \$67.50$.

19. Let $x$ = amount of money invested 8%.
Then $6000 - x$ = amount of money invested at 12%.
$$0.08x + 0.12(6000 - x) = 0.09(6000)$$
$$100[0.08x + 0.12(6000 - x)] = 100[0.09(6000)]$$
$$100(0.08x) + 100[0.12(6000 - x)] = 100[0.09(6000)]$$
$$8x + 12(6000 - x) = 9(6000)$$
$$8x + 72000 - 12x = 54000$$
$$-4x + 72000 = 54000$$
$$-4x = -18000$$
$$x = 4500$$

Then $6000 - x = 6000 - 4500 = 1500$.
Thus, \$1500 should be invested at 12%.

CHECK: The interest on the 8% investment is $0.08(\$4500) = \$360$. The interest on the 12% investment is $0.12(\$1500) = \$180$. The total interest $\$360 + \$180 = \$540$ is 9% of \$6000.

21. Let $x$ = number of ml of 30% hydrochloric acid solution.
$$0.30x + 0.50(30) = 0.45(x + 30)$$
$$100[0.30x + 0.50(30)] = 100[0.45(x + 30)]$$
$$100(0.30x) + 100[0.50(30)] = 100[0.45(x + 30)]$$
$$30x + 50(30) = 45(x + 30)$$
$$30x + 1500 = 45x + 1350$$
$$1500 = 15x + 1350$$
$$150 = 15x$$
$$10 = x$$

Thus, 10 ml of 30% hydrochloric acid solution should be used in the mixture.

CHECK: In 10 ml of a 30% solution, there are 3 ml of acid. In 30 ml of a 50% solution, there are 15 ml of acid. So there are $3 + 15 = 18$ ml of acid in the mixture, which is 45% of 40 ml.

23. Let $x$ = number of liters of 2.4% salt solution.
Then $90 - x$ = number of liters of 4.6% salt solution.
$$0.024x + 0.046(90 - x) = 0.03(90)$$
$$1000[0.024x + 0.046(90 - x)] = 1000[0.03(90)]$$
$$1000(0.024x) + 1000[0.046(90 - x)] = 1000[0.03(90)]$$
$$24x + 46(90 - x) = 30(90)$$
$$24x + 4140 - 46x = 2700$$
$$-22x + 4140 = 2700$$
$$-22x = -1440$$
$$x = 65.5$$

Then $90 - x = 90 - 65.5 = 24.5$.
Thus, 65.5 liters of the 2.4% solution should be mixed with 24.5 liters of the 4.6% solution.

CHECK: In 65.5 liters of a 2.4% salt solution, there are $0.024(65.5) = 1.57$ liters of salt. In 24.5 liters of a 4.6% salt solution, there are $0.046(24.5) = 1.13$ liters of salt. In the mixture, there are $1.57 + 1.13 = 2.70$ liters of salt, which is 3% of 90 liters.

25. Let $x$ = number of gallons of pure anti-freeze to be used.
$$1.00x + 0.30(10) = 0.50(x + 10)$$
$$100[1.00x + 0.30(10)] = 100[0.50(x + 10)]$$
$$100(1.00x) + 100[0.30(10)] = 100[0.50(x + 10)]$$
$$100x + 30(10) = 50(x + 10)$$
$$100x + 300 = 50x + 500$$
$$50x + 300 = 500$$
$$50x = 200$$
$$x = 4$$

Thus, 4 gallons of pure anti-freeze should be added to the radiator.

CHECK: 10 gallons of a 30% anti-freeze solution contains 3 gallons of anti-freeze. If we add 4 gallons of anti-freeze, we get 14 gallons in the solution, 7 of which are anti-freeze. This is 50%.

27. Let $x$ = number of pounds of candy that sells for $3.75/lb.
$$3.75x + 5.00(35) = 4.25(x + 35)$$
$$100[3.75x + 5.00(35)] = 100[4.25(x + 35)]$$
$$100(3.75x) + 100[5.00(35)] = 100[4.25(x + 35)]$$
$$375x + 500(35) = 425(x + 35)$$
$$375x + 17500 = 425x + 14875$$
$$17500 = 50x + 14875$$
$$2625 = 50x$$
$$52.5 = x$$

Thus, 52.5 pounds of candy that sells for $3.75/lb should be added to the mixture.

CHECK: 52.5 lbs. at $3.75/lb = $196.87$\frac{1}{2}$.

35 lbs. at $5/lb = $175. Total = $371.87$\frac{1}{2}$.

87.5 lbs at $4.25/lb = $371.87$\frac{1}{2}$.

29. Let $t$ = number of hours that John and Susan must travel before they meet.
$$4t + 8t = 9$$
$$12t = 9$$
$$t = \frac{9}{12} = \frac{3}{4}$$

Since each must travel $\frac{3}{4}$ of an hour or 45 minutes, they will meet at 8:45 a.m.

CHECK: John walks $4\left(\frac{3}{4}\right) = 3$ miles and Susan jogs $8\left(\frac{3}{4}\right) = 6$ miles. $3 + 6 = 9$.

31. Let $t$ = number of hours that John must travel before they meet.

Then $t - \frac{1}{4}$ = number of hours that Susan must travel before they meet.
$$4t + 8\left(t - \frac{1}{4}\right) = 9$$
$$4t + 8t - \overset{2}{\cancel{8}} \cdot \frac{1}{\cancel{4}} = 9$$
$$12t - 2 = 9$$
$$12t = 11$$
$$t = \frac{11}{12}$$

Since John must travel $\frac{11}{12}$ hours or 55 minutes before he meets Susan, they will meet at 8:40 a.m.

CHECK: John walks $4\left(\frac{11}{12}\right) = \frac{11}{3} = 3\frac{2}{3}$ miles and Susan jogs $8\left(\frac{11}{12} - \frac{1}{4}\right) = 8\left(\frac{8}{12}\right) = \frac{16}{3}$ $= 5\frac{1}{3}$ miles. $3\frac{2}{3} + 5\frac{1}{3} = 9$.

33. Let $t$ = number of hours that David spends jogging.
Then $2 - t$ = number of hours that David spends walking.

$5(2 - t) + 9t = 16$

$10 - 5t + 9t = 16$

$10 + 4t = 16$

$4t = 6$

$t = \dfrac{6}{4} = \dfrac{3}{2}$

Thus, David jogs for $1\frac{1}{2}$ hours.

CHECK: David jogs $9\left(\dfrac{3}{2}\right) = \dfrac{27}{2} = 13\dfrac{1}{2}$

miles and walks $5\left(\dfrac{1}{2}\right) = \dfrac{5}{2} = 2\dfrac{1}{2}$ miles.

$13\dfrac{1}{2} + 2\dfrac{1}{2} = 16.$

35. Let $r$ = rate at which additional money is invested.

$0.072(3200) + r(2800) = 390$

$230.4 + 2800r = 390$

$2800r = 159.6$

$r = \dfrac{159.6}{2800}$

$r = 0.057$

Sal must invest the additional $\$2,800$ at a 5.7% rate.

CHECK: An investment of $\$3,200$ at 7.2% pays $\$230.40$ investment. An investment of $\$2,800$ at 5.7% pays $\$159.60$ interest. $\$230.40 + \$159.60 = \$390.$

37. Let $p$ = percent solution for additional hydrochloric acid.

$0.20(30) + p(60) = 0.40(90)$

$6 + 60p = 36$

$60p = 30$

$p = \dfrac{30}{60}$

$p = 0.50$

The chemist must use a 50% hydrochloric acid solution.

CHECK: 30 ml of a 20% solution contains 6 ml of acid. 60 ml of a 50% solution contains 30 ml of acid. 90 ml of a 40% solution contains 36 ml of acid and $6 + 30 = 36.$

39. Let $w$ = width of the rectangle.
Then $2w - 1$ = length of the rectangle.

$w = \dfrac{1}{7}[2(w) + 2(2w - 1)] + 1$

$w = \dfrac{1}{7}(2w + 4w - 2) + 1$

$7w = 7\left[\dfrac{1}{7}(6w - 2) + 1\right]$

$7w = 6w - 2 + 7$

$7w = 6w + 5$

$w = 5$

Then $2w - 1 = 2(5) - 1 = 9.$
So the rectangle has a width of 5 and a length of 9.

CHECK: 9 is 1 less than twice 5. The perimeter of a $5 \times 9$ rectangle is $2(5) + 2(9) = 28$, and 5 is 1 more than $\dfrac{1}{7}(28) = 4.$

41. $C = \dfrac{5}{9}(F - 32)$, where $C$ is the temperature in degrees Celsius and $F$ is the temperature in degrees Fahrenheit.

$$25 < \quad C \quad < 40$$

$$25 < \quad \frac{5}{9}(F - 32) \quad < 40$$

$$9(25) < 9\left[\frac{5}{9}(F - 32)\right] < 9(40)$$

$$225 < \quad 5(F - 32) \quad < 360$$

$$45 < \quad F - 32 \quad < 72$$

$$77 < \quad F \quad < 104$$

Thus, the corresponding temperature range in degrees Fahrenheit is between 77°F and 104°F.

43. Let $x = $ original price of a skirt.
Then $0.80x = $ sale price of the skirt.

$$12.60 < \quad 0.80x \quad < 20.76$$

$$100(12.60) < 100(0.80x) < 100(20.76)$$

$$1260 < \quad 80x \quad < 2076$$

$$15.75 < \quad x \quad < 25.95$$

Thus, the original range of prices on the skirts was between \$15.75 and \$25.95.

## Chapter 4 Review Exercises

1. $\dfrac{-18}{42} = \dfrac{-3 \cdot \cancel{6}}{7 \cdot \cancel{6}} = \dfrac{-3}{7} = -\dfrac{3}{7}$

3. $\dfrac{15x^6}{6x^2} = \dfrac{5x^4 \cdot \cancel{3x^2}}{2 \cdot \cancel{3x^2}} = \dfrac{5x^4}{2}$

5. $\dfrac{-10x^3y^5}{4xy^{10}} = \dfrac{-5x^2 \cdot \cancel{2xy^5}}{2y^5 \cdot \cancel{2xy^5}} = \dfrac{-5x^2}{2y^5}$
   $$= -\dfrac{5x^2}{2y^5}$$

7. $\dfrac{3t - 7t - t}{-2t^2 - 3t^2} = \dfrac{-5t}{-5t^2} = \dfrac{1(\cancel{-5t})}{t(\cancel{-5t})} = \dfrac{1}{t}$

9. $\dfrac{a}{4} \cdot \dfrac{a}{4} = \dfrac{a^2}{16}$

11. $\dfrac{7a}{6} - \dfrac{5a}{6} = \dfrac{7a - 5a}{6} = \dfrac{\cancel{2}a^{\,1}}{\cancel{6}_3} = \dfrac{a}{3}$

13. $\dfrac{4x - 3}{6x} - \dfrac{x - 1}{6x} = \dfrac{4x - 3 - (x - 1)}{6x}$
    $$= \dfrac{4x - 3 - x + 1}{6} = \dfrac{3x - 2}{6x}$$

**15.** $\dfrac{2y^2 - 3y}{4} - \dfrac{y^2 - 3y}{4} + \dfrac{y^2}{4} = \dfrac{2y^2 - 3y - (y^2 - 3y) + y^2}{4}$

$$= \dfrac{2y^2 - 3y - y^2 + 3y + y^2}{4}$$

$$= \dfrac{\overset{1}{\cancel{2}}y^2}{\underset{2}{\cancel{4}}} = \dfrac{y^2}{2}$$

**17.** $\left( \dfrac{\overset{x}{\cancel{x^2}}}{\underset{2}{\cancel{4}}} \cdot \dfrac{\overset{3}{\cancel{6}}}{\cancel{x}y^2} \right) \div (2xy) = \dfrac{3x}{2y^2} \div \dfrac{2xy}{1}$

$$= \dfrac{3\cancel{x}}{2y^2} \cdot \dfrac{1}{2\cancel{x}y} = \dfrac{3}{4y^3}$$

**19.** $\dfrac{a}{2} \cdot \dfrac{a}{4} = \dfrac{a^2}{8}$

**21.** $\dfrac{x^2}{2} - \dfrac{x^2}{6} + \dfrac{x^2}{3} = \dfrac{x^2(3)}{2(3)} - \dfrac{x^2}{6} + \dfrac{x^2(2)}{3(2)}$

$$= \dfrac{3x^2}{6} - \dfrac{x^2}{6} + \dfrac{2x^2}{6} = \dfrac{3x^2 - x^2 + 2x^2}{6}$$

$$= \dfrac{\overset{2}{\cancel{4}}x^2}{\underset{3}{\cancel{6}}} = \dfrac{2x^2}{3}$$

**23.** $\dfrac{4}{x^2} + \dfrac{3}{2x} = \dfrac{4(2)}{x^2(2)} + \dfrac{3(x)}{2x(x)} = \dfrac{8}{2x^2} + \dfrac{3x}{2x^2} = \dfrac{8 + 3x}{2x^2}$

**25.** $\dfrac{3}{4a^2b} - \dfrac{5}{6ab} + \dfrac{7}{8b^3} = \dfrac{3(6b^2)}{4a^2b(6b^2)} - \dfrac{5(4ab^2)}{6ab(4ab^2)} + \dfrac{7(3a^2)}{8b^3(3a^2)}$

$$= \dfrac{18b^2}{24a^2b^3} - \dfrac{20ab^2}{24a^2b^3} + \dfrac{21a^2}{24a^2b^3}$$

$$= \dfrac{18b^2 - 20ab^2 + 21a^2}{24a^2b^3}$$

**27.**
$$\dfrac{x}{6} - \dfrac{1}{4} = \dfrac{7}{12}$$
$$12\left( \dfrac{x}{6} - \dfrac{1}{4} \right) = 12\left( \dfrac{7}{12} \right)$$
$$\dfrac{\overset{2}{\cancel{12}}}{1} \cdot \dfrac{x}{\underset{1}{\cancel{6}}} - \dfrac{\overset{3}{\cancel{12}}}{1} \cdot \dfrac{1}{\underset{1}{\cancel{4}}} = \dfrac{\cancel{12}}{1} \cdot \dfrac{7}{\cancel{12}}$$
$$2x - 3 = 7$$
$$2x = 10$$
$$x = 5$$

**29.**
$$\dfrac{t+1}{2} + \dfrac{t+2}{3} < \dfrac{t+7}{6}$$
$$6\left( \dfrac{t+1}{2} + \dfrac{t+2}{3} \right) < 6\left( \dfrac{t+7}{6} \right)$$
$$\dfrac{\overset{3}{\cancel{6}}}{1} \cdot \dfrac{t+1}{\underset{1}{\cancel{2}}} + \dfrac{\overset{2}{\cancel{6}}}{1} \cdot \dfrac{t+2}{\underset{1}{\cancel{3}}} < \dfrac{\cancel{6}}{1} \cdot \dfrac{t+7}{\cancel{6}}$$
$$3(t+1) + 2(t+2) < t+7$$
$$3t + 3 + 2t + 4 < t + 7$$
$$5t + 7 < t + 7$$
$$4t < 0$$
$$t < 0$$

**31.**

$$\frac{y+3}{5} - \frac{y-2}{3} = 1$$

$$15\left(\frac{y+3}{5} - \frac{y-2}{3}\right) = 15(1)$$

$$\frac{\overset{3}{\cancel{15}}}{1} \cdot \frac{y+3}{\cancel{5}} - \frac{\overset{5}{\cancel{15}}}{1} \cdot \frac{y-2}{\cancel{3}} = 15$$

$$3(y+3) - 5(y-2) = 15$$
$$3y + 9 - 5y + 10 = 15$$
$$-2y + 19 = 15$$
$$-2y = -4$$
$$y = 2$$

**33.**

$$\frac{x}{3} = \frac{x+1}{6}$$

$$6 \cdot \frac{x}{3} = 6 \cdot \frac{x+1}{6}$$

$$\frac{\overset{2}{\cancel{6}}}{1} \cdot \frac{x}{\underset{1}{\cancel{3}}} = \frac{\cancel{6}}{1} \cdot \frac{x+1}{\cancel{6}}$$

$$2x = x + 1$$
$$x = 1$$

**35.**

$$2x + 0.2(x+6) = 10$$
$$10[2x + 0.2(x+6)] = 10 \cdot 10$$
$$10(2x) + 10[0.2(x+6)] = 100$$
$$20x + 2(x+6) = 100$$
$$20x + 2x + 12 = 100$$
$$22x + 12 = 100$$
$$22x = 88$$
$$x = 4$$

**37.** Let $x$ = the number of ounces in 1 kilogram (1000 grams).

$$\frac{1000}{x} = \frac{28.4}{1}$$

$$\cancel{x} \cdot \frac{1000}{\cancel{x}} = x \cdot \frac{28.4}{1}$$

$$1000 = 28.4x$$

$$\frac{1000}{28.4} = x$$

$35.21 = x$ (to the nearest hundredth).
Thus, there are 35.21 ounces in 1 kilogram.

**39.** Let $x$ = amount invested at 6%.
Then $2x$ = amount invested at 7%
and $7000 - 3x$ = amount invested at 8%.

$$0.06x + 0.07(2x) + 0.08(7000 - 3x) \geq 500$$

$$100[0.06x + 0.07(2x) + 0.08(7000 - 3x)] \geq 100(500)$$

$$100(0.06x) + 100[0.07(2x)] + 100[.08(7000 - 3x)] \geq 50000$$

$$6x + 7(2x) + 8(7000 - 3x) \geq 50000$$

$$6x + 14x + 56000 - 24x \geq 50000$$

$$-4x + 56000 \geq 50000$$

$$-4x \geq -6000$$

$$x \leq 1500$$

Thus, the most that can be invested at 6% is $1500.

**41.** Let $x$ = Bill's present speed.

$$5x = 3(x + 20)$$

$$5x = 3x + 60$$

$$2x = 60$$

$$x = 30$$

Thus, Bill's present speed is 30 mph.

**CHECK:** After 5 hours, Bill covered $(30)(5) = 150$ miles. In 3 hours, he would cover $(50)(3) = 150$ miles at the faster rate as well.

## Chapter 4 Practice Test

1.  (a) $\dfrac{-10}{24} = \dfrac{\cancel{2}(-5)}{\cancel{2}(12)} = \dfrac{-5}{12} = -\dfrac{5}{12}$

    (b) $\dfrac{\overset{x^8}{\cancel{x^{10}}}}{\underset{1}{\cancel{x^2}}} = \dfrac{x^8}{1} = x^8$

    (c) $\dfrac{\overset{2a^3}{\cancel{-6a^6}}}{\underset{1}{\cancel{-3a^3}}} = \dfrac{2a^3}{1} = 2a^3$

    (d) $\dfrac{\overset{5t^2}{\cancel{25r^2t^3}}}{\underset{-3r^2}{\cancel{-15r^4t}}} = \dfrac{5t^2}{-3r^2} = -\dfrac{5t^2}{3r^2}$

3.  (a)
$$\frac{x}{3} + \frac{x}{5} = 8$$
$$15\left(\frac{x}{3} + \frac{x}{5}\right) = 15 \cdot 8$$
$$\frac{\overset{5}{\cancel{15}}}{1} \cdot \frac{x}{\underset{1}{\cancel{3}}} + \frac{\overset{3}{\cancel{15}}}{1} \cdot \frac{x}{\underset{1}{\cancel{5}}} = 120$$
$$5x + 3x = 120$$
$$8x = 120$$
$$x = 15$$

    (b)
$$\frac{x-5}{2} + \frac{x}{5} \geq 8$$
$$10\left(\frac{x-5}{2} + \frac{x}{5}\right) \geq 10 \cdot 8$$
$$\frac{\overset{5}{\cancel{10}}}{1} \cdot \frac{x-5}{\underset{1}{\cancel{2}}} + \frac{\overset{2}{\cancel{10}}}{1} \cdot \frac{x}{\underset{1}{\cancel{5}}} \geq 80$$
$$5(x-5) + 2x \geq 80$$
$$5x - 25 + 2x \geq 80$$
$$7x - 25 \geq 80$$
$$\underline{\phantom{7x}+25 \quad +25}$$
$$7x \phantom{-25} \geq 105$$
$$\frac{\cancel{7}x}{\cancel{7}} \geq \frac{105}{7}$$
$$x \geq 15$$

    (c)
$$\frac{a+3}{5} - \frac{a-2}{4} = 1$$
$$20\left(\frac{a+3}{5} - \frac{a-2}{4}\right) = 20 \cdot 1$$
$$\frac{\overset{4}{\cancel{20}}}{1} \cdot \frac{a+3}{\underset{1}{\cancel{5}}} - \frac{\overset{5}{\cancel{20}}}{1} \cdot \frac{a-2}{\underset{1}{\cancel{4}}} = 20$$
$$4(a+3) - 5(a-2) = 20$$
$$4a + 12 - 5a + 10 = 20$$
$$-a + 22 = 20$$
$$\underline{\phantom{-a}-22 \quad -22}$$
$$-a \phantom{+22} = -2$$
$$\frac{-a}{-1} = \frac{-2}{-1}$$
$$a = 2$$

    (d)
$$0.03t + 0.5t = 10.6$$
$$100(0.03t + 0.5t) = 100(10.6)$$
$$100(0.03t) + 100(0.5t) = 100(10.6)$$
$$3t + 50t = 1060$$
$$53t = 1060$$
$$\frac{\cancel{53}t}{\cancel{53}} = \frac{1060}{53}$$
$$t = 20$$

5. Let $x =$ the number.

$$x + \frac{2}{3}x = 2x - 5$$

$$3\left(x + \frac{2}{3}x\right) = 3(2x - 5)$$

$$3x + \frac{\cancel{3}}{1} \cdot \frac{2}{\cancel{3}}x = 3(2x) - 3(5)$$

$$3x + 2x = 6x - 15$$

$$5x = 6x - 15$$

$$\underline{-6x \quad -6x}$$

$$-x = -15$$

$$\frac{-x}{-1} = \frac{-15}{-1}$$

$$x = 15$$

CHECK: $\frac{2}{3}$ of $15 = \frac{2}{\cancel{3}} \cdot \frac{\overset{5}{\cancel{15}}}{1} = \frac{10}{1} = 10.$

Then $15 + 10 = 25$, which is 5 less than twice 15.

7. Let $x =$ amount invested at 8%.
   Then $7000 - x =$ amount invested at 13%.

$$0.08x + 0.13(7000 - x) = 750$$
$$100[0.08x + 0.13(7000 - x)] = 100(750)$$
$$100(0.08x) + 100[0.13(7000 - x)] = 100(750)$$
$$8x + 13(7000 - x) = 75000$$
$$8x + 91000 - 13x = 75000$$
$$-5x + 91000 = 75000$$
$$\underline{-91000 \quad -91000}$$
$$-5x = -16000$$
$$\frac{\cancel{-5}x}{\cancel{-5}} = \frac{-16000}{-5}$$
$$x = 3200$$

Then $7000 - x = 7000 - 3200 = 3800$.
Thus, $3200 is invested at 8% and $3800 is invested at 13%.

CHECK: $3200 + 3800 = 7000$. The interest on $3200 at 8% is $256\,(= 0.08 \cdot 3200)$. The interest on $3800 at 13% is $494\,(= 0.13 \cdot 3800)$. Then $256 + $494 = $750$, as required.

9. Let $x =$ number of hours that the first person drives.
   Then $x - 4 =$ number of hours that the second person drives.

$$48x + 55(x - 4) = 604$$
$$48x + 55x - 220 = 604$$
$$103x - 220 = 604$$
$$\underline{+220 \quad +220}$$
$$103x = 824$$
$$\frac{\cancel{103}x}{\cancel{103}} = \frac{824}{103}$$
$$x = 8$$

Thus, they will be 604 kilometers apart after the first person drives for 8 hours, or at 7:00 p.m.

CHECK: In 8 hours, the first person covers $48(8) = 384$ km. In $8 - 4 = 4$ hours, the second person covers $55(4) = 220$ km. Together they cover $384 + 220 = 604$ km.

# Chapter 5
# Graphing Straight Lines

## Exercises 5.1

1-19.  (odd)

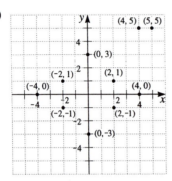

21.  Quadrant I

23.  Quadrant III

25.  on $y$-axis

27.  Quadrant II

29.  on both $x$-axis and $y$-axis (origin)

31.  Quadrant IV

33.  $(1, 7)$ satisfies the equation $2y - 4x = 10$, since $2(7) - 4(1) = 14 - 4 = 10$.

35.  $(0, -1)$ satisfies the equation $y = 4x - 1$, since $4(0) - 1 = 0 - 1 = -1$.

37.  All three points satisfy the equation $6x - 4y - 8 = 0$, since
$6(-2) - 4(-5) - 8 = -12 + 20 - 8 = 0;$
$6(6) - 4(7) - 8 = 36 - 28 - 8 = 0;$
$6(-10) - 4(-17) - 8 = -60 + 68 - 8 = 0.$

39.  $(-6, -10)$ satisfies the equation $\dfrac{2}{3}x - \dfrac{1}{2}y = 1$, since $\dfrac{2}{3}(-6) - \dfrac{1}{2}(-10) = -4 + 5 = 1$.

41.  $(1, -6)$ satisfies the equation $y = x^2 - 3x - 4$, since $(1)^2 - 3(1) - 4 = 1 - 3 - 4 = -6$.

43.  The last two ordered pairs satisfy the equation $y = \dfrac{1}{x+2}$, since $\dfrac{1}{1+2} = \dfrac{1}{3}$ and $\dfrac{1}{-1+2} = 1$.

45.  $\{(x, y) \mid y = x + 2 \text{ and } x = -3, 0, 4\}$

| $x$ | $y$ | $(x, y)$ |
|-----|-----|----------|
| $-3$ | $-1$ | $(-3, -1)$ |
| $0$ | $2$ | $(0, 2)$ |
| $4$ | $6$ | $(4, 6)$ |

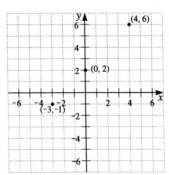

46. $\{(x,y)\,|\,x = y + 2 \text{ and } x = -3, 0, 4\}$

| $x$ | $y$ | $(x,y)$ |
|---|---|---|
| $-3$ | $-5$ | $(-3,-5)$ |
| $0$ | $-2$ | $(0,-2)$ |
| $4$ | $2$ | $(4,2)$ |

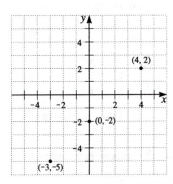

47. $(2,3), (0,6), (6,-3), (4,0)$

48. $(-2,-16), (0,-8), (3,4), (2,0)$

49. An ordered pair $(x,y)$ is a pair of numbers in which the order of appearance matters. For instance, $(3,4)$ and $(4,3)$ are different ordered pairs, even though both involve the same two numbers.

50. The notation $\{x,y\}$ stands for the set containing the two elements $x$ and $y$. Here, order does not matter. That is, $\{3,4\}$ and $\{4,3\}$ are equal sets. When we write $(x,y)$, we mean that $x$ comes first and $y$ second.

51.

| $x$ | $y$ | $(x,y)$ |
|---|---|---|
| $-1$ | $1$ | $(-1,1)$ |
| $0$ | $0$ | $(0,0)$ |
| $1$ | $1$ | $(1,1)$ |
| $2$ | $4$ | $(2,4)$ |

The graph of $y = x^2$ is not a straight line.

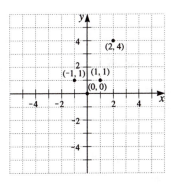

52.

| $x$ | $y$ | $(x,y)$ |
|---|---|---|
| $-1$ | $-1$ | $(-1,-1)$ |
| $0$ | $0$ | $(0,0)$ |
| $1$ | $1$ | $(1,1)$ |
| $2$ | $8$ | $(2,8)$ |

The graph of $y = x^3$ is not a straight line.

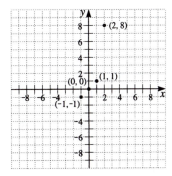

53. Let $x =$ number of HW problems Marina had.

$$\frac{1}{3}x + 8 + \frac{2}{5}x = x$$

$$15\left(\frac{1}{3}x + 8 + \frac{2}{5}x\right) = 15x$$

$$\overset{5}{15}\left(\frac{1}{3}x\right) + 15(8) + \overset{3}{15}\left(\frac{2}{5}x\right) = 15x$$

$$5x + 120 + 6x = 15x$$

$$11x + 120 = 15x$$

$$120 = 4x$$

$$30 = x$$

Thus, Marina had 30 homework problems.

## Exercises 5.2

1. To find the $x$-intercept, we set $y = 0$ and solve for $x$.

$$x - y = 7$$
$$x - (0) = 7$$
$$x - 0 = 7$$
$$x = 7$$

Therefore, the $x$-intercept is 7.

To find the $y$-intercept, we set $x = 0$ and solve for $y$.

$$x - y = 7$$
$$(0) - y = 7$$
$$0 - y = 7$$
$$-y = 7$$
$$y = -7$$

Therefore, the $y$-intercept is $-7$.

5. To find the $x$-intercept, we set $y = 0$ and solve for $x$.

$$2x + 3y = 12$$
$$2x + 3(0) = 12$$
$$2x + 0 = 12$$
$$2x = 12$$
$$x = 6$$

Therefore, the $x$-intercept is 6.

To find the $y$-intercept, we set $x = 0$ and solve for $y$.

$$2x + 3y = 12$$
$$2(0) + 3y = 12$$
$$0 + 3y = 12$$
$$3y = 12$$
$$y = 4$$

Therefore, the $y$-intercept is 4.

9. To find the $x$-intercept, we set $y = 0$ and solve for $x$.

$$2x = 5y$$
$$2x = 5(0)$$
$$2x = 0$$
$$x = 0$$

Therefore, the $x$-intercept is 0.

To find the $y$-intercept, we set $x = 0$ and solve for $y$.

$$2x = 5y$$
$$2(0) = 5y$$
$$0 = 5y$$
$$0 = y$$

Therefore, the $y$-intercept is 0.

11. To find the $x$-intercept, we set $y = 0$ and solve for $x$. Since the equation $x = 5$ does not contain $y$, its solution is just $x = 5$. So the $x$-intercept is 5. To find the $y$-intercept, we set $x = 0$ and solve for $y$. But this gives $0 = 5$, which is a contradiction. So the equation $x = 5$ has no $y$-intercept.

15. $x + y = -5$

    $x$-intercept:        $y$-intercept:

    $x + (0) = -5$    $(0) + y = -5$

    $x + 0\ \ = -5$     $0 + y = -5$

    $x\ \ \ \ \ \ \ = -5$        $y = -5$

    Plot $(-5, 0)$       Plot $(0, -5)$

    check point: choose $x = 1$

    $(1) + y = -5$

     $1 + y = -5$

         $y = -6$

    Plot $(1, -6)$

The graph of $x + y = -5$

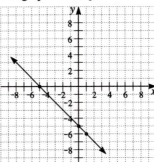

19. $3x - 4y = 12$

    $x$-intercept:        $y$-intercept:

    $3x - 4(0) = 12$    $3(0) - 4y = 12$

    $3x - 0\ \ \ = 12$     $0 - 4y = 12$

    $3x\ \ \ \ \ \ \ = 12$       $-4y = 12$

    $x\ \ \ \ \ \ \ \ = 4$          $y = -3$

    Plot $(4, 0)$        Plot $(0, -3)$

    check point: choose $x = 8$

    $3(8) - 4y =\ \ \ 12$

      $24 - 4y =\ \ \ 12$

         $-4y = -12$

            $y =\ \ \ \ 3$

    Plot $(8, 3)$

The graph of $3x - 4y = 12$

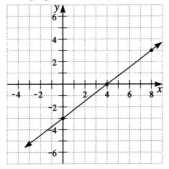

23. $y = -4x$

    $x$-intercept:

    $(0) = -4x$

    $\ 0 = -4x$

    $\ 0 =\ \ \ \ x$

    Plot $(0, 0)$

    second point: choose $x = 1$

    $y = -4(1)$

    $y = -4$

    Plot $(1, -4)$

    check point: choose $x = -1$

    $y = -4(-1)$

    $y = 4$

    Plot $(-1, 4)$

The graph of $y = -4x$

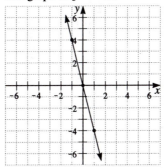

27. $y = 5$
Its graph is a line parallel to the $x$-axis and 5 units above it.

The graph of $y = 5$

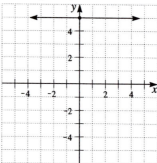

29. $x = -4$
Its graph is a line parallel to the $y$-axis and 4 units to the left of it.

The graph of $x = -4$

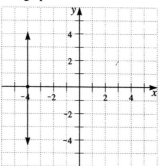

33. $2(x - 3) = 4(y + 2)$

$x$-intercept:
$2(x - 3) = 4[(0) + 2]$
$2(x - 3) = 4(0 + 2)$
$2x - 6 = 8$
$2x = 14$
$x = 7$

Plot $(7, 0)$

$y$-intercept:
$2[(0) - 3] = 4(y + 2)$
$2(0 - 3) = 4(y + 2)$
$-6 = 4y + 8$
$-14 = 4y$
$-\dfrac{7}{2} = y$

Plot $\left(0, -\dfrac{7}{2}\right)$

check point: choose $x = 1$
$2[(1) - 3] = 4(y + 2)$
$2(1 - 3) = 4(y + 2)$
$2(-2) = 4(y + 2)$
$-4 = 4y + 8$
$-12 = 4y$
$-3 = y$
Plot $(1, -3)$

The graph of $2(x - 3) = 4(y + 2)$

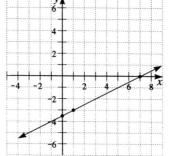

37. $d = 3t + 4$

    $t$-intercept:        $d$-intercept:

      $0 = 3t + 4$      $d = 3(0) + 4$

     $-4 = 3t$        $d = 0 + 4$

    $-\dfrac{4}{3} = t$         $d = 4$

    Plot $\left(-\dfrac{4}{3}, 0\right)$    Plot $(0, 4)$

    check point: choose $t = -1$
    $d = 3(-1) + 4$
    $d = -3 + 4$
    $d = 1$
    Plot $(-1, 1)$

The graph of $d = 3t + 4$

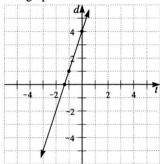

41. $u - 4v = 8$

    $u$-intercept:        $v$-intercept:

    $u - 4(0) = 8$    $(0) - 4v = 8$

      $u - 0 = 8$       $-4v = 8$

         $u = 8$         $v = -2$

    Plot $(8, 0)$        Plot $(0, -2)$

    check point: choose $v = -1$
    $u - 4(-1) = 8$
      $u + 4 = 8$
         $u = 4$
    Plot $(4, -1)$

The graph of $u - 4v = 8$

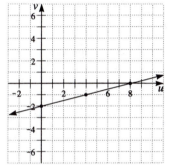

43.  (a)  $125v + 165t = 9110$

    (b)  $v$-intercept:           $t$-intercept:

       $125v + 165(0) = 9110$    $125(0) + 165t = 9110$

           $125v + 0 = 9110$         $0 + 165t = 9110$

              $125v = 9110$           $165t = 9110$

                 $v \approx 72.88$              $t \approx 55.21$

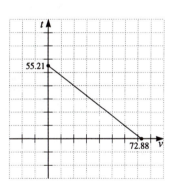

    (c)  $125(28) + 165t = 9110$

          $3500 + 165t = 9110$

              $165t = 5610$

                  $t = 34$

    The store orders 34 TVs.

45. (a) $12a + 8c = 360$

(b)  $a$-intercept:             $c$-intercept:
$$12a + 8(0) = 360 \qquad 12(0) + 8c = 360$$
$$12a + 0 = 360 \qquad 0 + 8c = 360$$
$$12a = 360 \qquad 8c = 360$$
$$a = 30 \qquad c = 45$$

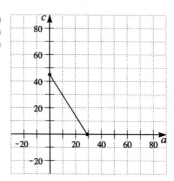

(c)  $$12(25) + 8c = 360$$
$$300 + 8c = 360$$
$$8c = 60$$
$$c = 7.5$$
Mary checked 7.5 credit reports.

47.

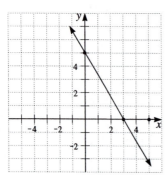

48. Some points on this line are $(-4, -4)$, $(-3, -3)$, $(0, 0)$, $(2, 2)$, and $(5, 5)$. Since the $y$-coordinate of any such point is equal to its $x$-coordinate, an equation of this line would be $y = x$.

The graph of $y = x$

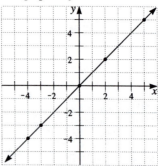

49. Some points on this line are $(-4, 4)$, $(-3, 3)$, $(0, 0)$, $(2, -2)$, and $(5, -5)$. Since the $y$-coordinate of any such point is equal to the negative of its $x$-coordinate, an equation of this line would be $y = -x$.

The graph of $y = -x$

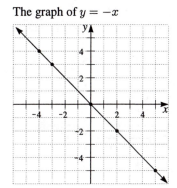

50. The $x$-coordinate of a point where a graph intersects the $x$-axis is called an $x$-intercept of the graph. To find an $x$-intercept, set $y = 0$ and solve for $x$. The $y$-coordinate of a point where a graph intersects the $y$-axis is called a $y$-intercept of the graph. To find a $y$-intercept, set $x = 0$ and solve for $y$.

51. $x^2(x^3)(x^2)^3 = x^2 x^3 x^6 = x^{2+3+6} = x^{11}$

53. $2x^2(5xy - y^2) - x^2y^2 = 2x^2(5xy) - 2x^2(y^2) - x^2y^2$
$$= 10x^3y - 2x^2y^2 - x^2y^2$$
$$= 10x^3y - 3x^2y^2$$

## Exercises 5.3

1. $m = \dfrac{y_2 - y_1}{x_2 - x_1} = \dfrac{9 - 5}{6 - 3} = \dfrac{4}{3}$

7. $m = \dfrac{y_2 - y_1}{x_2 - x_1} = \dfrac{-4 - (-2)}{-3 - (-1)} = \dfrac{-4 + 2}{-3 + 1}$
$$= \dfrac{-2}{-2} = 1$$

11. $m = \dfrac{y_2 - y_1}{x_2 - x_1} = \dfrac{7 - 7}{-3 - 4} = \dfrac{0}{-7} = 0$

13. $m$ is undefined. (If we tried to use the formula, we would have obtained $m = \dfrac{9 - 6}{2 - 2} = \dfrac{3}{0}$, and division by zero is undefined.)

17. $m = \dfrac{y_2 - y_1}{x_2 - x_1} = \dfrac{\frac{1}{4} - \frac{1}{5}}{\frac{3}{2} - \left(-\frac{1}{3}\right)} = \dfrac{\frac{1}{4} - \frac{1}{5}}{\frac{3}{2} + \frac{1}{3}}$

$= \dfrac{\frac{5}{20} - \frac{4}{20}}{\frac{9}{6} + \frac{2}{6}} = \dfrac{\frac{1}{20}}{\frac{11}{6}} = \dfrac{1}{20} \cdot \dfrac{\overset{3}{\cancel{6}}}{11} = \dfrac{3}{110}$

23. $m = \dfrac{y_2 - y_1}{x_2 - x_1} = \dfrac{0 - a}{a - 0} = \dfrac{-a}{a} = -1$

27. $m = \dfrac{y_2 - y_1}{x_2 - x_1} = \dfrac{4.72 - 2.65}{1.3 - 0.8} = \dfrac{2.07}{0.5}$
$$= 4.14$$

29. $m = \dfrac{y_2 - y_1}{x_2 - x_1} = \dfrac{4.9 - (-1.05)}{-2.16 - 3.7}$
$$= \dfrac{4.9 + 1.05}{-2.16 - 3.7} = \dfrac{5.95}{-5.86} \approx -1.02$$

31. $m = \dfrac{y_2 - y_1}{x_2 - x_1} = \dfrac{8.77 - 8.77}{-1.4 - 9.62} = \dfrac{0}{-11.02} = 0$

37.

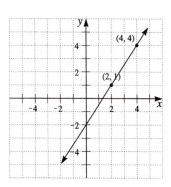

41.

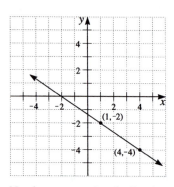

43. A slope of 0 means that the line is horizontal.   45. No slope means that the line is vertical.

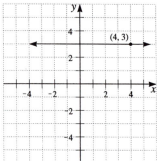

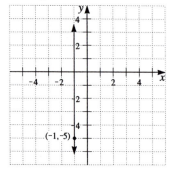

47. $\dfrac{-5}{4}$      49. 5      51. $\dfrac{7}{3}$      53. 0

55. positive      57. zero      59. $M_3, M_2, M_1$

61. (a) 40 calories

(b) Consider the points $(2, 10)$ and $(4, 20)$: slope $= \dfrac{c_2 - c_1}{m_2 - m_1} = \dfrac{20 - 10}{4 - 2} = \dfrac{10}{2} = 5$.

(c) 5 calories per minute are burned.

63. First, compute the slope of the line through $(-1, -2)$ and $(2, 0)$: $m = \dfrac{0 - (-2)}{2 - (-1)} = \dfrac{2}{3}$.

Next, compute the slope of the line through $(2, 0)$ and $(5, 2)$: $m = \dfrac{2 - 0}{5 - 2} = \dfrac{2}{3}$.

Since these two lines have equal slopes, they are either parallel or actually one line. But the two lines in question cannot be parallel, since $(2, 0)$ is a point on each and parallel lines have no point in common. We can then conclude that the three points $(-1, -2)$, $(2, 0)$, and $(5, 2)$ all lie on a straight line. (Points with this property are called <u>collinear</u>.)

65. $LCD = 12$

$$\frac{x}{3} - \frac{x-1}{4} + \frac{x+1}{2} = \frac{(4)x}{(4)3} - \frac{(3)(x-1)}{(3)4} + \frac{(6)(x+1)}{(6)2}$$

$$= \frac{4x}{12} - \frac{3(x-1)}{12} + \frac{6(x+1)}{12}$$

$$= \frac{4x - 3(x-1) + 6(x+1)}{12}$$

$$= \frac{4x - 3x + 3 + 6x + 6}{12}$$

$$= \frac{7x + 9}{12}$$

## Exercises 5.4

1. $y - y_1 = m(x - x_1)$
   $(x_1, y_1) = (1, 5), m = 3$
   $y - 5 = 3(x - 1)$ or
   $\qquad y = 3x + 2$

5. $y - y_1 = m(x - x_1)$
   $(x_1, y_1) = (-3, -5), m = -\dfrac{2}{3}$
   $y - (-5) = -\dfrac{2}{3}[x - (-3)]$
   $\qquad y + 5 = -\dfrac{2}{3}(x + 3)$ or
   $\qquad\quad y = -\dfrac{2}{3}x - 7$

9. $y = mx + b$
   $m = \dfrac{1}{4}, b = -2$
   $y = \dfrac{1}{4}x + (-2)$ or
   $y = \dfrac{1}{4}x - 2$

11. $y - y_1 = m(x - x_1)$
    $(x_1, y_1) = (-2, 0), m = -\dfrac{3}{4}$
    $y - 0 = -\dfrac{3}{4}[x - (-2)]$
    $\qquad y = -\dfrac{3}{4}(x + 2)$ or
    $\qquad y = -\dfrac{3}{4}x - \dfrac{3}{2}$

15. $y - y_1 = m(x - x_1)$
    $(x_1, y_1) = (5, 6), m = 0$
    $y - 6 = 0(x - 5)$
    $y - 6 = 0$ or
    $\qquad y = 6$

19. $m = \dfrac{y_2 - y_1}{x_2 - x_1} = \dfrac{-2 - 4}{2 - (-1)} = \dfrac{-6}{3} = -2$
    $(x_1, y_1) = (-1, 4)$
    $y - y_1 = m(x - x_1)$
    $\quad y - 4 = -2[x - (-1)]$
    $\quad y - 4 = -2(x + 1)$ or $y = -2x + 2$

23. $m = \dfrac{y_2 - y_1}{x_2 - x_1} = \dfrac{-1 - 0}{0 - (-1)} = \dfrac{-1}{1} = -1$

$b = -1$

$y = mx + b$

$y = (-1)x + (-1)$ or

$y = -x - 1$

27. Since the line is vertical, it has no slope. Therefore, we cannot use either the point-slope form or the slope-intercept form of an equation of a line. We know that all points on a vertical line have the same $x$-coordinate. Since our line passes through $(4, -3)$, its equation is $x = 4$.

31. An $x$-intercept of $-3$ means $(-3, 0)$ is on the line. A $y$-intercept of 4 means $(0, 4)$ is on the line.
Then $m = \dfrac{y_2 - y_1}{x_2 - x_1} = \dfrac{4 - 0}{0 - (-3)} = \dfrac{4}{3}$ and $b = 4$, so that an equation (in slope-intercept form) is

$y = \dfrac{4}{3}x + 4.$

33. $m = \dfrac{y_2 - y_1}{x_2 - x_1} = \dfrac{82 - 75}{3 - 2} = \dfrac{7}{1} = 7$

$y - y_1 = m(x - x_1)$

$y - 75 = 7(x - 2)$

$y - 75 = 7x - 14$

$y = 7x + 61$

When $x = 5$, $y = 7(5) + 61 = 35 + 61 = 96$. The grade would be 96 if a student studies for 5 hours.

35. $m = \dfrac{y_2 - y_1}{x_2 - x_1} = \dfrac{5 - 8}{8 - 6} = \dfrac{-3}{2} = -\dfrac{3}{2}$

$y - y_1 = m(x - x_1)$

$y - 8 = -\dfrac{3}{2}(x - 6)$

$y - 8 = -\dfrac{3}{2}x + 9$

$y = -\dfrac{3}{2}x + 17$

When $x = 4$, $y = -\dfrac{3}{2}(4) + 17 = -6 + 17 = 11$. The waiting time would be 11 minutes if 4 clerks are working.

37. $y = 5x + 7$

$y = mx + b$. So $m = 5$.

41. $\begin{aligned} x + y &= \quad 7 \\ -x \quad\quad &\quad -x \\ \hline y &= -x + 7 \end{aligned}$

$y = mx + b$. So $m = -1$.

**45.** $2x - 5y + 7 = 0$

$$\underline{-2x \qquad\qquad -2x}$$
$$-5y + 7 = -2x$$
$$\underline{\quad -7 \qquad\qquad -7}$$
$$-5y \quad = -2x - 7$$
$$\frac{-5y}{-5} = \frac{-2x - 7}{-5}$$
$$y = \frac{-2x}{-5} + \frac{-7}{-5}$$
$$y = \frac{2}{5}x + \frac{7}{5}$$
$$y = mx + b. \text{ So } m = \frac{2}{5}.$$

**47.** $y = mx + b$

$$m = \frac{2 - 0}{3 - 0} = \frac{2}{3}$$
$$b = 0$$
$$\text{So } y = \frac{2}{3}x.$$

**51.** $y - y_1 = m(x - x_1)$

$$m = \frac{3 - (-1)}{6 - (-6)} = \frac{4}{12} = \frac{1}{3}$$
$$y - 3 = \frac{1}{3}(x - 6)$$
$$y - 3 = \frac{1}{3}x - 2$$
$$y = \frac{1}{3}x + 1$$

**55.** By comparison with $y = mx + b$, the line whose equation is $y = 3x - 8$ has slope $= 3$. Since parallel lines have equal slopes, the line in question also has a slope of 3.

**57.** The line whose equation is $y = -x + 4$ has slope $= -1$. So the line in question also has a slope of $-1$, and it passes through $(-3, 0)$.

$$y - y_1 = m(x - x_1)$$
$$y - 0 = -1[x - (-3)]$$
$$y = -1(x + 3)$$
$$y = -x - 3$$

**59.** (a) $\quad m = \dfrac{c_2 - c_1}{f_2 - f_1} = \dfrac{80 - 73}{100 - 80} = \dfrac{7}{20}$

$$c - c_1 = m(f - f_1)$$
$$c - 73 = \frac{7}{20}(f - 80)$$
$$c - 73 = \frac{7}{20}f - 28$$
$$c = \frac{7}{20}f + 45$$

(b) When $f = 90$, $c = \dfrac{7}{20}(90) + 45$

$$= 31.5 + 45 = 76.5$$

The cost of selling 90 franks is \$76.50.

(c)  When $c = 90.50$, $90.50 = \dfrac{7}{20}f + 45$

$$45.50 = \dfrac{7}{20}f$$

$$\dfrac{20}{7}(45.50) = f$$

$$130 = f$$

130 franks can be sold for $90.50.

(d)  When $f = 0$, $c = \dfrac{7}{20}(0) + 45 = 0 + 45 = 45$.

So the vendor's fixed costs are $45.

61.  (a)  $m = \dfrac{N_2 - N_1}{s_2 - s_1} = \dfrac{82 - 80}{18 - 15} = \dfrac{2}{3}$

$N - N_1 = m(s - s_1)$

$N - 80 = \dfrac{2}{3}(s - 15)$

$N - 80 = \dfrac{2}{3}s - 10$

$N = \dfrac{2}{3}s + 70$

(b)  When $s = 20$, $N = \dfrac{2}{3}(20) + 70 = 13.33 + 70 = 83.33$.

The jogger's heart rate will be 83.33 beats per minute.

(c)  When $s = 0$, $N = \dfrac{2}{3}(0) + 70 = 0 + 70 = 70$.

The jogger's heart rate at rest is 70 beats per minute.

63.  (a)  $m = \dfrac{c_2 - c_1}{n_2 - n_1} = \dfrac{3.20 - 1.70}{2 - 1} = \dfrac{1.50}{1} = 1.50$

$c - c_1 = m(n - n_1)$

$c - 1.70 = 1.50(n - 1)$

$c - 1.70 = 1.50n - 1.50$

$c = 1.50n + 0.20$

(b)  When $n = 3.5$, $c = 1.50(3.5) + 0.20$

$$= 5.25 + 0.20 = 5.45$$

The cost of 3.5 oz of coffee would be $5.45.

(c)  When $c = 6.50$, $6.50 = 1.50n + 0.20$

$$6.30 = 1.50n$$

$$4.2 = n$$

The package contains 4.2 oz of coffee.

(d)

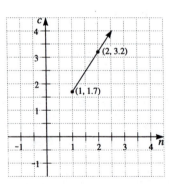

65. (a) $m = \dfrac{s_2 - s_1}{n_2 - n_1} = \dfrac{115 - 62}{-2 - 6} = \dfrac{53}{-8} = -6.625$

$s - s_1 = m(n - n_1)$

$s - 62 = -6.625(n - 6)$

$s - 62 = -6.625n + 39.75$

$s = -6.625 + 101.75$

(b) When $n = 4.5$, $s = -6.625(4.5) + 101.75$

$\approx -29.81 + 101.75$

$\approx 71.94$

On a 4.5° incline, the maximum speed would be 71.94 mph.

(c) When $n = -2.8$, $s = -6.625(-2.8) + 101.75$

$\approx 18.55 + 101.75$

$\approx 120.30$

On a 2.8° decline, the maximum speed would be 120.30 mph.

(d)

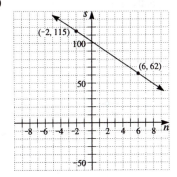

(e) When $n = 0$, $s = -6.625(0) + 101.75$

$= 0 + 101.75$

$= 101.75$

The $s$-intercept of this graph is 101.75. This means that the maximum speed on level ground is 101.75 mph.

(f) When $s = 0$, $0 = -6.625n + 101.75$

$6.625n = 101.75$

$n \approx 15.36$

The $n$-intercept of this graph is 15.36. This means that the car cannot make it up an incline which is 15.36° or more.

67. (a) $m = \dfrac{w_2 - w_1}{t_2 - t_1} = \dfrac{0.7 - 1.5}{33 - 24} = \dfrac{-0.8}{9} \approx -0.09$

$w - w_1 = m(t - t_1)$

$w - 1.5 = -0.09(t - 24)$

$w - 1.5 = -0.09t + 2.16$

$w = -0.09t + 3.66$

(b) When $t = 28$, $w = -0.09(28) + 3.66$

$\approx -2.52 + 3.66$

$\approx 1.14$

At 28° C, the width of a gap in this roadway would be 1.14 cm.

(c) When $w = 0$, $0 = -0.09t + 3.66$

$0.09t = 3.66$

$t \approx 40.67$

The gap would close completely when the temperature is 40.67° C.

(d) It is unlikely to occur, since 40.67° C is more than 105° F.

69.  The point-slope form is the most flexible since it accommodates *any point* and the slope, whereas the slope-intercept form is most useful for determining the slope of a line given its equation.

70.  There are many possible answers to this question. The following answers all depend on the "best fit" line we choose in part (a).

   (a)  $n = 0.85h - 14.5$

   (c)  If $h = 150$, we get $n = 133$ phone calls.

   (d)  The predicted value from the equation (133 calls) agrees very well with the actual number of calls (130).

   (e)  If $h = 225$ we get $n = 176.75$ which we would round off to 177 calls.

71.  (a)

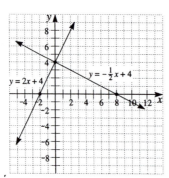

   (b)  From the graph, it appears that these lines are perpendicular.

   (c)  The line whose equation is $y = 2x + 4$ has slope $= 2$. The line whose equation is $y = -\frac{1}{2}x + 4$ has slope $= -\frac{1}{2}$. The product of these slopes is equal to $-1$. Put another way, the slopes are negative reciprocals of one another.

   (d)  The line whose equation is $y = \frac{2}{5}x + 7$ has slope $= \frac{2}{5}$. Therefore, any line that is perpendicular to this line must have slope $= -\frac{5}{2}$.

73.  Let $r =$ the alcohol percentage of the added solution.
$$0.60(40) + r(30) = 0.45(70)$$
$$100[0.60(40) + r(30)] = 100[0.45(70)]$$
$$100[0.60(40)] + 100[r(30)] = 100[0.45(70)]$$
$$60(40) + 3000r = 45(70)$$
$$2400 + 3000r = 3150$$
$$3000r = 750$$
$$r = 0.25$$
The added solution is 25% alcohol.

## Chapter 5 Review Exercises

1, 3, 5.

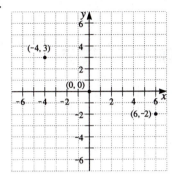

7.  $2x + 4y = 14$
    $2(3) + 4y = 14$
    $6 + 4y = 14$
    $4y = 8$
    $y = 2$
    Therefore, the ordered pair is $(3, 2)$.

9.  $x + 2y = 4$
    $x + 2(-3) = 4$
    $x - 6 = 4$
    $x = 10$
    Therefore, the ordered pair is $(10, -3)$.

11. $x - y = 0$
    $(0) - y = 0$
    $0 - y = 0$
    $-y = 0$
    $y = 0$
    Therefore, the ordered pair is $(0, 0)$.
    The graph of $x - y = 8$

13. $x - y = 8$
    x-intercept:       y-intercept:
    $x - (0) = 8$      $(0) - y = 8$
    $x - 0 = 8$        $0 - y = 8$
    $x = 8$            $-y = 8$
                       $y = -8$
    Plot $(8, 0)$      Plot $(0, -8)$

    check point: choose $x = 4$
    $(4) - y = 8$
    $4 - y = 8$
    $-y = 4$
    $y = -4$
    Plot $(4, -4)$

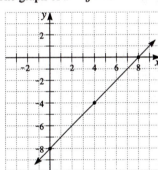

15. $3x + 7y = 21$
    x-intercept:         y-intercept:
    $3x + (0) = 21$      $3(0) + 7y = 21$
    $3x + 0 = 21$        $0 + 7y = 21$
    $3x = 21$            $7y = 21$
    $x = 7$              $y = 3$
    Plot $(7, 0)$        Plot $(0, 3)$

    check point: choose $y = 6$
    $3x + 7(6) = 21$
    $3x + 42 = 21$
    $3x = -21$
    $x = -7$
    Plot $(-7, 6)$

The graph of $3x + 7y = 21$

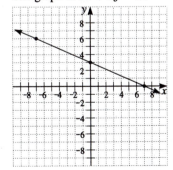

17. $y = 3x + 2$

$x$-intercept:     $y$-intercept:

$0 = 3x + 2$     $y = 3(0) + 2$

$-2 = 3x$     $y = 0 + 2$

$-\frac{2}{3} = x$     $y = 2$

Plot $\left(-\frac{2}{3}, 0\right)$     Plot $(0, 2)$

check point: choose $x = 1$
$y = 3(1) + 2$
$y = 3 + 2$
$y = 5$
Plot $(1, 5)$

The graph of $y = 3x + 2$

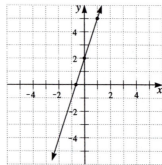

19. $x - 2y = 4$

$x$-intercept:     $y$-intercept:

$x - 2(0) = 4$     $(0) - 2y = 4$

$x - 0 = 4$     $0 - 2y = 4$

$x = 4$     $-2y = 4$

                  $y = -2$

Plot $(4, 0)$     Plot $(0, -2)$

check point: choose $y = -1$
$x - 2(-1) = 4$
$x + 2 = 4$
$x = 2$
Plot $(2, -1)$

The graph of $x - 2y = 4$

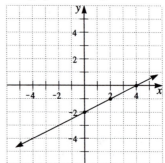

21. $x = 4$

Its graph is a line parallel to the $y$-axis and 4 units to the right of it.

The graph of $x = 4$

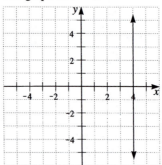

23. $3x + 2y = 12$

x-intercept:

$3x + 2(0) = 12$

$3x + 0 = 12$

$3x = 12$

$x = 4$

Plot $(4, 0)$

y-intercept:

$3(0) + 2y = 12$

$0 + 2y = 12$

$2y = 12$

$y = 6$

Plot $(0, 6)$

check point: choose $x = 2$

$3(2) + 2y = 12$

$6 + 2y = 12$

$2y = 6$

$y = 3$

Plot $(2, 3)$

The graph of $3x + 2y = 12$

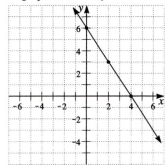

25. $3y = 6$

This equation is equivalent to $y = 2$ (divide both sides by 3), whose graph is a line parallel to the $x$-axis and 2 units above it.

The graph of $3y = 6$

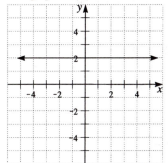

27. $m = \dfrac{y_2 - y_1}{x_2 - x_1} = \dfrac{7 - 5}{1 - 2} = \dfrac{2}{-1} = -2$

29. $m = \dfrac{y_2 - y_1}{x_2 - x_1} = \dfrac{2 - (-2)}{3 - (-3)}$

$= \dfrac{2 + 2}{3 + 3} = \dfrac{4}{6} = \dfrac{2}{3}$

31. $m = \dfrac{y_2 - y_1}{x_2 - x_1} = \dfrac{8 - 8}{-1 - 1} = \dfrac{0}{-2} = 0$

33. $(x_1, y_1) = (3, 2), m = 4$

$y - y_1 = m(x - x_1)$

$y - 2 = 4(x - 3)$ or $y = 4x - 10$

35. $m = \dfrac{y_2 - y_1}{x_2 - x_1} = \dfrac{0 - (-6)}{1 - 2} = \dfrac{0 + 6}{1 - 2} = \dfrac{6}{-1} = -6$

$(x_1, y_1) = (2, -6)$

$y - y_1 = m(x - x_1)$

$y - (-6) = -6(x - 2)$

$y + 6 = -6(x - 2)$ or $y = -6x + 6$

37. All points on a horizontal line have the same $y$-coordinate. Since the point $(3, 8)$ is on our line, an equation for this horizontal line must be $y = 8$.

39. $-\dfrac{3}{4}$

41. The line whose equation is $y = 5x - 1$ has slope $= 5$. So the line in question has slope $= 5$, and it passes through $(-3, 5)$.

$$y - y_1 = m(x - x_1)$$
$$y - 5 = 5[x - (-3)]$$
$$y - 5 = 5(x + 3)$$
$$y - 5 = 5x + 15$$
$$y = 5x + 20$$

43. 
$$y = mx + b$$
$$m = \frac{4 - 0}{3 - 0} = \frac{4}{3}$$
$$b = 0$$
So $y = \frac{4}{3}x$

45. Let $h$ = number of overtime hours.
$d$ = number of defective items.
$$m = \frac{d_2 - d_1}{h_2 - h_1} = \frac{17 - 12}{10 - 8} = \frac{5}{2}$$
$$d - d_1 = m(h - h_1)$$
$$d - 12 = \frac{5}{2}(h - 8)$$
$$d - 12 = \frac{5}{2}x - 20$$
$$d = \frac{5}{2}x - 8$$

When $h = 20$, $d = \frac{5}{2}(20) - 8 = 50 - 8 = 42$.

If the worker puts in 20 overtime hours, you would expect to find 42 defective items.

## Chapter 5 Practice Test

1. 
$$\frac{2y - x}{5} = x - y$$
$$\frac{2(-4) - (-3)}{5} \overset{?}{=} (-3) - (-4)$$
$$\frac{-8 + 3}{5} \overset{?}{=} -3 + 4$$
$$\frac{-5}{5} \overset{?}{=} 1$$
$$-1 \neq 1$$

So the point $(-3, -4)$ does not satisfy the given equation.

3. (a) $3x - 5y = 15$

x-intercept:     y-intercept:

$3x - 5(0) = 15$    $3(0) - 5y = 15$

$3x - 0 = 15$       $0 - 5y = 15$

$3x = 15$          $-5y = 15$

$x = 5$           $y = -3$

Plot $(5, 0)$      Plot $(0, -3)$

check point: choose $x = -5$

$3(-5) - 5y = 15$

$-15 - 5y = 15$

$-5y = 30$

$y = -6$

Plot $(-5, -6)$

The graph of $3x - 5y = 15$

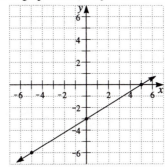

(b) $y = 3x - 7$

x-intercept:     y-intercept:

$0 = 3x - 7$      $y = 3(0) - 7$

$7 = 3x$           $y = 0 - 7$

$\frac{7}{3} = x$         $y = -7$

Plot $\left(\frac{7}{3}, 0\right)$     Plot $(0, -7)$

check point: choose $x = 2$

$y = 3(2) - 7$

$y = 6 - 7$

$y = -1$

Plot $(2, -1)$

The graph of $y = 3x - 7$

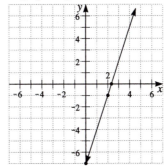

(c) $y + 2x = 0$

x-intercept:     (This implies that

$0 + 2x = 0$     $y = 0$ is the

$2x = 0$       y-intercept.)

$x = 0$

Plot $(0, 0)$

second point: choose $x = 1$

$y + 2(1) = 0$

$y + 2 = 0$

$y = -2$

Plot $(1, -2)$

check point: choose $x = -1$

$y + 2(-1) = 0$

$y - 2 = 0$

$y = 2$

Plot $(-1, 2)$

The graph of $y + 2x = 0$

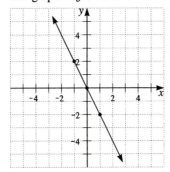

(d)  $x = 4$

This is an equation of a line parallel to the $y$-axis and 4 units to the right of it.

The graph of $x = 4$

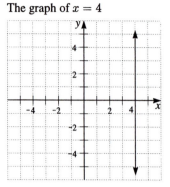

(e)  $y = -3$

This is an equation of a line parallel to the $x$-axis and 3 units below it.

The graph of $y = -3$

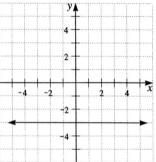

5.  $m = \dfrac{4}{3}$, $(x_1, y_1) = (4, -1)$

$$y - y_1 = m(x - x_1)$$
$$y - (-1) = \frac{4}{3}(x - 4)$$
$$y + 1 = \frac{4}{3}(x - 4) \text{ or}$$
$$y = \frac{4}{3}x - \frac{19}{3}$$

7.  $y = 5$ is the horizontal line passing through $(3, 5)$.

$x = 3$ is the vertical line passing through $(3, 5)$.

9.  The line whose equation is $y = -3x + 4$ has slope $= -3$. So the line in equation has slope $= -3$ and it passes through $(-4, 0)$.

$$y - y_1 = m(x - x_1)$$
$$y - 0 = -3[x - (-4)]$$
$$y - 0 = -3(x + 4)$$
$$y = -3x - 12$$

11.  Let $C$ = number of phone calls.
     $A$ = amount of charity pledged.

$$m = \frac{A_2 - A_1}{C_2 - C_1} = \frac{115 - 80}{21 - 15} = \frac{35}{6}$$

$$A - A_1 = m(C - C_1)$$

$$A - 80 = \frac{35}{6}(C - 15)$$

$$A - 80 = \frac{35}{6}C - \frac{175}{2}$$

$$A = \frac{35}{6}C - \frac{15}{2}$$

When $C = 30$, $A = \dfrac{35}{6}(30) - \dfrac{15}{2}$

$$= 175 - 7.50 = 167.50.$$

If they make 30 phone calls, the amount of charity pledged will be $167.50.

# Chapter 6
# Interpreting Graphs and Systems of Equations

## Exercises 6.1

1. (a) B      (b) B      (c) A      (d) B

3. (a) 2,000 calls      (b) 4:00 a.m.      (c) 10,000 calls
   (d) From 4:00 p.m. to 4:00 a.m.      (e) 10:00 a.m. to 4:00 p.m.
   (e) 10:00 a.m. to 4:00 p.m.

5. (a) This graph denies the belief that temperature drops steadily as altitude increases. Notice that the temperature increases as the altitude increases from 12 to 50 miles and again when the altitude is greater than 80 miles.

   (b) The temperature decreases when the altitude is between 0 and 12 miles and also when it is between 50 and 80 miles.

   (c) For the first 5 miles that the balloon rises, the temperature will increase; for the last 5 miles, it will decrease.

7. (a) Immediately after the students memorized the list of 20 words, the group remembered all twenty.

   (b) after 4 hours

   (c) during the first two hours

   (d) 16 words are forgotten during the first 4 hours; 2 more words are forgotten during the next 4 hours.

9. (a) 150 deer

   (b) between 1996 and 1998

   (c) 800 deer

11. (a) The purchase price of Machine A is $10,000; the purchase price of Machine B is $7,500.

    (b) It takes Machine A 8 years and it takes Machine B 10 years.

    (c) The machines have the same value 5 years after they were purchased.

    (d) The value of Machine A is greater than the value of Machine B for the first five years. The opposite is true during the next five years.

13. Let $m$ = number of minutes that elapse from the beginning of the experiment.
    $T$ = temperature of the metal bar (in $°C$).

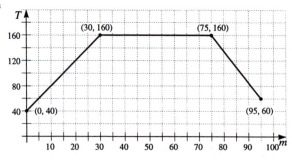

15. $\dfrac{6x^2}{25yz^3} \div 15xyz = \dfrac{6x^2}{25yz^3} \div \dfrac{15xyz}{1}$

$= \dfrac{\overset{2x}{\cancel{6}x^2}}{25yz^3} \cdot \dfrac{1}{\underset{5}{\cancel{15}}\cancel{x}yz} = \dfrac{2x}{125y^2z^4}$

17. Let $x$ = number of pairs of dress slacks.
Then $36 - x$ = number of pairs of casual slacks.
$$40x + 25(36 - x) = 1185$$
$$40x + 900 - 25x = 1185$$
$$15x + 900 = 1185$$
$$15x = 285$$
$$x = 19$$

Then $36 - x = 36 - 19 = 17$. So 19 pairs of dress slacks and 17 pairs of casual slacks were bought.

## Exercises 6.2

1. (a)

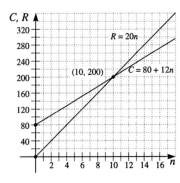

(b) $(10, 200)$

(c) In order to break even, the potter must sell 10 pieces.

3. $x + y = 4$
Set $y = 0$ and solve for $x$ to get an $x$-intercept of 4.
Set $x = 0$ and solve for $y$ to get a $y$-intercept of 4.

$x - y = 2$
Set $y = 0$ and solve for $x$ to get an $x$-intercept of 2.
Set $x = 0$ and solve for $y$ to get a $y$-intercept of $-2$.

The lines cross at the point $(3, 1)$. So the system
$\begin{cases} x + y = 4 \\ x - y = 2 \end{cases}$ is satisfied by the point $(3, 1)$.

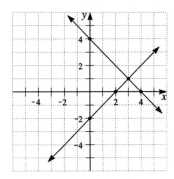

7. $3x + y = 6$
Set $y = 0$ and solve for $x$ to get an $x$-intercept of 2.
Set $x = 0$ and solve for $y$ to get a $y$-intercept of 6.

$6x + 2y = 12$
Set $y = 0$ and solve for $x$ to get an $x$-intercept of 2.
Set $x = 0$ and solve for $y$ to get a $y$-intercept of 6.

Since these lines have the same $x$-intercept and the same $y$-intercept, the lines coincide. Therefore, there are infinitely many solutions to the system.
$$\begin{cases} 3x + y = 6 \\ 6x + 2y = 12 \end{cases} : \{(x,\ y)|3x + y = 6\}.$$

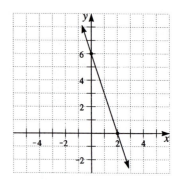

11. $3x - 2y = 6$
Set $y = 0$ and solve for $x$ to get an $x$-intercept of 2.
Set $x = 0$ and solve for $y$ to get a $y$-intercept of $-3$.

$x + y = 2$
Set $y = 0$ and solve for $x$ to get an $x$-intercept of 2.
Set $x = 0$ and solve for $y$ to get a $y$-intercept of 2.

The lines cross at the point $(2,\ 0)$. So the system
$$\begin{cases} 3x - 2y = 6 \\ x + y = 2 \end{cases}$$ is satisfied by the point $(2,\ 0)$.

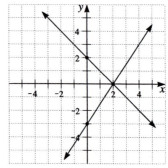

15. $5x - 3y = 15$
Set $y = 0$ and solve for $x$ to get an $x$-intercept of 3.
Set $x = 0$ and solve for $y$ to get a $y$-intercept of $-5$.

$2x - y = 4$
Set $y = 0$ and solve for $x$ to get an $x$-intercept of 2.
Set $x = 0$ and solve for $y$ to get a $y$-intercept of $-4$.

The lines cross at the point $(-3,\ -10)$. So the system
$$\begin{cases} 5x - 3y = 15 \\ 2x - y = 4 \end{cases}$$ is satisfied by the point $(-3,\ -10)$.

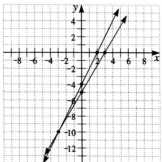

19. $y = x - 3$
Set $y = 0$ and solve for $x$ to get an $x$-intercept of 3.
Set $x = 0$ and solve for $y$ to get a $y$-intercept of $-3$.

$y = x + 4$
Set $y = 0$ and solve for $x$ to get an $x$-intercept of $-4$.
Set $x = 0$ and solve for $y$ to get a $y$-intercept of 4.

These lines appear to be parallel. Therefore, the system
$$\begin{cases} y = x - 3 \\ y = x + 4 \end{cases}$$ has no solution.

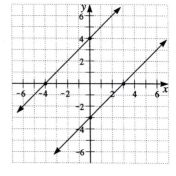

23. $3x + y = 3$
Set $y = 0$ and solve for $x$ to get an $x$-intercept of 1.
Set $x = 0$ and solve for $y$ to get a $y$-intercept of 3.

$y = x + 5$
Set $y = 0$ and solve for $x$ to get an $x$-intercept of $-5$.
Set $x = 0$ and solve for $y$ to get a $y$-intercept of 5.

The lines cross at the point $\left(-\dfrac{1}{2}, \dfrac{9}{2}\right)$. So the system

$\begin{cases} 3x + y = 3 \\ y = x + 5 \end{cases}$ is satisfied by the point $\left(-\dfrac{1}{2}, \dfrac{9}{2}\right)$.

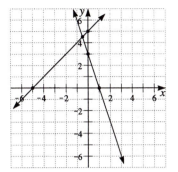

27. $2x + y = 10$
Set $y = 0$ and solve for $x$ to get an $x$-intercept of 5.
Set $x = 0$ and solve for $y$ to get a $y$-intercept of 10.

$2y = 20 - 4x$
Set $y = 0$ and solve for $x$ to get an $x$-intercept of 5.
Set $x = 0$ and solve for $y$ to get a $y$-intercept of 10.

Since these lines have the same $x$-intercept and the same $y$-intercept, the lines coincide. Therefore, there are infinitely many solutions to the system.

$\begin{cases} 2x + y = 10 \\ 2y = 20 - 4x \end{cases}$ : $\{(x, y)\,|\,2x + y = 10\}$.

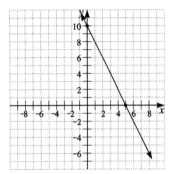

31. $x + y = 6$
Set $y = 0$ and solve for $x$ to get an $x$-intercept of 6.
Set $x = 0$ and solve for $y$ to get a $y$-intercept of 6.

$y = -2$
This is a horizontal line two units below the $x$-axis.

The lines cross at the point $(8, -2)$. So the system
$\begin{cases} x + y = 6 \\ y = -2 \end{cases}$ is satisfied by the point $(8, -2)$.

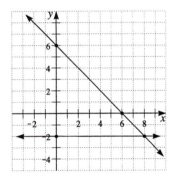

35. $y = 4$
This is a horizontal line four units above the $x$-axis.

$x = -1$
This is a vertical line one unit to the left of the $y$-axis.

The lines cross at the point $(-1, 4)$. So the system
$\begin{cases} y = 4 \\ x = -1 \end{cases}$ is satisfied by the point $(-1, 4)$.

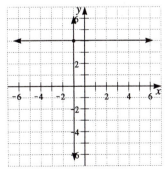

37. A solution to a system of equations is an ordered pair that satisfies both equations simultaneously.

38. It is impossible for a system of two linear equations to have exactly two solutions. It this were possible, then we would be able to draw two straight lines that intersect exactly twice. But two straight lines either do not intersect at all (when they are parallel), intersect in one point, or intersect in infinitely many points (when they coincide).

39. 1-b, 2-e, 3-a, 4-c, 5-d, 6-f

41.
$$9 - 5(x + 4) > 9$$
$$9 - 5x - 20 > 9$$
$$-5x - 11 > 9$$
$$-5x > 20$$
$$x < -4$$

43.
$$1 \le 7 - 2x < 13$$
$$-6 \le -2x < 6$$
$$3 \ge x > -3$$
or
$$-3 < x \le 3$$

45. Let $8x$ = number of members who voted in favor.
Then $7x$ = number of members who voted against.
$$8x + 7x = 435$$
$$15x = 435$$
$$x = 29$$

When $x = 29, 8x = 8(29) = 232$. So 232 members were in favor of the bill.

# Exercises 6.3

3.
$$\begin{cases} 2x + y = 5 \\ x - y = 4 \end{cases} \text{Add.}$$
$$3x = 9$$
$$x = 3$$

$$2x + y = 5$$
$$2(3) + y = 5$$
$$y = -1$$

Solution: $(3, -1)$
CHECK: $x - y = 4$
$$(3) - (-1) \overset{?}{=} 4$$
$$3 + 1 \overset{?}{=} 4$$
$$4 \overset{\checkmark}{=} 4$$

7.
$$\begin{cases} 2x + y = 15 \\ x - 2y = 0 \end{cases} \xrightarrow[\text{multiply by } -2]{\text{as is}}$$

$$2x + y = 15$$
$$-2x + 4y = 0 \text{ Add.}$$
$$5y = 15$$
$$y = 3$$

$$2x + y = 15$$
$$2x + (3) = 15$$
$$2x + 3 = 15$$
$$2x = 12$$
$$x = 6$$

Solution: $(6, 3)$
CHECK: $x - 2y = 0$
$$(6) - 2(3) \overset{?}{=} 0$$
$$6 - 6 \overset{?}{=} 0$$
$$0 \overset{\checkmark}{=} 0$$

11. 
$$\begin{cases} x + 2y = 9 \\ y = 3x + 1 \end{cases}$$
Substitute the value of $y$ given in the second equation into the first.

$x + 2(3x + 1) = 9$

$x + 6x + 2 = 9$ 

Solution: $(1, 4)$

$7x + 2 = 9$ 

CHECK: $x + 2y = 9$

$7x = 7$

$(1) + 2(4) \stackrel{?}{=} 9$

$x = 1$  When $x = 1, y = 3(1) + 1 = 3 + 1 = 4$

$1 + 8 \stackrel{?}{=} 9$

$9 \stackrel{\checkmark}{=} 9$

17. 
$$\begin{cases} 2r - 5s = 9 \\ s = 1 - r \end{cases}$$
Substitute the value of $s$ given in the second equation into the first.

$2r - 5(1 - r) = 9$

$2r - 5 + 5r = 9$ 

Solution: $(2, -1)$

$7r - 5 = 9$ 

CHECK: $2r - 5s = 9$

$7r = 14$

$2(2) - 5(-1) \stackrel{?}{=} 9$

$r = 2$  When $r = 2, s = 1 - 2 = -1$

$4 + 5 \stackrel{?}{=} 9$

$9 \stackrel{\checkmark}{=} 9$

21. 
$$\begin{cases} r + 2t = 10 \\ 3r + t = -15 \end{cases}$$
$\xrightarrow[\text{multiply by } -2]{\text{as is}}$

$r + 2t = 10$

$-6r - 2t = 30$  Add

$-5r \quad = 40$

$r = -8$

$r + 2t = 10$

$(-8) + 2t = 10$

$-8 + 2t = 10$

$2t = 18$

$t = 9$

Solution: $(-8, 9)$

CHECK: $3r + t = -15$

$3(-8) + (9) \stackrel{?}{=} -15$

$-24 + 9 \stackrel{?}{=} -15$

$-15 \stackrel{\checkmark}{=} -15$

23. 
$$\begin{cases} 6x + y = 6 \\ 4x + 1 = y \end{cases}$$
Rewrite the second equation in the system as $4x - y = -1$.

$6x + y = 6$

$4x - y = -1$  Add

$10x \quad = 5$

$x = \dfrac{5}{10} = \dfrac{1}{2}$

$6x + y = 6$

$6\left(\dfrac{1}{2}\right) + y = 6$

$3 + y = 6$

$y = 3$

Solution: $\left(\dfrac{1}{2}, 3\right)$

CHECK: $4x + 1 = y$

$4\left(\dfrac{1}{2}\right) + 1 \stackrel{?}{=} (3)$

$2 + 1 \stackrel{?}{=} 3$

$3 \stackrel{\checkmark}{=} 3$

27. $\begin{cases} 11a - 2b = 30 \\ 3a + 3b = -6 \end{cases}$ $\xrightarrow[\text{multiply by 2}]{\text{multiply by 3}}$ $\begin{aligned} 33a - 6b &= 90 \\ 6a + 6b &= -12 \quad \text{Add} \\ \hline 39a \phantom{{}-6b} &= 78 \\ a &= 2 \end{aligned}$     $\begin{aligned} 11a - 2b &= 30 \\ 11(2) - 2b &= 30 \\ 22 - 2b &= 30 \\ -2b &= 8 \\ b &= -4 \end{aligned}$

Solution: $(2, -4)$
CHECK: $3a + 3b = -6$
$\qquad 3(2) + 3(-4) \overset{?}{=} -6$
$\qquad\qquad 6 - 12 \overset{?}{=} -6$
$\qquad\qquad\quad -6 \overset{\checkmark}{=} -6$

31. $\begin{cases} 5x + 2y = 4y + 9 \\ y = x - 3 \end{cases}$ $\xrightarrow[\text{subtract } x \text{ from both sides}]{\text{subtract } 4y \text{ from both sides}}$ $\begin{aligned} 5x - 2y &= 9 \\ -x + y &= -3 \end{aligned}$ $\xrightarrow[\text{multiply by 2}]{\text{as is}}$ $\begin{aligned} 5x - 2y &= 9 \\ -2x + 2y &= -6 \quad \text{Add} \\ \hline 3x \phantom{{}- 2y} &= 3 \\ x &= 1 \end{aligned}$

$y = x - 3$       Solution: $(1, -2)$
$y = (1) - 3$     CHECK: $5x + 2y = 4y + 9$
$y = 1 - 3$       $\qquad 5(1) + 2(-2) \overset{?}{=} 4(-2) + 9$
$y = -2$          $\qquad\qquad 5 - 4 \overset{?}{=} -8 + 9$
$\qquad\qquad\qquad\qquad\qquad 1 \overset{\checkmark}{=} 1$

33. $\begin{cases} \dfrac{x}{2} + \dfrac{y}{3} = 1 \\[2mm] \dfrac{x}{4} - y = 11 \end{cases}$ $\xrightarrow[\text{multiply by 4}]{\text{multiply by 6}}$ $\begin{aligned} 3x + 2y &= 6 \\ x - 4y &= 44 \end{aligned}$ $\xrightarrow[\text{as is}]{\text{multiply by 2}}$ $\begin{aligned} 6x + 4y &= 12 \\ x - 4y &= 44 \quad \text{Add} \\ \hline 7x \phantom{{}+ 4y} &= 56 \\ x &= 8 \end{aligned}$

$\dfrac{x}{2} + \dfrac{y}{3} = 1$          Solution: $(8, -9)$
$\dfrac{(8)}{2} + \dfrac{y}{3} = 1$          CHECK: $\dfrac{x}{4} - y = 11$
$4 + \dfrac{y}{3} = 1$             $\dfrac{(8)}{4} - (-9) \overset{?}{=} 11$
$\dfrac{y}{3} = -3$                $2 + 9 \overset{?}{=} 11$
$y = -9$                   $11 \overset{\checkmark}{=} 11$

37. $\begin{cases} 0.4x + 0.2y = 8 \\ 0.7x - 0.3y = 1 \end{cases}$ $\xrightarrow[\text{multiply by 10}]{\text{multiply by 10}}$ $\begin{aligned} 4x + 2y &= 80 \\ 7x - 3y &= 10 \end{aligned}$ $\xrightarrow[\text{multiply by 2}]{\text{multiply by 3}}$ $\begin{aligned} 12x + 6y &= 240 \\ 14x - 6y &= 20 \quad \text{Add} \\ \hline 26x \phantom{{}- 6y} &= 260 \\ x &= 10 \end{aligned}$

$$0.4x + 0.2y = 8$$
$$0.4(10) + 0.2y = 8$$
$$4 + 0.2y = 8$$
$$0.2y = 4$$
$$2y = 40$$
$$y = 20$$

Solution: $(10, 20)$

CHECK: $0.7x - 0.3y = 1$
$$0.7(10) - 0.3(20) \overset{?}{=} 1$$
$$7 - 6 \overset{?}{=} 1$$
$$1 \overset{\vee}{=} 1$$

41.  $(5.7, 1.1)$

43.  $(1.9, -0.6)$

45.  The student forgot to multiply the right hand side of the second equation by 2.

46.  The student mistakenly multiplied 0 times 2 in the second equation and got a product of 2 rather than 0.

## Exercises 6.4

3.  Let $n$ = number of nickels that Sam has.
    Let $q$ = number of quarters that Sam has.

$$\begin{cases} n + q = 80 \\ 5n + 25q = 1360 \end{cases} \quad \xrightarrow[\text{as is}]{\text{multiply by } -5} \quad \begin{array}{c} -5n - 5q = -400 \\ 5n + 25q = 1360 \ \text{Add} \\ \hline 20q = 960 \\ q = 48 \end{array} \qquad \begin{array}{c} n + q = 80 \\ n + 48 = 80 \\ n = 32 \end{array}$$

Thus, Sam has 32 nickels and 48 quarters.
CHECK: $32 + 48 \overset{\vee}{=} 80$ and $5(32) + 25(48) = 160 + 1200 \overset{\vee}{=} 1360$.

5.  Let $x$ = speed of the slower car.
    Let $y$ = speed of the faster car.

$$\begin{cases} y = x + 15 \\ 5x + 5y = 275 \end{cases} \quad \xrightarrow[\text{as is}]{\text{subtract } x \text{ from both sides}} \quad \begin{array}{c} -x + y = 15 \\ 5x + 5y = 275 \end{array} \quad \xrightarrow[\text{as is}]{\text{multiply by } 5} \quad \begin{array}{c} -5x + 5y = 75 \\ 5x + 5y = 275 \ \text{Add} \\ \hline 10y = 350 \\ y = 35 \end{array}$$

$$35 = x + 15$$
$$20 = x \qquad \text{Thus, the speed of the slower car is 20 kph and the speed of the faster car is 35 kph.}$$
CHECK: The difference of the speeds is $35 - 20 = 15$ kph. The slower car travels $5(20) = 100$ km and the faster car travels $5(35) = 175$ kph. The distance between the cars after 5 hours is $100 + 175 = 275$ km.

7.  Let $x$ = amount of money that Carmen invested at 6%.
    Let $y$ = amount of money that Carmen invested at 7%.

$$\begin{cases} x + y = 1700 \\ 0.06x + 0.07y = 110 \end{cases} \quad \xrightarrow[\text{multiply by } 100]{\text{as is}} \quad \begin{array}{c} x + y = 1700 \\ 6x + 7y = 11000 \end{array} \quad \xrightarrow[\text{as is}]{\text{multiply by } -6} \quad \begin{array}{c} -6x - 6y = -10200 \\ 6x + 7y = 11000 \\ \text{Add:} \ \hline y = 800 \end{array}$$

$$x + y = 1700$$
$$x + 800 = 1700$$
$$x = 900 \qquad \text{Thus, Carmen invested \$900 at 6\% and \$800 at 7\%.}$$
CHECK: $\$900 + \$800 \overset{\vee}{=} \$1700$ and $0.06(\$900) + 0.07(\$800) = \$54 + \$56 \overset{\vee}{=} \$110$

9. Let $x$ = price of a cassette.
   Let $y$ = price of a CD.

   $$\begin{cases} 4x + 6y = 107.66 \\ 5x + 3y = 76.30 \end{cases} \xrightarrow[\text{multiply by } -2]{as\ is}$$

   $$\begin{array}{r} 4x + 6y = \phantom{-}107.66 \\ -10x - 6y = -152.60 \\ \hline -6x \phantom{aaaaa} = -44.94 \\ x = \phantom{-}7.49 \end{array} \text{Add}$$

   $$\begin{array}{r} 4x + 6y = 107.66 \\ 4(7.49) + 6y = 107.66 \\ 29.96 + 6y = 107.66 \\ 6y = 77.70 \\ y = 12.95 \end{array}$$

   Thus, a cassette costs \$7.49 and a CD costs \$12.95.
   CHECK: $4(\$7.49) + 6(\$12.95) = \$29.96 + \$77.70 \overset{\checkmark}{=} \$107.66$ and
   $5(\$7.49) + 3(\$12.95) = \$37.45 + \$38.85 \overset{\checkmark}{=} \$76.30$.

11. Let $L$ = length of the rectangle.
    Let $W$ = width of the rectangle.

    $$\begin{cases} L = 2W \\ 2L + 2W = 28 \end{cases}$$ Substitute the value of $L$ from the first equation into the second to get

    $$2(2W) + 2W = 28$$
    $$4W + 2W = 28$$
    $$6W = 28$$
    $$W = \frac{28}{6} = \frac{14}{3} \qquad \text{So } L = 2W = 2\left(\frac{14}{3}\right) = \frac{28}{3}$$

    Thus, the width of the rectangle is $\frac{14}{3}$ inches and the length is $\frac{28}{3}$ inches.
    CHECK: $\frac{28}{3}$ is twice as large as $\frac{14}{3}$.
    The perimeter of the rectangle is $2\left(\frac{28}{3}\right) + 2\left(\frac{14}{3}\right) = \frac{56}{3} + \frac{28}{3} = \frac{84}{3} = 28$, as required.

15. Let $x$ = larger number.
    Let $y$ = smaller number.

    $$\begin{cases} \dfrac{x}{y} = \dfrac{6}{5} \\ x - y = 8 \end{cases}$$

    Solve the second equation for $x$, obtaining $x = y + 8$. Then substitute this result into the first equation.

    $$\frac{y + 8}{y} = \frac{6}{5}$$
    $$\frac{5\cancel{y}}{1} \cdot \frac{y + 8}{\cancel{y}} = \frac{\cancel{5}y}{1} \cdot \frac{6}{\cancel{5}}$$
    $$5(y + 8) = 6y$$
    $$5y + 40 = 6y$$
    $$40 = y$$

    Then $x = y + 8 = 40 + 8 = 48$. Thus, the numbers are 48 and 40.
    CHECK: $\frac{48}{40} = \frac{6(8)}{5(8)} \overset{\checkmark}{=} \frac{6}{5}$ and $48 - 40 \overset{\checkmark}{=} 8$.

19.　Let $x$ = cost of a receiver.
　　Let $y$ = cost of a turntable.

$$\begin{cases} 8x + 4y = 2060 \\ 5x + 6y = 1690 \end{cases} \xrightarrow[\text{multiply by } -2]{\text{multiply by } 3}$$

$$\begin{array}{rl} 24x + 12y = & 6180 \\ -10x - 12y = & -3380 \quad \text{Add} \\ \hline 14x \quad\quad = & 2800 \\ x = & 200 \end{array}$$

$$\begin{array}{rl} 8x + 4y = & 2060 \\ 8(200) + 4y = & 2060 \\ 1600 + 4y = & 2060 \\ 4y = & 460 \\ y = & 115 \end{array}$$

　　Thus, a receiver costs \$200 and a turntable costs \$115.
　　CHECK: $8(\$200) + 4(\$115) = \$1600 + \$460 \overset{\checkmark}{=} \$2060$ and
　　$5(\$200) + 6(\$115) = \$1000 + \$690 \overset{\checkmark}{=} \$1690.$

23.　Let $x$ = number of \$7 books bought.
　　Let $y$ = number of \$9 books bought.

$$\begin{cases} x + y = 35 \\ 7x + 9y = 271 \end{cases} \xrightarrow[\text{as is}]{\text{multiply by } -7}$$

$$\begin{array}{rl} -7x - 7y = & -245 \\ 7x + 9y = & 271 \quad \text{Add} \\ \hline 2y = & 26 \\ y = & 13 \end{array}$$

$$\begin{array}{rl} x + y = & 35 \\ x + 13 = & 35 \\ x = & 22 \end{array}$$

　　Thus, the bookstore bought 22 books at \$7 each and 13 books at \$9 each.
　　CHECK: $22 + 13 \overset{\checkmark}{=} 35$ and $(\$7)(22) + (\$9)(13) = \$154 + \$117 \overset{\checkmark}{=} \$271.$

27.　Let $p$ = speed (in mph) of the plane in still air.
　　Let $w$ = speed (in mph) of the wind.

$$\begin{array}{rl} \begin{cases} p + w = 150 \\ \underline{p - w = \phantom{0}90} \quad \text{Add} \\ 2p \phantom{ - w} = 240 \\ p = 120 \end{cases} & \end{array} \qquad \begin{array}{l} p + w = 150 \\ 120 + w = 150 \\ w = 30 \end{array}$$

　　Thus, the speed of the plane in still air is 120 mph and the speed of the wind is 30 mph.
　　CHECK: With the tailwind, the speed of the plane is increased by the speed of the wind, giving
　　$120 + 30 \overset{\checkmark}{=} 150$ mph. With the headwind, the speed of the plane is decreased by the speed of the
　　wind, giving $120 - 30 \overset{\checkmark}{=} 90$ mph.

31.　Let $d$ = number of desktop setups.
　　Let $C$ = total cost of system.

$$\begin{cases} C = 100,000 + 800d \\ C = \phantom{0}16,000 + 1200d \end{cases} \quad \text{Substitute the value of } C \text{ from the first equation into the second.}$$

$$\begin{array}{rl} 100,000 + 800d = & 16,000 + 1200d \\ 100,000 = & 16,000 + 400d \\ 84,000 = & 400d \\ 210 = & d \end{array}$$

　　The two systems will cost the same when there are 210 desktops setups.
　　CHECK: $100,000 + 800(210) = 100,000 + 168,000 \overset{\checkmark}{=} \$268,000.$
　　$16,000 + 1200(210) = 16,000 + 252,000 \overset{\checkmark}{=} \$268,000.$

33. Let $c =$ cost of a computer.
Let $p =$ cost of a printer.

$$\begin{cases} 10c + 10p = 10,000 \\ 12c + 2p = 10,000 \end{cases} \xrightarrow[\text{multiply by } -5]{\text{as is}}$$

$$\begin{aligned} 10c + 10p &= 10,000 \\ -60c - 10p &= -50,000 \\ \hline -50c &= -40,000 \\ c &= 800 \end{aligned} \text{Add}$$

$$\begin{aligned} 10c + 10p &= 10,000 \\ 10(800) + 10p &= 10,000 \\ 8,000 + 10p &= 10,000 \\ 10p &= 2000 \\ p &= 200 \end{aligned}$$

Thus, a computer costs \$800 and a printer costs \$200.

CHECK: $10(\$800) + 10(\$200) = \$8,000 + \$2,000 \overset{\checkmark}{=} \$10,000$ and
$12(\$800) + 2(\$200) = \$9,600 + \$400 \overset{\checkmark}{=} \$10,000.$

## Chapter 6 Review Exercises

1. $x + y = 4$
Set $y = 0$ and solve for $x$ to get an $x$-intercept of 4.
Set $x = 0$ and solve for $y$ to get a $y$-intercept of 4.

$x - y = 0$
Here, both the $x$ and $y$ intercepts are 0. To find a
second point on this line, choose $y = 1$ and find
$x = 1$. This gives $(1, 1)$.

The lines cross at the point $(2, 2)$. So the system
$\begin{cases} x + y = 4 \\ x - y = 0 \end{cases}$ is satisfied by the point $(2, 2)$.

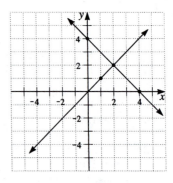

3. $x - 2y = 8$
Set $y = 0$ and solve for $x$ to get an $x$-intercept of 8.
Set $x = 0$ and solve for $y$ to get a $y$-intercept of $-4$.

$y = x - 5$
Set $y = 0$ and solve for $x$ to get an $x$-intercept of 5.
Set $x = 0$ and solve for $y$ to get a $y$-intercept of $-5$.

The lines cross at the point $(2, -3)$. So the system
$\begin{cases} x - 2y = 8 \\ y = x - 5 \end{cases}$ is satisfied by the point $(2, -3)$.

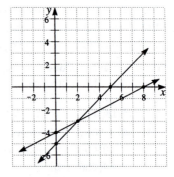

5. $y = x$

Here, both the $x$ and $y$ intercepts are 0. To find a second point on this line, choose $x = 1$ and find $y = 1$. This gives $(1, 1)$.

$3x - 2y = 6$

Set $y = 0$ and solve for $x$ to get an $x$-intercept of 2.
Set $x = 0$ and solve for $y$ to get a $y$-intercept of $-3$.

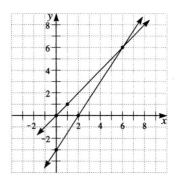

The lines cross at the point $(6, 6)$. So the system
$\begin{cases} y = x \\ 3x - 2y = 6 \end{cases}$ is satisfied by the point $(6, 6)$.

7. $\begin{cases} x + y = 4 \\ x - y = 6 \end{cases}$

$\begin{aligned} x + y &= 4 \\ x - y &= 6 \\ \hline 2x &= 10 \quad \text{Add} \\ x &= 5 \end{aligned}$

$\begin{aligned} x + y &= 4 \\ 5 + y &= 4 \\ y &= -1 \end{aligned}$

Solution: $(5, -1)$
CHECK: $x - y = 6$
$(5) - (-1) \overset{?}{=} 6$
$5 + 1 \overset{?}{=} 6$
$6 \overset{\checkmark}{=} 6$

9. $\begin{cases} y = 2x - 3 \\ x = 3y - 2 \end{cases}$ Substitute the value of $x$ given in the second equation into the first equation.

$\begin{aligned} y &= 2(3y - 2) - 3 \\ y &= 6y - 4 - 3 \\ y &= 6y - 7 \\ -5y &= -7 \\ y &= \frac{7}{5} \end{aligned}$

When $y = \frac{7}{5}$, $x = 3\left(\frac{7}{5}\right) - 2 = \frac{11}{5}$.

Solution: $\left(\frac{11}{5}, \frac{7}{5}\right)$
CHECK: $y = 2x - 3$
$\frac{7}{5} \overset{?}{=} 2\left(\frac{11}{5}\right) - 3$
$\frac{7}{5} \overset{?}{=} \frac{22}{5} - 3$
$\frac{7}{5} \overset{\checkmark}{=} \frac{7}{5}$

11. $\begin{cases} 2x - y = 10 \\ x + 3y = -16 \end{cases}$ $\xrightarrow{\text{multiply by 3}}$ $\xrightarrow{\text{as is}}$

$\begin{aligned} 6x - 3y &= 30 \\ x + 3y &= -16 \quad \text{Add} \\ \hline 7x &= 14 \\ x &= 2 \end{aligned}$

$\begin{aligned} 2x - y &= 10 \\ 2(2) - y &= 10 \\ 4 - y &= 10 \\ -y &= 6 \\ y &= -6 \end{aligned}$

Solution: $(2, -6)$
CHECK: $x + 3y = -16$
$(2) + 3(-6) \overset{?}{=} -16$
$2 - 18 \overset{?}{=} -16$
$-16 \overset{\checkmark}{=} -16$

13. $\begin{cases} x - 5y = 1 \\ 3x - 2y = 3 \end{cases}$ $\xrightarrow[\text{as is}]{\text{multiply by } -3}$ $\begin{array}{r} -3x + 15y = -3 \\ 3x - 2y = 3 \end{array}$ Add $\qquad$ $\begin{array}{r} x - 5y = 1 \\ x - 5(0) = 1 \\ x - 0 = 1 \\ x = 1 \end{array}$

$\begin{array}{r} \underline{\phantom{3x - 2y = 3}} \\ 13y = 0 \\ y = 0 \end{array}$

Solution: $(1, 0)$

CHECK: $3x - 2y = 3$

$3(1) - 2(0) \overset{?}{=} 3$

$3 - 0 \overset{?}{=} 3$

$3 \overset{\checkmark}{=} 3$

15. $\begin{cases} 4x - 3y = 10 \\ 9x + 2y = 5 \end{cases}$ $\xrightarrow[\text{multiply by } 3]{\text{multiply by } 2}$ $\begin{array}{r} 8x - 6y = 20 \\ 27x + 6y = 15 \end{array}$ Add $\qquad$ $\begin{array}{r} 4x - 3y = 10 \\ 4(1) - 3y = 10 \\ 4 - 3y = 10 \\ -3y = 6 \\ y = -2 \end{array}$

$\begin{array}{r} \underline{\phantom{27x + 6y = 15}} \\ 35x = 35 \\ x = 1 \end{array}$

Solution: $(1, -2)$

CHECK: $9x + 2y = 5$

$9(1) + 2(-2) \overset{?}{=} 5$

$9 - 4 \overset{?}{=} 5$

$5 \overset{\checkmark}{=} 5$

17. $\begin{cases} \dfrac{x}{2} + y = 5 \\ 2y = 8 - x \end{cases}$ $\xrightarrow[\text{add } x \text{ to both sides}]{\text{multiply by } 2}$ $\begin{array}{r} x + 2y = 10 \\ x + 2y = 8 \end{array}$ Subtract

$\begin{array}{r} \underline{\phantom{x + 2y = 8}} \\ 0 = 2 \end{array}$

This is a contradiction.

Therefore, the system of equations has no solution.

19. $\begin{cases} x + y - 8 = 2x - 4 \\ 2(y - x) = 8 \end{cases}$ $\xrightarrow[\text{divide both sides by } 2]{\text{subtract } 2x \text{ from both sides}}$ $\begin{array}{r} y - x - 8 = -4 \\ y - x = 4 \end{array}$ $\xrightarrow[\text{as is}]{\text{add } 8 \text{ to both sides}}$ $\begin{array}{r} y - x = 4 \\ y - x = 4 \end{array}$

Subtract: $\overline{\phantom{y - x = 4}}$
$0 = 0$

This is an identity.

Therefore, the system of equations has infinitely many solutions: $\{(x, y) \mid y - x = 4\}$.

21. Let $x =$ number of gallons pure water in the mixture.
Let $y =$ number of gallons of 30% alcohol solution in the mixture.

$\begin{cases} x + y = 30 \\ 0x + 0.30y = 0.25(30) \end{cases}$ $\xrightarrow[\text{multiply by } 100]{\text{as is}}$ $\begin{array}{r} x + y = 30 \\ 30y = 750 \end{array}$

From the second equation, find $y = 25$. Then $30 = x + y = x + 25$, so $x = 5$. Thus, 5 gallons of water should be added to 25 gallons of a 30% alcohol solution to produce 30 gallons of 25% alcohol solution.

CHECK: $0.30(25) = 7.5$ gallons of pure alcohol and $0.25(30) = 7.5$ gallons of pure alcohol.

23.  Let $x$ = walking speed (in kph).
Let $y$ = jogging speed (in kph).

$$\begin{cases} x + y = 17 \\ 2x + \dfrac{1}{2}y = 16 \end{cases} \xrightarrow[\text{multiply by } -2]{\text{as is}} \quad \begin{aligned} x + y &= \ \ 17 \\ -4x - y &= -32 \end{aligned} \ \text{Add} \quad \begin{aligned} x + y &= 17 \\ 5 + y &= 17 \\ y &= 12 \end{aligned}$$

$$\overline{\begin{aligned} -3x &= -15 \\ x &= \ \ 5 \end{aligned}}$$

Thus, their walking speed is 5 kph and their jogging speed is 12 kph.

CHECK:  $1(5) + 1(12) \overset{\checkmark}{=} 17$ and $2(5) + \dfrac{1}{2}(12) = 10 + 6 \overset{\checkmark}{=} 16.$

25.  (a)  7
     (b)  decreasing
     (c)  0; occurs when $w = 9$
     (d)  9; occurs when $w = 7$
     (e)  the value of $R$ when $w = 5$.

## Chapter 6 Practice Test

1.  (a)  \$32 per share
    (b)  between 10:00 a.m. and 11:00 a.m. and between 12:00 noon and 1:00 p.m.
    (c)  \$30 per share
    (d)  Yes.  This occurs between 11:00 a.m. and 12:00 noon.
    (e)  rising

3.  $\begin{cases} 3x - 4 = y - 1 \\ 9 + 3y = x \end{cases}$  Substitute the value of $x$ given in the second equation into the first equation.

$3(9 + 3y) - 4 = y - 1$  When $y = -3,$  Solution:  $(0, -3)$
$27 + 9y - 4 = y - 1$  $x = 9 + 3(-3) = 9 - 9 = 0$  CHECK: $3x - 4 = y - 1$
$9y + 23 = y - 1$  $3(0) - 4 \overset{?}{=} (-3) - 1$
$8y + 23 = -1$  $0 - 4 \overset{?}{=} -3 - 1$
$8y = -24$  $-4 \overset{\checkmark}{=} -4$
$y = -3$

5.  $\begin{cases} \dfrac{3x}{2} - y = 6 \\ x - \dfrac{2y}{3} = 5 \end{cases} \begin{aligned} &\xrightarrow{\text{multiply by } 2} \\ &\xrightarrow{\text{multiply by } 3} \end{aligned} \quad \begin{aligned} 3x - 2y &= 12 \\ 3x - 2y &= 15 \end{aligned} \ \text{Subtract}$

$$\overline{\quad\quad 0 = -3 \quad}$$

This is a contradiction.

Therefore, the system of equations has no solution.

7.  $2x - y = -8$
    Set $y = 0$ and solve for $x$ to get an $x$-intercept of $-4$.
    Set $x = 0$ and solve for $y$ to get a $y$-intercept of $8$.

    $x + 2y = 6$
    Set $y = 0$ and solve for $x$ to get an $x$-intercept of $6$.
    Set $x = 0$ and solve for $y$ to get a $y$-intercept of $3$.

    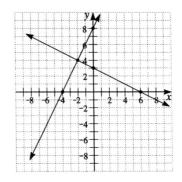

    The lines cross at the point $(-2, 4)$. So the system
    $\begin{cases} 2x - y = -8 \\ x + 2y = 6 \end{cases}$ is satisfied by the point $(-2, 4)$.

9.  Let $x =$ amount invested at 4%.
    Let $y =$ amount invested at 5%.

    $\begin{cases} x + y = 5550 \\ 0.04x + 0.05y = 259 \end{cases}$ $\xrightarrow{\substack{as\ is}}$ $x + y = 5550$ $\xrightarrow{\substack{multiply\ by\ -4}}$ $-4x - 4y = -22200$

    $\xrightarrow{\substack{multiply\ by\ 100}}$ $4x + 5y = 25900$ $\xrightarrow{\substack{as\ is}}$ $4x + 5y = \phantom{0}25900$

    Add: $\overline{\phantom{4x + 5y = \ } y = \phantom{00}3700}$

    $x + y = 5550$
    $x + 3700 = 5550$
    $x = 1850$
    Thus, \$1850 is invested at 4% and \$3700 is invested at 5%.
    CHECK: $\$1850 + \$3700 \overset{\checkmark}{=} \$5500$
    $0.04(\$1850) + 0.05(\$3700) = \$74 + \$185 \overset{\checkmark}{=} \$259.$

# Chapters 4 - 6
# Cumulative Review

1. $\dfrac{-24}{42} = \dfrac{(-4)(6)}{(7)(6)} = \dfrac{-4}{7} = -\dfrac{4}{7}$

3. $\dfrac{36 \overset{9}{s^8} \overset{t}{t^9}}{\underset{5}{20} s^9 \underset{s}{t^8}} = \dfrac{9t}{5s}$

5. $\dfrac{6\overset{}{\cancel{t}}}{\underset{5}{25}} \cdot \dfrac{\overset{2}{10}}{\cancel{t}} = \dfrac{12}{5}$

7. $\dfrac{6x}{25} + \dfrac{10}{x} = \dfrac{6x(x)}{25(x)} + \dfrac{10(25)}{x(25)}$

$$= \dfrac{6x^2}{25x} + \dfrac{250}{25x}$$

$$= \dfrac{6x^2 + 250}{25x}$$

9. $\dfrac{3t-5}{6t^2} + \dfrac{9t+5}{6t^2} = \dfrac{3t-5+9t+5}{6t^2}$

$$= \dfrac{\overset{2}{\cancel{12t}}}{\underset{t}{6t^2}} = \dfrac{2}{t}$$

11. $\dfrac{12x^3y^2}{35z^2} \div \dfrac{20xy}{14z} = \dfrac{\overset{3x^2y}{\cancel{12x^3y^2}}}{\underset{5z}{\cancel{35z^2}}} \cdot \dfrac{\overset{2}{\cancel{14z}}}{\underset{5}{\cancel{20xy}}} = \dfrac{6x^2y}{25z}$

13. $\dfrac{5}{3x} - \dfrac{7}{2x} = \dfrac{5(2)}{3x(2)} - \dfrac{7(3)}{2x(3)}$

$$= \dfrac{10}{6x} - \dfrac{21}{6x} = \dfrac{10-21}{6x}$$

$$= \dfrac{-11}{6x} = -\dfrac{11}{6x}$$

15. $\dfrac{5}{6x^2y} - \dfrac{9}{10xy^3} = \dfrac{5(5y^2)}{6x^2y(5y^2)} - \dfrac{9(3x)}{10xy^3(3x)}$

$$= \dfrac{25y^2}{30x^2y^3} - \dfrac{27x}{30x^2y^3}$$

$$= \dfrac{25y^2 - 27x}{30x^2y^3}$$

17. $\left(8 \cdot \dfrac{4}{x}\right) \div \dfrac{16}{x^2} = \dfrac{32}{x} \div \dfrac{16}{x^2}$

$$= \dfrac{\overset{2}{32}}{\underset{1}{\cancel{x}}} \cdot \dfrac{\overset{x}{x^2}}{\underset{1}{16}}$$

$$= \dfrac{2x}{1} = 2x$$

19. $2 + \dfrac{3}{x} - \dfrac{1}{x^2} = \dfrac{2}{1} + \dfrac{3}{x} - \dfrac{1}{x^2}$

$$= \dfrac{2(x^2)}{1(x^2)} + \dfrac{3(x)}{x(x)} - \dfrac{1}{x^2}$$

$$= \dfrac{2x^2}{x^2} + \dfrac{3x}{x^2} - \dfrac{1}{x^2}$$

$$= \dfrac{2x^2 + 3x - 1}{x^2}$$

21.
$$\frac{x}{3} - \frac{x}{4} = \frac{x-4}{6} \qquad \text{LCD} = 12$$

$$12\left(\frac{x}{3} - \frac{x}{4}\right) = 12\left(\frac{x-4}{6}\right)$$

$$\frac{\overset{4}{\cancel{12}}}{1} \cdot \frac{x}{\underset{1}{\cancel{3}}} - \frac{\overset{3}{\cancel{12}}}{1} \cdot \frac{x}{\underset{1}{\cancel{4}}} = \frac{\overset{2}{\cancel{12}}}{1} \cdot \frac{x-4}{\underset{1}{\cancel{6}}}$$

$$4x - 3x = 2(x-4)$$

$$x = 2x - 8$$

$$-x = -8$$

$$x = 8$$

23.
$$\frac{a}{5} - \frac{a}{6} = \frac{a}{30} \qquad \text{LCD} = 30$$

$$30\left(\frac{a}{5} - \frac{a}{6}\right) = 30\left(\frac{a}{30}\right)$$

$$\frac{\overset{6}{\cancel{30}}}{1} \cdot \frac{a}{\underset{1}{\cancel{5}}} - \frac{\overset{5}{\cancel{30}}}{1} \cdot \frac{a}{\underset{1}{\cancel{6}}} = \frac{\overset{1}{\cancel{30}}}{1} \cdot \frac{a}{\underset{1}{\cancel{30}}}$$

$$6a - 5a = a$$

$$a = a \qquad \text{Identity}$$

25.
$$\frac{7-2y}{4} - \frac{5-4y}{6} = \frac{8y+5}{9} \qquad \text{LCD} = 36$$

$$36\left(\frac{7-2y}{4} - \frac{5-4y}{6}\right) = 36\left(\frac{8y+5}{9}\right)$$

$$\frac{\overset{9}{\cancel{36}}}{1} \cdot \frac{7-2y}{\underset{1}{\cancel{4}}} - \frac{\overset{6}{\cancel{36}}}{1} \cdot \frac{5-4y}{\underset{1}{\cancel{6}}} = \frac{\overset{4}{\cancel{36}}}{1} \cdot \frac{8y+5}{\underset{1}{\cancel{9}}}$$

$$9(7-2y) - 6(5-4y) = 4(8y+5)$$

$$63 - 18y - 30 + 24y = 32y + 20$$

$$6y + 33 = 32y + 20$$

$$33 = 26y + 20$$

$$13 = 26y$$

$$\frac{\overset{1}{\cancel{13}}}{\underset{2}{\cancel{26}}} = y$$

$$\frac{1}{2} = y$$

27.
$$0.8x - 0.07(x-5) = 58.75 \qquad \text{LCD} = 100$$

$$100[0.8x - 0.07(x-5)] = 100(58.75)$$

$$100(0.8x) - 100[0.07(x-5)] = 100(58.75)$$

$$80x - 7(x-5) = 5875$$

$$80x - 7x + 35 = 5875$$

$$73x + 35 = 5875$$

$$73x = 5840$$

$$x = 80$$

29. $y = 2x - 6$

x-intercept:  y-intercept:

$0 = 2x - 6$  $y = 2(0) - 6$

$6 = 2x$  $y = 0 - 6$

$3 = x$  $y = -6$

Plot $(3, 0)$  Plot $(0, -6)$

check point: choose $x = 2$

$y = 2(2) - 6$

$y = 4 - 6$

$y = -2$

Plot $(2, -2)$

The graph of $y = 2x - 6$

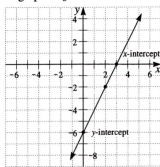

31. $3y - 6x = 12$

x-intercept:  y-intercept:

$3(0) - 6x = 12$  $3y - 6(0) = 12$

$0 - 6x = 12$  $3y - 0 = 12$

$-6x = 12$  $3y = 12$

$x = -2$  $y = 4$

Plot $(-2, 0)$  Plot $(0, 4)$

check point: choose $y = 2$

$3(2) - 6x = 12$

$6 - 6x = 12$

$-6x = 6$

$x = -1$

Plot $(-1, 2)$

The graph of $3y - 6x = 12$

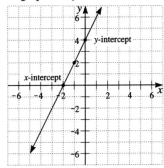

33. $x - 2 = 0$

Add 2 to both sides of this equation to get the equivalent equation $x = 2$. The graph of the equation is a line parallel to the $y$-axis and 2 units to the right of it.

The graph of $x - 2 = 0$

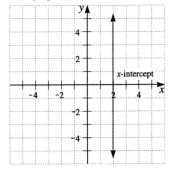

35. $y = 5x$
x-intercept:
$(0) = 5x$
$\quad 0 = 5x$
$\quad 0 = x$
Plot $(0, 0)$
(This implies that the $y$-intercept is 0.)

second point: choose $x = 1$
$y = 5(1)$
$y = 5$
Plot $(1, 5)$

check point: choose $x = -1$
$y = 5(-1)$
$y = -5$
Plot $(-1, -5)$

The graph of $y = 5x$

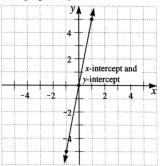

37. $m = \dfrac{y_2 - y_1}{x_2 - x_1} = \dfrac{4 - (-1)}{-3 - 2} = \dfrac{4 + 1}{-3 - 2}$

$\quad = \dfrac{5}{-5} = -1$

39. $m = \dfrac{y_2 - y_1}{x_2 - x_1} = \dfrac{4 - 4}{-1 - 2} = \dfrac{0}{-3} = 0$

41. $m = 4, (x_1, y_1) = (2, 3)$

$y - y_1 = m(x - x_1)$

$y - 3 = 4(x - 2)$ or $y = 4x - 5$

43. $m = -\dfrac{3}{4}, b = 3$

$y = mx + b$

$y = -\dfrac{3}{4}x + 3$

45. $m = \dfrac{y_2 - y_1}{x_2 - x_1} = \dfrac{-2 - 5}{2 - (-3)} = \dfrac{-2 - 5}{2 + 3}$

$\quad = \dfrac{-7}{5} = -\dfrac{7}{5}$

$(x_1, y_1) = (-3, 5)$

$y - y_1 = m(x - x_1)$

$y - 5 = -\dfrac{7}{5}[x - (-3)]$

$y - 5 = -\dfrac{7}{5}(x + 3)$ or

$\quad y = -\dfrac{7}{5}x + \dfrac{4}{5}$

47. $2x - y = 7$

Set $y = 0$ and solve for $x$ to get an $x$-intercept of $\dfrac{7}{2}$.

Set $x = 0$ and solve for $y$ to get a $y$-intercept of $-7$.

$x + 2y = 6$
Set $y = 0$ and solve for $x$ to get an $x$-intercept of 6.
Set $x = 0$ and solve for $y$ to get a $y$-intercept of 3.

The lines cross at the point $(4,\ 1)$. So the system

$\begin{cases} 2x - y = 7 \\ x + 2y = 6 \end{cases}$ is satisfied by the point $(4,\ 1)$.

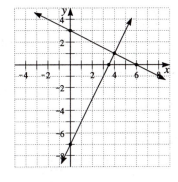

49. $\begin{cases} 2x - y = 7 \\ x + 2y = 6 \end{cases}$ $\xrightarrow[\text{as is}]{\text{multiply by 2}}$ $\begin{aligned} 4x - 2y &= 14 \\ x + 2y &= \phantom{0}6 \quad \text{Add} \\ \hline 5x \phantom{ - 2y} &= 20 \\ x &= \phantom{0}4 \end{aligned}$ $\qquad \begin{aligned} 2x - y &= 7 \\ 2(4) - y &= 7 \\ 8 - y &= 7 \\ -y &= -1 \\ y &= 1 \end{aligned}$

Solution: $(4,\ 1)$
CHECK: $x + 2y = 6$

$(4) + 2(1) \overset{?}{=} 6$

$4 + 2 \overset{?}{=} 6$

$6 \overset{\checkmark}{=} 6$

51. $\begin{cases} 4x - 3y = 0 \\ 2x - y = \dfrac{1}{3} \end{cases}$ $\begin{array}{l}\xrightarrow{\text{as is}} \\ \xrightarrow{\text{multiply by } -3}\end{array}$ $\begin{aligned} 4x - 3y &= \phantom{-}0 \\ -6x + 3y &= -1 \quad \text{Add} \\ \hline -2x \phantom{+ 3y} &= -1 \\ x &= \dfrac{1}{2} \end{aligned}$ $\qquad \begin{aligned} 4x - 3y &= 0 \\ 4\left(\dfrac{1}{2}\right) - 3y &= 0 \\ 2 - 3y &= 0 \\ 2 &= 3y \\ \dfrac{2}{3} &= y \end{aligned}$

Solution: $\left(\dfrac{1}{2},\ \dfrac{2}{3}\right)$

CHECK: $2x - y = \dfrac{1}{3}$

$2\left(\dfrac{1}{2}\right) - \left(\dfrac{2}{3}\right) \overset{?}{=} \dfrac{1}{3}$

$1 - \dfrac{2}{3} \overset{?}{=} \dfrac{1}{3}$

$\dfrac{1}{3} \overset{\checkmark}{=} \dfrac{1}{3}$

53. $\begin{cases} y = 5x - 4 \\ x = 3y + 12 \end{cases}$

Substitute the value of $y$ given in the first equation into the second. We get

$x = 3(5x - 4) + 12$                             Solution: $(0, -4)$

$x = 15x - 12 + 12$        Then $y = 5x - 4$       CHECK: $x = 3y + 12$

$x = 15x$                   $= 5(0) - 4$           $(0) \overset{?}{=} 3(-4) + 12$

$0 = 14x$                   $= -4$              $0 \overset{?}{=} -12 + 12$

$0 = x$                                     $0 \overset{\checkmark}{=} 0$

55. $\begin{cases} x + \dfrac{y}{2} = 5 & \xrightarrow{\text{multiply by 2}} & 2x + y = 10 \\ 2x + y = 10 & \xrightarrow{\text{as is}} & 2x + y = 10 \quad \text{Subtract} \end{cases}$

$$\frac{}{\qquad 0 = 0 \qquad}$$

This is an identity.

Therefore, the system has infinitely many solutions: $\{(x, y) \mid 2x + y = 10\}$.

57. Let $x =$ number of cheaper tickets sold.

Then $360 - x =$ number of more expensive tickets sold.

$$6.25x + 8.75(360 - x) = 2850$$
$$100[6.25x + 8.75(360 - x)] = 100 \cdot 2850$$
$$100(6.25x) + 100[8.75(360 - x)] = 100 \cdot 2850$$
$$625x + 875(360 - x) = 285000$$
$$625x + 315000 - 875x = 285000$$
$$-250x + 315000 = 285000$$
$$-250x = -30000$$
$$x = 120$$

Then $360 - x = 360 - 120 = 240$. Thus, 120 tickets at \$6.25 each and 240 tickets at \$8.75 were sold.

59. Let $x =$ number of votes that Party A received.

$$\frac{x}{15700} = \frac{8}{5}$$
$$15700 \cdot \frac{x}{15700} = 15700 \cdot \frac{8}{5}$$
$$\cancel{15700} \cdot \frac{x}{\cancel{15700}} = \overset{3140}{\cancel{15700}} \cdot \frac{8}{\cancel{5}}$$
$$x = 25,120$$

Thus, Party A received $25,120$ votes.

61. Let $t =$ number of hours needed for the faster car to overtake the slower one.

$$80t = 65\left(t + \frac{1}{4}\right)$$
$$80t = 65t + 65 \cdot \frac{1}{4}$$
$$15t = \frac{65}{4}$$
$$t = \frac{\left(\frac{65}{4}\right)}{15} = \frac{65}{4} \cdot \frac{1}{15} = \frac{65}{60} = 1\frac{5}{60}$$

Thus, the faster car to overtakes the slower one after 1 hour and 5 minutes.

63. (a) 18

(b) No. It decreases from 1960 to 1970 and from 1970 to 1980.

(c) 1900 to 1910; 6 people per square mile.

(d) 1910 to 1920.

# Chapters 4 - 6
# Cumulative Practice Test

1. $\dfrac{\overset{4s}{\cancel{12s^2}}\ \cancel{t^3}}{\underset{1}{\cancel{5d^2}}} \cdot \dfrac{\overset{\frac{1}{3}d^3}{\cancel{15d^5}}}{\underset{\underset{1}{3t}}{\cancel{9st^4}}} = \dfrac{4sd^3}{t}$

3. $\dfrac{11a}{9x} - \dfrac{a}{9x} + \dfrac{5a}{9x} = \dfrac{11a - a + 5a}{9x}$

$$= \dfrac{\overset{5}{\cancel{15}}a}{\underset{3}{\cancel{9}}x} = \dfrac{5a}{3x}$$

5. $\dfrac{5}{6ab^2} + \dfrac{4}{9b} = \dfrac{5(3)}{6ab^2(3)} + \dfrac{4(2ab)}{9b(2ab)}$

$$= \dfrac{15}{18ab^2} + \dfrac{8ab}{18ab^2}$$

$$= \dfrac{15 + 8ab}{18ab^2}$$

7. $\dfrac{a}{6} - \dfrac{a}{9} = 18$

$18\left(\dfrac{a}{6} - \dfrac{a}{9}\right) = 18(18)$

$\dfrac{\overset{3}{\cancel{18}}}{1} \cdot \dfrac{a}{\underset{1}{\cancel{6}}} - \dfrac{\overset{2}{\cancel{18}}}{1} \cdot \dfrac{a}{\underset{1}{\cancel{9}}} = 324$

$3a - 2a = 324$

$a = 324$

9. $x - 3y = 0$

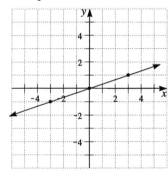

11. $x - 3 = 6$

(This is equivalent to the equation $x = 9$.)

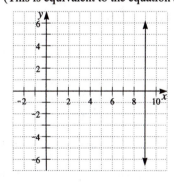

13. $m = \dfrac{y_2 - y_1}{x_2 - x_1} = \dfrac{3 - (-4)}{-1 - 2} = \dfrac{7}{-3} = -\dfrac{7}{3}$

$(x_1, y_1) = (2, -4)$

$y - y_1 = m(x - x_1)$

$y - (-4) = -\dfrac{7}{3}(x - 2)$

$y + 4 = -\dfrac{7}{3}(x - 2)$  or

$y = -\dfrac{7}{3}x + \dfrac{2}{3}$

15. $\begin{cases} \dfrac{x}{2} - y = 5 \\ -x + 2y = 8 \end{cases}$ $\xrightarrow{\text{multiply by 2}}$ $x - 2y = 10$

$\xrightarrow{\text{as is}}$ $\dfrac{-x + 2y = \ 8}{0 = 18}$ Add

This is a contradiction.

So the system has no solution.

17. Let $x$ = number of 22¢ stamps Jamie bought.
Let $y$ = number of 13¢ stamps Jamie bought.

$\begin{cases} x + y = 28 \\ 22x + 13y = 535 \end{cases}$ $\xrightarrow{\text{multiply by } -13}$ $-13x - 13y = -364$

$\xrightarrow{\text{as is}}$ $\dfrac{22x + 13y = \ \ \ 535}{9x \ \ \ \ \ \ \ \ \ = \ \ \ 171}$ Add

$x = \ \ \ 19$

$x + y = 28$
$19 + y = 28$
$y = 9$

Thus, Jamie bought nineteen 22¢ stamps and nine 13¢ stamps.
CHECK: $19 + 9 \overset{\checkmark}{=} 28$ and $22(19) + 13(9) = 418 + 117 \overset{\checkmark}{=} 535$

19. Let $t$ = amount of time Terry walks (in hours).

$6t + 8\left(t - \dfrac{1}{3}\right) = 9$

$6t + 8t - \dfrac{8}{3} = 9$

$14t - \dfrac{8}{3} = 9$

$14t = \dfrac{35}{3}$

$t = \dfrac{\left(\frac{35}{3}\right)}{14} = \dfrac{\overset{5}{\cancel{35}}}{3} \cdot \dfrac{1}{\underset{2}{\cancel{14}}} = \dfrac{5}{6}$

Since Terry walks for $\dfrac{5}{6}$ hours or for 50 minutes, Terry and Tom will be 9 km apart at 11:50 a.m.

CHECK: From 11:00 a.m. to 11:50 a.m., Terry walks $6\left(\dfrac{5}{6}\right) = 5$ km. From 11:20 a.m. to 11:50

a.m., Tom walks $8\left(\dfrac{1}{2}\right) = 4$ km. $5$ km $+ 4$ km $\overset{\checkmark}{=} 9$ km

21. Graph A.

# Chapter 7
# Exponents and Polynomials

## Exercises 7.1

1. $x^3x^2 = x^{3+2} = x^5$

3. $\left(x^3\right)^2 = x^{3\cdot2} = x^6$

5. $x^3xx^5 = x^{3+1+5} = x^9$

7. $10^410^5 = 10^{4+5} = 10^9$

9. $2^33^4$ cannot be simplified $(2^3 \cdot 3^4 = 8 \cdot 81 = 648)$

11. $\dfrac{y^3y^5}{y^2y^4} = \dfrac{y^{3+5}}{y^{2+4}} = \dfrac{y^8}{y^6} = y^{8-6} = y^2$

13. $\dfrac{9u^9v^8}{3u^3v^4} = \dfrac{9}{3}u^{9-3}v^{8-4} = 3u^6v^4$

15. $\dfrac{(a^3)^5}{(a^4)^2} = \dfrac{a^{3\cdot5}}{a^{4\cdot2}} = \dfrac{a^{15}}{a^8} = a^{15-8} = a^7$

17. $\left(-x^2\right)^4 = (-1)^4(x^2)^4 = 1 \cdot x^{2\cdot4} = x^8$

19. $\left(x^2y\right)^2 = \left(x^2\right)^2y^2 = x^{2\cdot2}y^2 = x^4y^2$

21. $\left(x^2y^3\right)^5 = \left(x^2\right)^5\left(y^3\right)^5 = x^{2\cdot5}y^{3\cdot5} = x^{10}y^{15}$

23. $\left(2r^3s^5\right)^4 = 2^4\left(r^3\right)^4\left(s^5\right)^4$
$= 16r^{3\cdot4}s^{5\cdot4} = 16r^{12}s^{20}$

25. $\left(-x^3y\right)^3 = (-1)^3\left(x^3\right)^3y^3$
$= -1x^{3\cdot3}y^3 = -x^9y^3$

27. $\left(\dfrac{x^3}{y^2}\right)^4 = \dfrac{\left(x^3\right)^4}{\left(y^2\right)^4} = \dfrac{x^{3\cdot4}}{y^{2\cdot4}} = \dfrac{x^{12}}{y^8}$

29. $\left(2x^3\right)^4\left(3x^2\right)^2 = 2^4\left(x^3\right)^4 \cdot 3^2\left(x^2\right)^2 = 16x^{3\cdot4} \cdot 9x^{2\cdot2}$
$= (16 \cdot 9)x^{12}x^4 = 144x^{12+4} = 144x^{16}$

31. $\dfrac{\left(x^4y^2\right)^3}{x^5\left(y^3\right)^2} = \dfrac{\left(x^4\right)^3\left(y^2\right)^3}{x^5\left(y^3\right)^2} = \dfrac{x^{4\cdot3}y^{2\cdot3}}{x^5y^{3\cdot2}}$
$= \dfrac{x^{12}y^6}{x^5y^6} = \dfrac{x^{12}}{x^5} = x^{12-5} = x^7$

33. $\dfrac{\left(3x^5y^4\right)^2}{9\left(x^3y\right)^3} = \dfrac{3^2\left(x^5\right)^2\left(y^4\right)^2}{9\left(x^3\right)^3y^3} = \dfrac{9x^{5\cdot2}y^{4\cdot2}}{9x^{3\cdot3}y^3}$
$= \dfrac{9x^{10}y^8}{9x^9y^3} = \dfrac{x^{10}}{x^9} \cdot \dfrac{y^8}{y^3}$
$= x^{10-9}y^{8-3} = xy^5$

35. $\left(\dfrac{x^2y}{4u}\right)^4 = \dfrac{\left(x^2y\right)^4}{(4u)^4} = \dfrac{\left(x^2\right)^4y^4}{4^4u^4}$
$= \dfrac{x^{2\cdot4}y^4}{256u^4} = \dfrac{x^8y^4}{256u^4}$

37. $\left(\dfrac{2x^3y^4}{xy^6}\right)^5 = \left(2 \cdot \dfrac{x^3}{x} \cdot \dfrac{y^4}{y^6}\right)$
$= \left(2x^{3-1} \cdot \dfrac{1}{y^2}\right)^5 = \left(\dfrac{2x^2}{y^2}\right)^5$
$= \dfrac{\left(2x^2\right)^5}{\left(y^2\right)^5} = \dfrac{2^5\left(x^2\right)^5}{\left(y^2\right)^5}$
$= \dfrac{32x^{2\cdot5}}{y^{2\cdot5}} = \dfrac{32x^{10}}{y^{10}}$

39. 
$$\left(\frac{-3a^2b^3}{2c}\right)^3 = \frac{(-3a^2b^3)^3}{(2c)^3} = \frac{(-3)^3(a^2)^3(b^3)^3}{2^3c^3}$$
$$= \frac{-27a^{2\cdot3}b^{3\cdot3}}{8c^3} = \frac{-27a^6b^9}{8c^3} = -\frac{27a^6b^9}{8c^3}$$

41. 
$$\frac{-3^2}{(-3)^2} = \frac{\overset{-1}{\cancel{-9}}}{\underset{1}{\cancel{9}}} = \frac{-1}{1} = -1$$

43. 
$$\frac{-x^2}{(-x)^2} = \frac{\overset{-1}{\cancel{-x^2}}}{\underset{1}{(-1)^2\cancel{x^2}}} = \frac{-1}{1} = -1$$

45. 
$$\frac{-u^3}{(-u)^3} = \frac{-u^3}{(-1)^3u^3} = \frac{\overset{1}{\cancel{-u^3}}}{\underset{1}{\cancel{-u^3}}} = \frac{1}{1} = 1$$

47. 
$$\frac{(-x^2)^4}{-(x^3)^2} = \frac{(-1)^4(x^2)^4}{-(x^3)^2} = \frac{1\cdot x^{2\cdot4}}{-x^{3\cdot2}}$$
$$= \frac{x^8}{-x^6} = -x^{8-6} = -x^2$$

49. 
$$\frac{-2^4+3^2}{(-4+3)^2} = \frac{-16+9}{(-1)^2} = \frac{-7}{1} = -7$$

51. 
$$\frac{-w^2}{(-w)^2} = \frac{-2^2}{(-2)^2} = \frac{-4}{4} = -1$$

53. 
$$\frac{-w^4+x^2}{(-y+x)^2} = \frac{-2^4+(-3)^2}{[-4+(-3)]^2} = \frac{-16+9}{(-7)^2}$$
$$= \frac{-7}{49} = -\frac{1}{7}$$

55. 
$$(w-x+y-z)^2 = [2-(-3)+4-(-5)]^2$$
$$= (14)^2 = 196$$

57. 
$$\frac{x(-y)}{x-y} = \frac{-3(-4)}{-3-4} = \frac{12}{-7} = -\frac{12}{7}$$

59. When we multiply two powers of the same base, we keep the base and add the exponents. When we raise a power to a power, we keep the base and multiply the exponents.

60. $x^8$ means '$x$ multiplied by itself 8 times'. $x^5 + x^3$ means 'multiply $x$ by itself 5 times and then add $x$ multiplied by itself 3 times'. These are not the same.

61. (a) $\dfrac{x^8}{x^8} = x^{8-8} = x^0$        (b) $\dfrac{x^4}{x^7} = x^{4-7} = x^{-3}$

62. (a) According to Exponent Rule 2, we must multiply the exponents, not add them.

    (b) According to Exponent Rule 1, we must add the exponents, not multiply them.

    (c) According to Exponent Rule 3, we must subtract the exponents, not divide them.

    (d) Since $x^2$ and $x^3$ are unlike terms, we cannot combine them when they are added.

63. $\dfrac{2}{x} \cdot \dfrac{8}{y} = \dfrac{2\cdot8}{x\cdot y} = \dfrac{16}{xy}$

65. $\dfrac{2}{x} + \dfrac{8}{y} = \dfrac{2\cdot y}{x\cdot y} + \dfrac{8\cdot x}{y\cdot x} = \dfrac{2y}{xy} + \dfrac{8x}{xy} = \dfrac{2y+8x}{xy}$

67. $\dfrac{x}{3} - \dfrac{x}{4} = \dfrac{x\cdot4}{3\cdot4} - \dfrac{x\cdot3}{4\cdot3} = \dfrac{4x}{12} - \dfrac{3x}{12} = \dfrac{x}{12}$

69. Let $m$ = number of miles driven.

$$0.20m + 3(15) = 72$$
$$0.20m + 45 = 72$$
$$0.20m = 27$$
$$10(0.20m) = 10(27)$$
$$2m = 270$$
$$m = 135$$

Thus, 135 miles were driven.

## Exercises 7.2

1. (a) $-3(2) = -6$

   (c) $\left(x^2\right)^{-3} = x^{2(-3)} = x^{-6} = \dfrac{1}{x^6}$

   (b) $x^2x^{-3} = x^{2+(-3)} = x^{-1} = \dfrac{1}{x}$

   (d) $2^{-3} = \dfrac{1}{2^3} = \dfrac{1}{8}$

3. (a) $-4(3) = -12$

   (c) $\left(x^3\right)^{-4} = x^{3(-4)} = x^{-12} = \dfrac{1}{x^{12}}$

   (b) $x^3x^{-4} = x^{3+(-4)} = x^{-1} = \dfrac{1}{x}$

   (d) $3^{-4} = \dfrac{1}{3^4} = \dfrac{1}{81}$

5. $8^0 = 1$

7. $5 \cdot 4^0 = 5 \cdot 1 = 5$

9. $xy^0 = x \cdot 1 = x$

11. $5^{-2} = \dfrac{1}{5^2} = \dfrac{1}{25}$

13. $\dfrac{1}{5^{-2}} = \dfrac{1}{\frac{1}{5^2}} = \dfrac{1}{\frac{1}{25}} = 1 \cdot \dfrac{25}{1} = 25$

15. $x^{-4}x^4 = x^{-4+4} = x^0 = 1$

17. $x^{-4}x^{-6} = x^{-4+(-6)}$
    $$= x^{-10} = \dfrac{1}{x^{10}}$$

19. $a^2a^{-4}aa^{-7} = a^{2+(-4)+1+(-7)}$
    $$= a^{-8} = \dfrac{1}{a^8}$$

21. $10^{-3}10^8 = 10^{-3+8}$
    $$= 10^5 = 100,000$$

23. $10^6 10^{-5} 10^{-4} = 10^{6+(-5)+(-4)}$
    $$= 10^{-3} = \dfrac{1}{10^3}$$
    $$= \dfrac{1}{1000} = 0.001$$

25. $10^7 10^{-7} = 10^{7+(-7)} = 10^0 = 1$

27. $(xy)^4 = x^4y^4$

29. $2a^{-3} = \dfrac{2}{1} \cdot \dfrac{1}{a^3} = \dfrac{2}{a^3}$

31. $-3y^{-2} = \dfrac{-3}{1} \cdot \dfrac{1}{y^2} = -\dfrac{3}{y^2}$

33. $-\left(3y^{-2}\right) = -\left(\dfrac{3}{1} \cdot \dfrac{1}{y^2}\right)$
    $$= -\left(\dfrac{3}{y^2}\right) = -\dfrac{3}{y^2}$$

35. $xy^{-1} = \dfrac{x}{1} \cdot \dfrac{1}{y} = \dfrac{x}{y}$

**37.** $\left(a^4 b^3\right)^{-2} = \left(a^4\right)^{-2}\left(b^3\right)^{-2}$

$\qquad = a^{4(-2)} b^{3(-2)} = a^{-8} b^{-6}$

$\qquad = \dfrac{1}{a^8} \cdot \dfrac{1}{b^6} = \dfrac{1}{a^8 b^6}$

**39.** $\left(a^{-4} b^3\right)^{-2} = \left(a^{-4}\right)^{-2}\left(b^3\right)^{-2}$

$\qquad = a^{-4(-2)} b^{3(-2)} = a^8 b^{-6}$

$\qquad = \dfrac{a^8}{1} \cdot \dfrac{1}{b^6} = \dfrac{a^8}{b^6}$

**41.** $\left(3x^{-2} y^3 z^{-4}\right)^2 = 3^2\left(x^{-2}\right)^2\left(y^3\right)^2\left(z^{-4}\right)^2$

$\qquad = 9x^{-2(2)} y^{3(2)} z^{-4(2)} = 9x^{-4} y^6 z^{-8}$

$\qquad = \dfrac{9}{1} \cdot \dfrac{1}{x^4} \cdot \dfrac{y^6}{1} \cdot \dfrac{1}{z^8} = \dfrac{9y^6}{x^4 z^8}$

**43..** $4\left(x^{-1} y\right)^{-3} = 4\left(x^{-1}\right)^{-3} y^{-3} = 4x^{-1(-3)} y^{-3}$

$\qquad = 4x^3 y^{-3} = \dfrac{4}{1} \cdot \dfrac{x^3}{1} \cdot \dfrac{1}{y^3} = \dfrac{4x^3}{y^3}$

**45.** $\dfrac{x^5}{x^2} = x^{5-2} = x^3$

**47.** $\dfrac{-3a^{-3}}{9a^9} = \dfrac{\overset{1}{\cancel{3}}}{\underset{3}{\cancel{9}}} \cdot \dfrac{a^{-3}}{a^9} = -\dfrac{1}{3} a^{-3-9}$

$\qquad = -\dfrac{1}{3} a^{-12} = -\dfrac{1}{3} \cdot \dfrac{1}{a^{12}} = -\dfrac{1}{3a^{12}}$

**49.** $x^{-2} + y^{-1} = \dfrac{1}{x^2} + \dfrac{1}{y}$

$\qquad = \dfrac{y}{x^2 y} + \dfrac{x^2}{x^2 y} = \dfrac{y + x^2}{x^2 y}$

**51.** $\dfrac{x^4 x^{-10}}{x^{-2} x^{-5}} = \dfrac{x^{4+(-10)}}{x^{-2+(-5)}} = \dfrac{x^{-6}}{x^{-7}}$

$\qquad = x^{-6-(-7)} = x^1 = x$

**53.** $\dfrac{x^4 y^{-10}}{x^{-2} y^{-5}} = \dfrac{x^4}{x^{-2}} \cdot \dfrac{y^{-10}}{y^{-5}}$

$\qquad = x^{4-(-2)} y^{-10-(-5)}$

$\qquad = x^6 y^{-5} = \dfrac{x^6}{1} \cdot \dfrac{1}{y^5} = \dfrac{x^6}{y^5}$

**55.** $\dfrac{10^{-3} 10^5}{10^6 10^{-10}} = \dfrac{10^{-3+5}}{10^{6+(-10)}}$

$\qquad = \dfrac{10^2}{10^{-4}} = 10^{2-(-4)}$

$\qquad = 10^6 = 1,000,000$

**57.** $\dfrac{12(10^{-3})}{4(10^{-7})} = \dfrac{\overset{3}{\cancel{12}}}{\underset{1}{\cancel{4}}} \cdot \dfrac{10^{-3}}{10^{-7}}$

$\qquad = 3 \cdot 10^{-3-(-7)} = 3 \cdot 10^4$

$\qquad = 3 \cdot 10,000 = 30,000$

**59.** $\left(\dfrac{a^{-2}}{a^3}\right)^{-3} = \left(a^{-2-3}\right)^{-3} = \left(a^{-5}\right)^{-3}$

$\qquad = a^{-5(-3)} = a^{15}$

**61.** $\dfrac{\left(x^2 y^{-1}\right)^{-1}}{\left(x^3 y^{-2}\right)^2} = \dfrac{\left(x^2\right)^{-1}\left(y^{-1}\right)^{-1}}{\left(x^3\right)^2\left(y^{-2}\right)^2} = \dfrac{x^{2(-1)} y^{-1(-1)}}{x^{3(2)} y^{-2(2)}}$

$\qquad = \dfrac{x^{-2} y^1}{x^6 y^{-4}} = \dfrac{x^{-2}}{x^6} \cdot \dfrac{y^1}{y^{-4}} = x^{-2-6} y^{1-(-4)}$

$\qquad = x^{-8} y^5 = \dfrac{1}{x^8} \cdot \dfrac{y^5}{1} = \dfrac{y^5}{x^8}$

**63.** $\left(\dfrac{2m^{-2}n^{-3}}{m^{-6}n^{-1}}\right)^{-2} = \left(\dfrac{2}{1} \cdot \dfrac{m^{-2}}{m^{-6}} \cdot \dfrac{n^{-3}}{n^{-1}}\right)^{-2}$

$$= \left(2m^{-2-(-6)}n^{-3-(-1)}\right)^{-2} = \left(2m^4n^{-2}\right)^{-2}$$

$$= 2^{-2}\left(m^4\right)^{-2}(n^{-2})^{-2} = \dfrac{1}{2^2} \cdot m^{4(-2)}n^{-2(-2)}$$

$$= \dfrac{1}{4}m^{-8}n^4 = \dfrac{1}{4} \cdot \dfrac{1}{m^8} \cdot \dfrac{n^4}{1} = \dfrac{n^4}{4m^8}$$

**65.** $\left(\dfrac{x^{-1}y^{-2}}{3x^{-2}y^{-3}}\right)^{-1} = \left(\dfrac{1}{3} \cdot \dfrac{x^{-1}}{x^{-2}} \cdot \dfrac{y^{-2}}{y^{-3}}\right)^{-1}$

$$= \left(\dfrac{1}{3}x^{-1-(-2)}y^{-2-(-3)}\right)^{-1} = \left(\dfrac{1}{3}xy\right)^{-1}$$

$$= \left(\dfrac{1}{3}\right)^{-1}x^{-1}y^{-1} = \dfrac{1}{\frac{1}{3}}x^{-1}y^{-1}$$

$$= \dfrac{3}{1} \cdot \dfrac{1}{x} \cdot \dfrac{1}{y} = \dfrac{3}{xy}$$

**67.** $\dfrac{(2m^{-2}n^{-3})^{-4}}{(m^{-6}n^{-1})^{-2}} = \dfrac{2^{-4}(m^{-2})^{-4}(n^{-3})^{-4}}{(m^{-6})^{-2}(n^{-1})^{-2}}$

$$= \dfrac{2^{-4}m^{-2(-4)}n^{-3(-4)}}{m^{-6(-2)}n^{-1(-2)}} = \dfrac{2^{-4}m^8n^{12}}{m^{12}n^2}$$

$$= \dfrac{1}{2^4} \cdot \dfrac{m^8}{m^{12}} \cdot \dfrac{n^{12}}{n^2} = \dfrac{1}{16}m^{8-12}n^{12-2}$$

$$= \dfrac{1}{16}m^{-4}n^{10} = \dfrac{1}{16} \cdot \dfrac{1}{m^4} \cdot \dfrac{n^{10}}{1} = \dfrac{n^{10}}{16m^4}$$

**69.** $\dfrac{(x^5y)^{-2}(x^{-2}y^3)^2}{(x^{-3}y^{-4})^{-2}} = \dfrac{(x^5)^{-2}y^{-2} \cdot (x^{-2})^2(y^3)^2}{(x^{-3})^{-2}(y^{-4})^{-2}}$

$$= \dfrac{x^{5(-2)}y^{-2}x^{-2(2)}y^{3(2)}}{x^{-3(-2)}y^{-4(-2)}} = \dfrac{x^{-10}y^{-2}x^{-4}y^6}{x^6y^8}$$

$$= \dfrac{x^{-10+(-4)}y^{-2+6}}{x^6y^8} = \dfrac{x^{-14}y^4}{x^6y^8}$$

$$= \dfrac{x^{-14}}{x^6} \cdot \dfrac{y^4}{y^8} = x^{-14-6}y^{4-8} = x^{-20}y^{-4}$$

$$= \dfrac{1}{x^{20}} \cdot \dfrac{1}{y^4} = \dfrac{1}{x^{20}y^4}$$

**71.** $x^{-3} = 2^{-3} = \dfrac{1}{2^3} = \dfrac{1}{8}$ 　　　　　　**73.** $8x^{-1} = 8 \cdot 2^{-1} = 8 \cdot \dfrac{1}{2} = 4$

75. $x^{-1} + y^{-1} = 2^{-1} + (-3)^{-1} = \dfrac{1}{2} + \dfrac{1}{-3}$

$\qquad\qquad = \dfrac{1}{2} - \dfrac{1}{3} = \dfrac{1}{6}$

77. $\dfrac{x^6}{x^4}$ requires us to divide $x^6$ by $x^4$, whereas $\dfrac{x^6}{x^{-4}} = \dfrac{x^6}{\left(\dfrac{1}{x^4}\right)} = x^6 \cdot \dfrac{x^4}{1}$ asks us to multiply $x^6$ by $x^4$.

78. When $-1$ appears in the exponent, it tells us to take the reciprocal of the base. Thus, $3^{-1} = \dfrac{1}{3}$.
When the minus sign appears in front of the 3, it tells us to take the opposite of 3. Put another way, $3^{-1}$ is the multiplicative inverse of 3, while $-3$ is the additive inverse of 3.

79. Let $t =$ number of hours Maria works.
Then $t - 2 =$ number of hours Francis works.
$$5t + 7(t - 2) = 70$$
$$5t + 7t - 14 = 70$$
$$12t - 14 = 70$$
$$12t = 84$$
$$t = 7$$
Seven hours after Maria started, it is 4 P.M.

## Exercises 7.3

1. $4,530 = 4.53 \times 10^3$

3. $0.0453 = 4.53 \times 10^{-2}$

5. $0.00007 = 7 \times 10^{-5}$

7. $7,000,000 = 7 \times 10^6$

9. $85,370 = 8.537 \times 10^4$

11. $0.0085370 = 8.537 \times 10^{-3}$

13. $90 = 9 \times 10^1$ or $9 \times 10$

15. $9 = 9 \times 10^0$ or $9 \times 1$

17. $0.9 = 9 \times 10^{-1}$

19. $0.09 = 9 \times 10^{-2}$

21. $0.00000003 = 3 \times 10^{-8}$

23. $28 = 2.8 \times 10^1$ or $2.8 \times 10$

25. $47.5 = 4.75 \times 10^1$ or $4.75 \times 10$

27. $9,7273 = 9.7273 \times 10^3$

29. $56 \times 10^{-2} = (5.6 \times 10^1) \times 10^{-2}$
$\qquad\qquad = 5.6 \times (10^1 \cdot 10^{-2})$
$\qquad\qquad = 5.6 \times 10^{-1}$

31. $0.154 \times 10^4 = (1.54 \times 10^{-1}) \times 10^4$
$\qquad\qquad = 1.54 \times (10^{-1} \cdot 10^4)$
$\qquad\qquad = 1.54 \times 10^3$

33. $28.40 \times 10^6 = (2.84 \times 10^1) \times 10^6$
$\qquad\qquad = 2.84 \times (10^1 \cdot 10^6)$
$\qquad\qquad = 2.84 \times 10^7$

35. $2.8 \times 10^4 = 28,000$

37. $2.8 \times 10^{-4} = 0.00028$

39. $4.29 \times 10^7 = 42,900,000$

41. $4.29 \times 10^{-7} = 0.000000429$

43. $3.52 \times 10^{-3} = 0.00352$

45. $3.5286 \times 10^5 = 352,860$

47. $0.026 \times 10^{-3} = 0.000026$

49. $(0.004)(250) = (4 \times 10^{-3})(2.5 \times 10^2)$

$$= (4)(2.5) \times (10^{-3} \cdot 10^2) = 10 \times 10^{-3+2}$$

$$= 10^1 \times 10^{-1} = 10^{1+(-1)} = 10^0 = 1$$

51. $\dfrac{0.003}{6,000} = \dfrac{3 \times 10^{-3}}{6 \times 10^3} = \dfrac{3}{6} \times \dfrac{10^{-3}}{10^3}$

$$= 0.5 \times 10^{-3-3} = 0.5 \times 10^{-6} = 5 \times 10^{-7}$$

53. $\dfrac{(480)(0.008)}{(0.24)(4,000)} = \dfrac{(4.8 \times 10^2)(8 \times 10^{-3})}{(2.4 \times 10^{-1})(4 \times 10^3)}$

$$= \dfrac{(\overset{2}{\cancel{4.8}})(\overset{2}{\cancel{8}})}{(\underset{1}{\cancel{2.4}})(\underset{1}{\cancel{4}})} \times \dfrac{10^2 10^{-3}}{10^{-1} 10^3} = 4 \times \dfrac{10^{2+(-3)}}{10^{-1+3}}$$

$$= 4 \times \dfrac{10^{-1}}{10^2} = 4 \times 10^{-1-2} = 4 \times 10^{-3} = 0.004$$

55. $\dfrac{(0.0036)(0.005)}{(0.01)(0.06)} = \dfrac{(3.6 \times 10^{-3})(5 \times 10^{-3})}{(1 \times 10^{-2})(6 \times 10^{-2})}$

$$= \dfrac{(\overset{0.6}{\cancel{3.6}})(5)}{(1)(\cancel{6})_{1}} \times \dfrac{10^{-3} 10^{-3}}{10^{-2} 10^{-2}} = 3 \times \dfrac{10^{-3+(-3)}}{10^{-2+(-2)}}$$

$$= 3 \times \dfrac{10^{-6}}{10^{-4}} = 3 \times 10^{-6-(-4)} = 3 \times 10^{-2} = 0.03$$

57. $5.98 \times 10^{24}$ kg

59. $(80,000)(9.3 \times 10^{-23}) = (8 \times 10^4)(9.3 \times 10^{-23})$

$$= (8)(9.3) \times (10^4 \cdot 10^{-23})$$

$$= 74.4 \times 10^{4+(-23)} = 74.4 \times 10^{-19}$$

$$= 7.44 \times 10^{-18} \text{ grams}$$

61. $0.00000001 = 1 \times 10^{-8}$ cm

63. $(153)(1 \times 10^{-8}) = 153 \times 10^{-8}$

$$= 1.53 \times 10^{-6} \text{ cm}$$

65. $4,250$ million $= (4,250)(1,000,000)$

$$= (4.25 \times 10^3)(1 \times 10^6)$$

$$= (4.25)(1) \times 10^3 10^6 = 4.25 \times 10^{3+6}$$

$$= 4.25 \times 10^9 \text{ miles}$$

67. Let $w$ = weight of the Earth in tons.

$$\frac{5.98 \times 10^{24}}{w} = \frac{888.9}{1}$$

$$5.98 \times 10^{24} = (8.889 \times 10^2)w$$

$$\frac{5.98 \times 10^{24}}{8.889 \times 10^2} = w$$

$$\left(\frac{5.98}{8.889}\right) \times \frac{10^{24}}{10^2} = w$$

$$0.6727 \times 10^{24-2} = w$$

$$0.6727 \times 10^{22} = 6.727 \times 10^{21} = w$$

Thus, the weight of the Earth is $6.727 \times 10^{21}$ tons.

69. There are $(365)(24)(60)(60) = 31,536,000$ seconds in one year. Then one light year equals $(186,000)(31,536,000)$ miles.

$$(186,000)(31,536,000) = (1.86 \times 10^5)(3.1536 \times 10^7)$$
$$= (1.86)(3.1536) \times 10^5 10^7$$
$$= 5.865696 \times 10^{5+7}$$
$$= 5.865696 \times 10^{12} \text{ miles.}$$

71. $(5 \times 10^9)(5.865696 \times 10^{12})(1.6) = (5)(5.865696)(1.6) \times 10^9 10^{12}$
$$= 46.925568 \times 10^{9+12}$$
$$= 46.925568 \times 10^{21}$$
$$= 4.6925568 \times 10^{22} \text{ kilometers}$$

73. First multiply 3.74 by 6.38; then multiply $10^{-5}$ by $10^4$. Take the product of these two results and express this product in scientific notation.

$$(3.74)(6.38) = 23.8612$$
$$10^{-5} 10^4 = 10^{-5+4} = 10^{-1}$$
$$23.8612 \times 10^{-1} = 2.38612 \times 10^0$$
$$= 2.38612$$

74. If the number is bigger than 1, the exponent cannot be negative; if the number is smaller than 1, the exponent must be negative.

## Exercises 7.4

1. (a) one term: $3x^5$
   (b) degree of $3x^5$: 5
   (c) degree of polynomial: 5

3. (a) two terms: $3x$, 4
   (b) degree of $3x$: 1
       degree of 4: 0
   (c) degree of polynomial: 1

5. (a) two terms: $x^2$, $y^3$
   (b) degree of $x^2$: 2
       degree of $y^3$: 3
   (c) degree of polynomial: 3

7. (a) one term: $x^2 y^3$
   (b) degree of $x^2 y^3$: $5(= 2+3)$
   (c) degree of polynomial: 5

9. (a) one term: 8
   (b) degree of 8: 0
   (c) degree of polynomial: 0

11. (a) three terms: $2x^3$, $-5x^2$, $x$
    (b) degree of $2x^3$: 3
        degree of $-5x^2$: 2
        degree of $x$: 1
    (c) degree of polynomial: 3

13. (a) two terms: $2x^3$, $y^5$
    (b) degree of $2x^3$: 3
        degree of $y^5$: 5
    (c) degree of polynomial: 5

15. (a) one term: $2x^3y^5$
    (b) degree of $2x^3y^5$: $8( = 3 + 5)$
    (c) degree of polynomial: 8

17. (a) four terms: $x^5$, $-x^3y^4$, $-2x^2y^3$, $y^6$
    (b) degree of $x^5$: 5
        degree of $-x^3y^4$: $7( = 3 + 4)$
        degree of $-2x^2y^3$: $5( = 2 + 3)$
        degree of $y^6$: 6
    (c) degree of polynomial: 7

19. (a) degree of $x^2$: 2
        degree of $-5x$: 1
        degree of 6: 0
    (b) degree of polynomial: 2
    (c) The coefficient of $x^2$ is 1.
        The coefficient of $-5x$ is $-5$.
        6 is both a term and a coefficient.

21. (a) degree of $x^2$: 2
        degree of 4: 0
    (b) degree of polynomial: 2
    (c) Write the polynomial as $x^2 + 0x + 4$. Then the coefficient of $x^2$ is 1 and the coefficient of $0x$ is 0. 4 is both a term and a coefficient.

23. (a) degree of $x^3$: 3
        degree of $-1$: 0
    (b) degree of polynomial: 3
    (c) Write the polynomial as $x^3 + 0x^2 + 0x - 1$. Then the coefficient of $x^3$ is 1 and the coefficients of $0x^2$ and $0x$ are 0. $-1$ is both a term and a coefficient.

25. (a) degree of 1: 0
        degree of $-x^5$: 5
    (b) degree of polynomial: 5
    (c) Write the polynomial as $-x^5 + 0x^4 + 0x^3 + 0x^2 + 0x + 1$. Then the coefficient of $-x^5$ is $-1$ and the coefficients of $0x^4$, $0x^3$, $0x^2$, and $0x$ are all 0. 1 is both a term and a coefficient.

27. $(2x^2 - 5) + (3x^2 - 5) = 2x^2 + 3x^2 - 5 - 5 = 5x^2 - 10$

29. $(3u^3 - 2u + 7) + (u^3 - u^2 + 7u) = 3u^3 + u^3 - u^2 - 2u + 7u + 7$
    $$= 4u^3 - u^2 + 5u + 7$$

31. $(3u^2 - 2u + 7) - (u^3 - u^2 + 7u) = 3u^2 - 2u + 7 - u^3 + u^2 - 7u$
    $$= -u^3 + 3u^2 + u^2 - 2u - 7u + 7$$
    $$= -u^3 + 4u^2 - 9u + 7$$

33. $(4t^3 - t) + (t^2 + t) - (t^3 - t^2) = 4t^3 - t + t^2 + t - t^3 + t^2$
$$= 4t^3 - t^3 + t^2 + t^2 - t + t$$
$$= 3t^3 + 2t^2$$

35. $(x^2 y + 3xy - x^2 y^2) + (x^2 y - 5x^2 y^2 - xy^2) = x^2 y + x^2 y + 3xy - x^2 y^2 - 5x^2 y^2 - xy^2$
$$= 2x^2 y + 3xy - 6x^2 y^2 - xy^2$$

37. $(x^2 y + 3xy - x^2 y^2) - (x^2 y - 5x^2 y^2 - xy^2) = x^2 y + 3xy - x^2 y^2 - x^2 y + 5x^2 y^2 + xy^2$
$$= x^2 y - x^2 y + 3xy - x^2 y^2 + 5x^2 y^2 + xy^2$$
$$= 3xy + 4x^2 y^2 + xy^2$$

39. $2(y^2 - 4y + 1) + 3(2y^2 - y - 1) = 2y^2 - 8y + 2 + 6y^2 - 3y - 3$
$$= 2y^2 + 6y^2 - 8y - 3y + 2 - 3$$
$$= 8y^2 - 11y - 1$$

41. $5(x^2 - 3x + 2) - 3(2x^2 - 5x - 2) = 5x^2 - 15x + 10 - 6x^2 + 15x + 6$
$$= 5x^2 - 6x^2 - 15x + 15x + 10 + 6$$
$$= -x^2 + 16$$

43. $(x^2 + 3x - 7) + (5x - x^2) + (3x^2 - x - 2) = x^2 - x^2 + 3x^2 + 3x + 5x - x - 7 - 2$
$$= 3x^2 + 7x - 9$$

45. $(2x^2 - 3x + 5) - (x^2 - 7x + 3) = 2x^2 - 3x + 5 - x^2 + 7x - 3$
$$= 2x^2 - x^2 - 3x + 7x + 5 - 3$$
$$= x^2 + 4x + 2$$

47. $(a^3 - b^2) + (a^2 b + 2b^2) - (a^3 - a^2 - b + b^2) = a^3 - b^2 + a^2 b + 2b^2 - a^3 + a^2 + b - b^2$
$$= a^3 - a^3 - b^2 + 2b^2 - b^2 + a^2 b + a^2 + b$$
$$= a^2 b + a^2 + b$$

49. $2x - 1 - [(3x + 6) + (5x - 8)] = 2x - 1 - (8x - 2)$
$$= 2x - 1 - 8x + 2$$
$$= 2x - 8x - 1 + 2$$
$$= -6x + 1$$

51. $x^2 - x + 3 = (-5)^2 - (-5) + 3$
$$= 25 + 5 + 3 = 33$$

53. $y^4 + y^3 + y^2 + y + 1 = (-3)^4 + (-3)^3 + (-3)^2 + (-3) + 1$
$$= 81 - 27 + 9 - 3 + 1 = 61$$

55. $-3x^2 y + 5xy^2 = -3(2)^2(-1) + 5(2)(-1)^2$
$$= -3(4)(-1) + 5(2)(1)$$
$$= 12 + 10 = 22$$

57. $5x - 12 - (3x + 8) = 5x - 12 - 3x - 8$
$$= 5x - 3x - 12 - 8 = 2x - 20$$

59. $P = 100x - x^2$

| When $x = 30$, | When $x = 50$, | When $x = 90$, |
|---|---|---|
| $P = 100(30) - (30)^2$ | $P = 100(50) - (50)^2$ | $P = 100(90) - (90)^2$ |
| $= 3,000 - 900$ | $= 5,000 - 2,500$ | $= 9,000 - 8,100$ |
| $= \$2,100.$ | $= \$2,500.$ | $= \$900.$ |

61. $h = 120 - 28t - 16t^2$

| When $t = 1$, | When $t = 1.5$, | When $t = 2$, |
|---|---|---|
| $h = 120 - 28(1) - 16(1)^2$ | $h = 120 - 28(1.5) - 16(1.5)^2$ | $h = 120 - 28(2) - 16(2)^2$ |
| $= 120 - 28 - 16$ | $= 120 - 42 - 36$ | $= 120 - 56 - 64$ |
| $= 76$ feet. | $= 42$ feet. | $= 0$ feet. |

63. In a sum, the expressions to be added are called terms; in a product, the expressions to be multiplied are called factors.

64. 3 is not a factor of the expression $6x + 8$ because 3 does not exactly divide 8.

65. 2 is a factor of the expression $6x + 8$ because 2 does exactly divide both $6x$ and 8. Here, we can write $6x + 8 = 2(3x + 4)$.

67. $2(x + 3) - 5(x + 4) = x - 2$
$$2x + 6 - 5x - 20 = x - 2$$
$$-3x - 14 = x - 2$$
$$-4x = 12$$
$$x = -3$$

69. $4(x - 1) - 3x = x - 4$
$$4x - 4 - 3x = x - 4$$
$$x - 4 = x - 4$$
Identity

71. Let $r$ = rate of interest on second investment.
$$0.08(1,800) + r(1,400) = 284$$
$$144 + 1,400r = 284$$
$$1,400r = 140$$
$$r = 0.10$$

The rate of interest on the second investment must be 10%.

## Exercises 7.5

1. $3x(5x^3)(4x^2) = (3)(5)(4)(x \cdot x^3 \cdot x^2)$
$$= 60x^6$$

3. $3x(5x^3 + 4x^2) = 3x \cdot 5x^3 + 3x \cdot 4x^2$
$$= 15x^4 + 12x^3$$

5. $4xy(3yz)(-5xz) = (4)(3)(-5)(xx)(yy)(zz)$
$$= -60x^2y^2z^2$$

7. $4xy(3yz - 5xz) = 4xy \cdot 3yz - 4xy \cdot 5xz$
$$= 12xy^2z - 20x^2yz$$

9.  $3x^2(x + 3y) + 4xy(x - 3y) = 3x^2 \cdot x + 3x^2 \cdot 3y + 4xy \cdot x - 4xy \cdot 3y$
$$= 3x^3 + 9x^2y + 4x^2y - 12xy^2$$
$$= 3x^3 + 13x^2y - 12xy^2$$

11.  $5xy^2(xy - y) - 2y(x^2y^2 - xy^2) = 5xy^2 \cdot xy - 5xy^2 \cdot y - 2y \cdot x^2y^2 + 2y \cdot xy^2$
$$= 5x^2y^3 - 5xy^3 - 2x^2y^3 + 2xy^3$$
$$= 3x^2y^3 - 3xy^3$$

13.  $(x + 2)(x^2 - x + 3) = x(x^2 - x + 3) + 2(x^2 - x + 3)$
$$= x^3 - x^2 + 3x + 2x^2 - 2x + 6$$
$$= x^3 + x^2 + x + 6$$

15.  $(y - 5)(y^2 + 2y - 6) = y(y^2 + 2y - 6) - 5(y^2 + 2y - 6)$
$$= y^3 + 2y^2 - 6y - 5y^2 - 10y + 30$$
$$= y^3 - 3y^2 - 16y + 30$$

17.  $(3x - 2)(x^2 + 3x - 5) = 3x(x^2 + 3x - 5) - 2(x^2 + 3x - 5)$
$$= 3x^3 + 9x^2 - 15x - 2x^2 - 6x + 10$$
$$= 3x^3 + 7x^2 - 21x + 10$$

19.  $(5z + 2)(3z^2 + 2z + 8) = 5z(3z^2 + 2z + 8) + 2(3z^2 + 2z + 8)$
$$= 15z^3 + 10z^2 + 40z + 6z^2 + 4z + 16$$
$$= 15z^3 + 16z^2 + 44z + 16$$

21.  $(x + y)(x^2 - xy + y^2) = x(x^2 - xy + y^2) + y(x^2 - xy + y^2)$
$$= x^3 - x^2y + xy^2 + x^2y - xy^2 + y^3$$
$$= x^3 + y^3$$

23.  $(x^2 + x + 1)(x^2 + x - 1) = x^2(x^2 + x - 1) + x(x^2 + x - 1) + 1(x^2 + x - 1)$
$$= x^4 + x^3 - x^2 + x^3 + x^2 - x + x^2 + x - 1$$
$$= x^4 + 2x^3 + x^2 - 1$$

25.  $(x^3 + xy - y^2)(x^3 - 3xy + y^2) = x^3(x^3 - 3xy + y^2) + xy(x^3 - 3xy + y^2) - y^2(x^3 - 3xy + y^2)$
$$= x^6 - 3x^4y + x^3y^2 + x^4y - 3x^2y^2 + xy^3 - x^3y^2 + 3xy^3 - y^4$$
$$= x^6 - 2x^4y - 3x^2y^2 + 4xy^3 - y^4$$

27.  $(x + 5)(x + 3) = x^2 + 3x + 5x + 15$
$$= x^2 + 8x + 15$$

29.  $(x - 5)(x - 3) = x^2 - 3x - 5x + 15$
$$= x^2 - 8x + 15$$

31.  $(x - 5)(x + 3) = x^2 + 3x - 5x - 15$
$$= x^2 - 2x - 15$$

33.  $(x + 5)(x - 3) = x^2 - 3x + 5x - 15$
$$= x^2 + 2x - 15$$

35.  $(x + 2y)(x + 3y) = x^2 + 3xy + 2xy + 6y^2$
$$= x^2 + 5xy + 6y^2$$

37.  $(a + 8b)(a - 5b) = a^2 - 5ab + 8ab - 40b^2$
$$= a^2 + 3ab - 40b^2$$

39.  $(3x - 4)(4x - 1) = 12x^2 - 3x - 16x + 4$
$$= 12x^2 - 19x + 4$$

41.  $(2r - s)(r + 3s) = 2r^2 + 6rs - rs - 3s^2$
$$= 2r^2 + 5rs - 3s^2$$

43.  $(x^2 + 3)(x^2 + 2) = x^4 + 2x^2 + 3x^2 + 6$
$$= x^4 + 5x^2 + 6$$

45.  $(x + 7)(x + 7) = x^2 + 7x + 7x + 49$
$$= x^2 + 14x + 49$$

47. $(x+7)(x-7) = x^2 - 7x + 7x - 49$
$\qquad = x^2 - 49$

49. $(x-4)^2 = (x-4)(x-4)$
$\qquad = x^2 - 4x - 4x + 16$
$\qquad = x^2 - 8x + 16$

51. $(x+2)^3 = (x+2)(x+2)(x+2)$
$\qquad = (x+2)(x^2 + 2x + 2x + 4)$
$\qquad = (x+2)(x^2 + 4x + 4)$
$\qquad = x(x^2 + 4x + 4) + 2(x^2 + 4x + 4)$
$\qquad = x^3 + 4x^2 + 4x + 2x^2 + 8x + 8$
$\qquad = x^3 + 6x^2 + 12x + 8$

53. $(3x-5)^2 = (3x-5)(3x-5)$
$\qquad = 9x^2 - 15x - 15x + 25$
$\qquad = 9x^2 - 30x + 25$

55. $(2a-9b)^2 = (2a-9b)(2a-9b)$
$\qquad = 4a^2 - 18ab - 18ab + 81b^2$
$\qquad = 4a^2 - 36ab + 81b^2$

57. $2x^2(x+4)(x-8) = 2x^2(x^2 - 8x + 4x - 32)$
$\qquad = 2x^2(x^2 - 4x - 32)$
$\qquad = 2x^4 - 8x^3 - 64x^2$

59. $3x(5x-6)(3x-2) = 3x(15x^2 - 10x - 18x + 12)$
$\qquad = 3x(15x^2 - 28x + 12)$
$\qquad = 45x^3 - 84x^2 + 36x$

61. $(x+4)(x-3) + (x-6)(x-2) = x^2 - 3x + 4x - 12 + x^2 - 2x - 6x + 12$
$\qquad = 2x^2 - 7x$

63. $(a-5)(a-4) - (a-3)(a-2) = a^2 - 4a - 5a + 20 - (a^2 - 2a - 3a + 6)$
$\qquad = a^2 - 9a + 20 - (a^2 - 5a + 6)$
$\qquad = a^2 - 9a + 20 - a^2 + 5a - 6$
$\qquad = -4a + 14$

65. $(x-6)^2 - (x+6)^2 = (x-6)(x-6) - (x+6)(x+6)$
$\qquad = x^2 - 6x - 6x + 36 - (x^2 + 6x + 6x + 36)$
$\qquad = x^2 - 12x + 36 - (x^2 + 12x + 36)$
$\qquad = x^2 - 12x + 36 - x^2 - 12x - 36$
$\qquad = -24x$

67. $(x+2)^3 - (x+2)^2 - (x+2) + 2 = (x+2)(x+2)(x+2) - (x+2)(x+2) - (x+2) + 2$
$\qquad = x^3 + 6x^2 + 12x + 8 - (x^2 + 4x + 4) - (x+2) + 2$
$\qquad$ (see Exercise 51 for details)
$\qquad = x^3 + 6x^2 + 12x + 8 - x^2 - 4x - 4 - x - 2 + 2$
$\qquad = x^3 + 5x^2 + 7x + 4$

69. Let $W =$ width of the rectangle.
Then $2W + 3 =$ length of the rectangle.
So $P = 2W + 2(2W + 3) = 2W + 4W + 6 = 6W + 6$ is the perimeter of the rectangle and
$A = W(2W + 3) = 2W^2 + 3W$ is the area of the rectangle.

71. Let $s =$ length of the side of the original square.
Then $s + 4.5 =$ length of the side of the new square.
So $s \cdot s = s^2$ is the area of the original square and $(s + 4.5)(s + 4.5) = s^2 + 4.5s + 4.5s + 20.25$
$= s^2 + 9s + 20.25$ is the area of the new square. Therefore, the change in the area is
$s^2 + 9s + 20.25 - s^2 = 9s + 20.25$.

73. Let $a =$ width of original rectangle.
Then $5a - 8 =$ length of original rectangle and $a + 3 =$ width of new rectangle.
So $a(5a - 8) = 5a^2 - 8a$ is the area of the original rectangle and
$(a + 3)(5a - 8) = 5a^2 - 8a + 15a - 24 = 5a^2 + 7a - 24$ is the area of the new rectangle.
Therefore, the change in area is
$5a^2 + 7a - 24 - (5a^2 - 8a) = 5a^2 + 7a - 24 - 5a^2 + 8a = 15a - 24$.

75. Let $x =$ length of side of the square to be cut out. The original rectangle has dimensions 8 by 10, so has an area of 80. The area of the square to be cut out is $x \cdot x = x^2$. Therefore, the remaining area is $80 - x^2$.

77. When $x$ is increased by 500,
$R = (x + 500)[50 - 0.004(x + 500)]$
$= (x + 500)(50 - 0.004x - 2) = (x + 500)(48 - 0.004x)$
$= 48x - 0.004x^2 + 24{,}000 - 2x$
$= 46x - 0.004x^2 + 24{,}000$
Then the increase in revenue is
$46x - 0.004x^2 + 24{,}000 - [x(50 - 0.004x)] = 46x - 0.004x^2 + 24{,}000 - (50x - 0.004x^2)$
$$= 46x - 0.004x^2 + 24{,}000 - 50x + 0.004x^2$$
$$= -4x + 24{,}000$$

79. (a) Both examples require us to find the square of an expression involving $x$ and $y$.

(b) The first example asks us to square a product; the second asks us to square a sum.

(c) $(xy)^2 = x^2y^2$, by Exponent Rule 4. $(x + y)^2 = x^2 + 2xy + y^2$ by the FOIL method.

80. $(x + y)^n = x^n + y^n$ only when $n = 1$.

81. (a) $-4x^2(2x - 7) = -8x^3 + 28x^2$
$-4x^2(2x - 7)(3x + 1) = (-8x^3 + 28x^2)(3x + 1)$
$$= -24x^4 - 8x^3 + 84x^3 + 28x^2$$
$$= -24x^4 + 76x^3 + 28x^2$$

(b) $-4x^2(3x + 1) = -12x^3 - 4x^2$
$-4x^2(2x - 7)(3x + 1) = -4x^2(3x + 1)(2x - 7)$
$$= (-12x^3 - 4x^2)(2x - 7)$$
$$= -24x^4 + 84x^3 - 8x^3 + 28x^2$$
$$= -24x^4 + 76x^3 + 28x^2$$

(c) $(2x - 7)(3x + 1) = 6x^2 + 2x - 21x - 7$
$$= 6x^2 - 19x - 7$$
$-4x^2(2x - 7)(3x + 1) = -4x^2(6x^2 - 19x - 7)$
$$= -24x^4 + 76x^3 + 28x^2$$

The three answers are the same, and should be because of the commutative and associative laws of multiplication.

83. $8 - 3(x + 1) < 32$
$8 - 3x - 3 < 32$
$-3x + 5 < 32$
$-3x < 27$
$x > -9$

85. $\begin{array}{rrr} -3 \leq & 9 - x < & 1 \\ -9 & -9 & -9 \\ \hline -12 \leq & -x & < -8 \\ 12 \geq & x & > 8 \end{array}$
or
$8 < \quad x \quad \leq 12$

87. Let $x =$ number of ounces of the 55% solution.

Then $40 - x =$ number of ounces of the 35% solution.

$0.55x + 0.35(40 - x) = 0.51(40)$

$0.55x + 14 - 0.35x = 20.4$

$0.20x + 14 = 20.4$

$0.20x = 6.4$

$2.0x = 64$

$x = 32$

Then $40 - x = 40 - 32 = 8$. So 32 ounces of the 55% solution should be mixed with 8 ounces of the 35% solution.


## Exercises 7.6

1. $(x + 4)(x + 3) = x^2 + 3x + 4x + 12$
$= x^2 + 7x + 12$

3. $(x - 4)(x - 3) = x^2 - 3x - 4x + 12$
$= x^2 - 7x + 12$

5. $(x + 4)(x - 3) = x^2 - 3x + 4x - 12$
$= x^2 + x - 12$

7. $(x - 4)(x + 3) = x^2 + 3x - 4x - 12$
$= x^2 - x - 12$

9. $(x + 6)(x + 2) = x^2 + 2x + 6x + 12$
$= x^2 + 8x + 12$

11. $(x - 6)(x - 2) = x^2 - 2x - 6x + 12$
$= x^2 - 8x + 12$

13. $(x + 6)(x - 2) = x^2 - 2x + 6x - 12$
$= x^2 + 4x - 12$

15. $(x - 6)(x + 2) = x^2 + 2x - 6x - 12$
$= x^2 - 4x - 12$

17. $(x + 12)(x + 1) = x^2 + x + 12x + 12$
$= x^2 + 13x + 12$

19. $(x - 12)(x - 1) = x^2 - x - 12x + 12$
$= x^2 - 13x + 12$

21. $(x + 12)(x - 1) = x^2 - x + 12x - 12$
$= x^2 + 11x - 12$

23. $(x - 12)(x + 1) = x^2 + x - 12x - 12$
$= x^2 - 11x - 12$

25. $(a + 8)(a + 8) = a^2 + 8a + 8a + 64$
$= a^2 + 16a + 64$

(Perfect square)

27. $(a - 8)(a - 8) = a^2 - 8a - 8a + 64$
$= a^2 - 16a + 64$

(Perfect square)

29. $(a+8)(a-8) = a^2 - 8a + 8a - 64$
$$= a^2 - 64$$
(Difference of two squares)

31. $(c-4)^2 = (c-4)(c-4)$
$$= c^2 - 4c - 4c + 16$$
$$= c^2 - 8a + 16$$
(Perfect square)

33. $(c+4)^2 = (c+4)(c+4)$
$$= c^2 + 4c + 4c + 16$$
$$= c^2 + 8c + 16$$
(Perfect square)

35. $(c+4)(c-4) = c^2 - 4c + 4c + 16$
$$= c^2 - 16$$
(Difference of two squares)

37. $(3x+4)(x+7) = 3x^2 + 21x + 4x + 28$
$$= 3x^2 + 25x + 28$$

39. $(3x+7)(x+4) = 3x^2 + 12x + 7x + 28$
$$= 3x^2 + 19x + 28$$

41. $(3x+4)(x-7) = 3x^2 - 21x + 4x - 28$
$$= 3x^2 - 17x - 28$$

43. $(3x-4)(x+7) = 3x^2 + 21x - 4x - 28$
$$= 3x^2 + 17x - 28$$

45. $(3x+4)(5x+7) = 15x^2 + 21x + 20x + 28$
$$= 15x^2 + 41x + 28$$

47. $(3x+7)(5x+4) = 15x^2 + 12x + 35x + 28$
$$= 15x^2 + 47x + 28$$

49. $(3x+4)(5x-7) = 15x^2 - 21x + 20x - 28$
$$= 15x^2 - x - 28$$

51. $(3x-4)(5x+7) = 15x^2 + 21x - 20x - 28$
$$= 15x^2 + x - 28$$

53. $(2a+5)^2 = (2a+5)(2a+5)$
$$= 4a^2 + 10a + 10a + 25$$
$$= 4a^2 + 20a + 25$$
(Perfect square)

55. $(2a+5)(2a-5) = 4a^2 - 10a + 10a - 25$
$$= 4a^2 - 25$$
(Difference of two squares)

57. $(3xy)^2 = 3^2 x^2 y^2 = 9x^2 y^2$

59. $(x^3 + y^2)^2 = (x^3 + y^2)(x^3 + y^2)$
$$= x^6 + x^3 y^2 + x^3 y^2 + y^4$$
$$= x^6 + 2x^3 y^2 + y^4$$
(Perfect square)

61. $(2a+5y)^2 = (2a+5y)(2a+5y)$
$$= 4a^2 + 10ay + 10ay + 25y^2$$
$$= 4a^2 + 20ay + 25y^2$$
(Perfect square)

63. $(2a+5y)(2a-5y) = 4a^2 - 10ay + 10ay - 25y^2$
$$= 4a^2 - 25y^2$$
(Difference of two squares)

65. $(x+6)(x+4) = x^2 + 4x + 6x + 24$     $(x-6)(x-4) = x^2 - 4x - 6x + 24$
$$= x^2 + 10x + 24 \qquad\qquad\qquad = x^2 - 10x + 24$$
The effect of switching both $+$ signs to $-$ signs is to change the sign of the middle term from $+$ to $-$ .

66. $(x+6)(x-4) = x^2 - 4x + 6x - 24$     $(x-6)(x+4) = x^2 + 4x - 6x - 24$
$$= x^2 + 2x - 24 \qquad\qquad\qquad = x^2 - 2x - 24$$
The middle terms of the resulting trinomials have opposite signs.

67.  (5) and (7)    (6) and (8)    (13) and (15)    (14) and (16)    (21) and (23)
     (22) and (24)   (41) and (43)   (42) and (44)    (49) and (51)    (50) and (52)

The middle terms will always have opposite signs, since

$(x + a)(x - b) = x^2 + (a - b)x - ab$

$(x - a)(x + b) = x^2 + (-a + b)x - ab$

and $(a - b)$ and $(-a + b)$ are always opposites of one another.

69.  $\dfrac{20xy}{9z} \div 18xyz = \dfrac{\overset{10}{\cancel{20}\cancel{x}\cancel{y}}}{9z} \cdot \dfrac{1}{\underset{9}{\cancel{18}\cancel{x}\cancel{y}z}} = \dfrac{10}{81z^2}$

71.  $\dfrac{a}{3} - \dfrac{a}{2} = \dfrac{a}{5}$    LCD $= 30$

$30\left(\dfrac{a}{3} - \dfrac{a}{2}\right) = 30\left(\dfrac{a}{5}\right)$

$10a - 15a = 6a$

$-5a = 6a$

$0 = 11a$

$0 = a$

73.  Let $r =$ rate of interest on second investment.

$0.09(1, 400) + r(1, 100) = 203$

$126 + 1, 100r = 203$

$1, 100r = 77$

$r = 0.07$

So the rate of interest on the second investment is 7%.

## Chapter 7 Review Exercises

1.  $3^{-4} = \dfrac{1}{3^4} = \dfrac{1}{81}$

3.  $\left(3^{-1} + 2^{-2}\right)^2 = \left(\dfrac{1}{3^1} + \dfrac{1}{2^2}\right)^2$

$= \left(\dfrac{1}{3} + \dfrac{1}{4}\right)^2 = \left(\dfrac{7}{12}\right)^2$

$= \dfrac{7^2}{12^2} = \dfrac{49}{144}$

5.  $\dfrac{(xy^2)^3}{(x^2y)^4} = \dfrac{x^3(y^2)^3}{(x^2)^4 y^4} = \dfrac{x^3 y^6}{x^8 y^4}$

$= \dfrac{x^3}{x^8} \cdot \dfrac{y^6}{y^4} = x^{-5}y^2$

$= \dfrac{1}{x^5} \cdot \dfrac{y^2}{1} = \dfrac{y^2}{x^5}$

7.  $\dfrac{(3x^3y^2)^4}{9(x^2y^4)^3} = \dfrac{3^4(x^3)^4(y^2)^4}{9(x^2)^3(y^4)^3}$

$= \dfrac{81x^{12}y^8}{9x^6y^{12}} = \dfrac{81}{9} \cdot \dfrac{x^{12}}{x^6} \cdot \dfrac{y^8}{y^{12}}$

$= 9 \cdot x^6 \cdot y^{-4} = \dfrac{9}{1} \cdot \dfrac{x^6}{1} \cdot \dfrac{1}{y^4} = \dfrac{9x^6}{y^4}$

9. $(x^{-2})^{-3} = x^6$

11. $\left(\dfrac{2x^{-2}x^3}{x^{-3}}\right)^{-2} = \left(\dfrac{2x}{x^{-3}}\right)^{-2}$

$$= (2x^4)^{-2} = 2^{-2}(x^4)^{-2}$$

$$= \dfrac{1}{2^2}x^{-8} = \dfrac{1}{4} \cdot \dfrac{1}{x^8} = \dfrac{1}{4x^8}$$

13. $58,700,000 = 5.87 \times 10^7$

15. $0.000002 = 2 \times 10^{-6}$

17. $2.56 \times 10^{-3} = 0.00256$

19. $5.773 \times 10^8 = 577,300,000$

21. $(0.008)(250,000) = (8 \times 10^{-3})(2.5 \times 10^5)$

$$= (8)(2.5) \times 10^{-3}10^5$$

$$= 20 \times 10^2 = 2 \times 10^3 = 2,000$$

23. $\dfrac{0.001}{0.000025} = \dfrac{1 \times 10^{-3}}{2.5 \times 10^{-5}}$

$$= \dfrac{1}{2.5} \times \dfrac{10^{-3}}{10^{-5}} = 0.4 \times 10^2 = 4 \times 10^1 = 40$$

25. (a) three terms: $x^2, 3x, -7$
    (b) degree of $x^2$: 2
    degree of $3x$: 1
    degree of $-7$: 0
    (c) degree of polynomial: 2

27. (a) three terms: $3x^3y, -5y^2, 6xy$
    (b) degree of $3x^3y$: $4(=3+1)$
    degree of $-5y^2$: 2
    degree of $6xy$: $2(=1+1)$
    (c) degree of polynomial: 4

29. (a) two terms: $8x, -5$
    (b) degree of $8x$: 1
    degree of $-5$: 0
    (c) degree of polynomial: 1

31. (a) one term: 9
    (b) degree of 9: 0
    (c) degree of polynomial: 0

33. (a) one term
    (b) degree of $(3x^5)(2x^3) = 6x^8$: 8
    (c) degree of polynomial: 8

35. $2x^3 - 7x^2 + 0x + 4$

37. $y^5 + 0y^4 + 0y^3 + y^2 - 2y - 1$

39. $(3x^2 - 5x + 7) + (5x - x^2 - 5) = 3x^2 - 5x + 7 + 5x - x^2 - 5$

$$= 2x^2 + 2$$

41. $(3x^2 - 5x + 7) - (5x - x^2 - 5) = 3x^2 - 5x + 7 - 5x + x^2 + 5$

$$= 4x^2 - 10x + 12$$

43. $2(x^2y - xy^2 - 5x^2y^2) + 3(xy^2 + x^2y + x^2y^2) = 2x^2y - 2xy^2 - 10x^2y^2 + 3xy^2 + 3x^2y + 3x^2y^2$

$$= 5x^2y + xy^2 - 7x^2y^2$$

45. $2(x^2y - xy^2) - 5x^2y^2 - 3(xy^2 - x^2y + x^2y^2) = 2x^2y - 2xy^2 - 5x^2y^2 - 3xy^2 + 3x^2y - 3x^2y^2$

$$= 5x^2y - 5xy^2 - 8x^2y^2$$

47. $2a^2(a - 3b) + 4a(a^2 + ab) - 2(a^3 - a^2b) = 2a^3 - 6a^2b + 4a^3 + 4a^2b - 2a^3 + 2a^2b$
$$= 4a^3$$

49. $x^2 + 4x - (x^2 - 4x) = x^2 + 4x - x^2 + 4x = 8x$

51. $(x^2 + 4x - 3) + (2x^2 - x - 2) - (3x - 5) = x^2 + 4x - 3 + 2x^2 - x - 2 - 3x + 5$
$$= 3x^2$$

53. $(x + 4)(x - 7) = x^2 - 7x + 4x - 28$
$$= x^2 - 3x - 28$$

55. $(2x - 3)(4x - 5) = 8x^2 - 10x - 12x + 15$
$$= 8x^2 - 22x + 15$$

57. $(3a - 4b)(2a + 5b) = 6a^2 + 15ab - 8ab - 20b^2$
$$= 6a^2 + 7ab - 20b^2$$

59. $(x + 2)(x - 3)(x + 1) = (x + 2)(x^2 + x - 3x - 3)$
$$= (x + 2)(x^2 - 2x - 3)$$
$$= x(x^2 - 2x - 3) + 2(x^2 - 2x - 3)$$
$$= x^3 - 2x^2 - 3x + 2x^2 - 4x - 6$$
$$= x^3 - 7x - 6$$

61. $(x + 5)(x + 7) = x^2 + 7x + 5x + 35$
$$= x^2 + 12x + 35$$

63. $(x - 5)(x - 7) = x^2 - 7x - 5x + 35$
$$= x^2 - 12x + 35$$

65. $(x + 5)(x - 7) = x^2 - 7x + 5x - 35$
$$= x^2 - 2x + 35$$

67. $(x - 5)(x + 7) = x^2 + 7x - 5x - 35$
$$= x^2 + 2x - 35$$

69. $(x - 5)(x - 5) = x^2 - 5x - 5x + 25$
$$= x^2 - 10x + 25$$

71. $(x - 5)(x + 5) = x^2 + 5x - 5x - 25$
$$= x^2 - 25$$

73. $(x + 9y)(x - 9y) = x^2 - 9xy + 9xy - 81y^2$
$$= x^2 - 81y^2$$

75. $(2x + 3)(x - 7) = 2x^2 - 14x + 3x - 21$
$$= 2x^2 - 11x - 21$$

77. $(5x - 2)(3x + 4) = 15x^2 + 20x - 6x - 8$
$$= 15x^2 + 14x - 8$$

79. $(x + 6)^2 = (x + 6)(x + 6)$
$$= x^2 + 6x + 6x + 36$$
$$= x^2 + 12x + 36$$

81. $(x - 5)^3 = (x - 5)(x - 5)(x - 5)$
$$= (x - 5)(x^2 - 5x - 5x + 25)$$
$$= (x - 5)(x^2 - 10x + 25)$$
$$= x(x^2 - 10x + 25) - 5(x^2 - 10x + 25)$$
$$= x^3 - 10x^2 + 25x - 5x^2 + 50x - 125$$
$$= x^3 - 15x^2 + 75x - 125$$

83. $3x^2(x-4)(x+2) = 3x^2(x^2 + 2x - 4x - 8)$
$$= 3x^2(x^2 - 2x - 8)$$
$$= 3x^4 - 6x^3 - 24x^2$$

85. $(x-5)(x+5) = x^2 + 5x - 5x - 25$
$$= x^2 - 25$$

87. $(x+2)(x^2 - 3x + 4) = x(x^2 - 3x + 4) + 2(x^2 - 3x + 4)$
$$= x^3 - 3x^2 + 4x + 2x^2 - 6x + 8$$
$$= x^3 - x^2 - 2x + 8$$

89. $(x^2 + 2x - 1)(x^2 + 2x + 1) = x^2(x^2 + 2x + 1) + 2x(x^2 + 2x + 1) - 1(x^2 + 2x + 1)$
$$= x^4 + 2x^3 + x^2 + 2x^3 + 4x^2 + 2x - x^2 - 2x - 1$$
$$= x^4 + 4x^3 + 4x^2 - 1$$

91. $(2x-3)(x+4) - (x-2)(x-1) = 2x^2 + 8x - 3x - 12 - (x^2 - x - 2x + 2)$
$$= 2x^2 + 5x - 12 - (x^2 - 3x + 2)$$
$$= 2x^2 + 5x - 12 - x^2 + 3x - 2$$
$$= x^2 + 8x - 14$$

93. Let $w$ = width of the rectangle .
Then $3w - 5$ = length of the rectangle.
Since the area of a rectangle, $A$, is the product of its length and width, we find
$A = (3w - 5)w = 3w^2 - 5w.$

95. $(60)(5,000)(10^{-6})$meters $= (6 \times 10^1)(5 \times 10^3)(10^{-6})$ meters
$$= 30 \times (10^1 \cdot 10^3 \cdot 10^{-6})\text{ meters}$$
$$= 30 \times 10^{-2}\text{ meters}$$
$$= 3.0 \times 10^{-1}\text{ or } 0.3\text{ meters}$$

## Chapter 7 Practice Test

1. $5^0 + 2^{-2} + 4^{-1} = 1 + \dfrac{1}{2^2} + \dfrac{1}{4^1}$
$$= 1 + \frac{1}{4} + \frac{1}{4}$$
$$= \frac{4}{4} + \frac{1}{4} + \frac{1}{4} = \frac{6}{4} = \frac{3}{2}$$

3. $x^{-4}x^{-5} = x^{-4+(-5)} = x^{-9} = \dfrac{1}{x^9}$

5. $\dfrac{\left(2x^{-3}y^4\right)^4}{4\left(x^{-2}y^{-1}\right)^3} = \dfrac{2^4\left(x^{-3}\right)^4\left(y^4\right)^4}{4\left(x^{-2}\right)^3\left(y^{-1}\right)^3}$

$\qquad = \dfrac{16x^{-3(4)}y^{4(4)}}{4x^{-2(3)}y^{-1(3)}}$

$\qquad = \dfrac{16x^{-12}y^{16}}{4x^{-6}y^{-3}}$

$\qquad = \left(\dfrac{16}{4}\right)\left(\dfrac{x^{-12}}{x^{-6}}\right)\left(\dfrac{y^{16}}{y^{-3}}\right)$

$\qquad = 4x^{-12-(-6)}y^{16-(-3)}$

$\qquad = 4x^{-6}y^{19}$

$\qquad = \dfrac{4}{1}\cdot\dfrac{1}{x^6}\cdot\dfrac{y^{19}}{1} = \dfrac{4y^{19}}{x^6}$

7. $-3x^2y(4x^2y)(-2x^3) = -3(4)(-2)x^2x^2x^3yy$

$\qquad = 24x^{2+2+3}y^{1+1}$

$\qquad = 24x^7y^2$

9. $2x(x^2-y) - 3(x-xy) - (2x^3-3x) = 2x^3 - 2xy - 3x + 3xy - 2x^3 + 3x$

$\qquad = 2x^3 - 2x^3 - 2xy + 3xy - 3x + 3x$

$\qquad = xy$

11. $3x^2(2x-y) - xy(x+y) = 6x^3 - 3x^2y - x^2y - xy^2$

$\qquad = 6x^3 - 4x^2y - xy^2$

13. $(a-1)^2 - (a+1)^2 = (a-1)(a-1) - (a+1)(a+1)$

$\qquad = a^2 - 2a + 1 - (a^2 + 2a + 1)$

$\qquad = a^2 - 2a + 1 - a^2 - 2a - 1$

$\qquad = -4a$

15. $\dfrac{(0.24)(5,000)}{0.006} = \dfrac{(2.4\times10^{-1})(5\times10^3)}{6\times10^{-3}}$

$\qquad = \dfrac{(2.4)(5)}{6}\times\dfrac{10^{-1}10^3}{10^{-3}}$

$\qquad = \dfrac{12}{6}\times\dfrac{10^2}{10^{-3}} = 2\times10^{2-(-3)}$

$\qquad = 2\times10^5 = 200,000$

17. $\dfrac{8.32\times10^{14}}{5.86\times10^{12}} = \dfrac{8.32}{5.86}\times\dfrac{10^{14}}{10^{12}}$

$\qquad = 1.4198\times10^{14-12}$

$\qquad = 1.4198\times10^2$

$\qquad = 141.98\text{ light years}$

# Chapter 8
# Factoring

## Exercises 8.1

1. $4x$

3. $5x$

5. $6mnp$

7. $5x + 20 = 5 \cdot x + 5 \cdot 4 = 5(x + 4)$

9. $6x - 18 = 6 \cdot x - 6 \cdot 3 = 6(x - 3)$

11. $4y + 27$ is not factorable.

13. $8a - 12 = 4 \cdot 2a - 4 \cdot 3 = 4(2a - 3)$

15. $12a + 9 = 3 \cdot 4a + 3 \cdot 3 = 3(4a + 3)$

17. $3a + 6b - 8c$ is not factorable.

19. $x^2 + 3x = x \cdot x + x \cdot 3 = x(x + 3)$

21. $2t^2 + 8t = 2 \cdot t \cdot t + 2 \cdot t \cdot 4 = 2t(t + 4)$

23. $6y^3 - 9y^2 = 3y^2 \cdot 2y - 3y^2 \cdot 3 = 3y^2(2y - 3)$

25. $4x^5 + 2x^2 - 8x = 2x \cdot 2x^4 + 2x \cdot x - 2x \cdot 4$
$$= 2x(2x^4 + x - 4)$$

27. $a^2 + a = a \cdot a + a \cdot 1 = a(a + 1)$

29. $x^2 - 5x + xy = x \cdot x - x \cdot 5 + x \cdot y$
$$= x(x - 5 + y)$$

31. $3c^6 - 6c^3 = 3c^3 \cdot c^3 - 3c^3 \cdot 2$
$$= 3c^3(c^3 - 2)$$

33. $x^2y - xy^2 = xy \cdot x - xy \cdot y$
$$= xy(x - y)$$

35. $6x^2 + 3x = 3x \cdot 2x + 3x \cdot 1$
$$= 3x(2x + 1)$$

37. $8x^3y^2 - 25z^4$ is not factorable.

39. $12c^3d^5 + 4c^2d^3 = 4c^2d^3 \cdot 3cd^2 + 4c^2d^3 \cdot 1$
$$= 4c^2d^3(3cd^2 + 1)$$

41. $x^2y^3 - y^2z^4 + x^3z^2$ is not factorable.

43. $2x^2yz^3 + 8xyz^2 - 10x^2y^2z^2 = 2xyz^2 \cdot xz + 2xyz^2 \cdot 4 - 2xyz^2 \cdot 5xy$
$$= 2xyz^2(xz + 4 - 5xy)$$

45. $6u^3v^2 + 18u^3v^3 - 12u^3v^5 = 6u^3v^2 \cdot 1 + 6u^3v^2 \cdot 3v - 6u^3v^2 \cdot 2v^3$
$$= 6u^3v^2(1 + 3v - 2v^3)$$

47. $x(x - 5) + 4(x - 5) = (x - 5)(x + 4)$

49. $y(y + 6) - 3(y + 6) = (y + 6)(y - 3)$

51. $x^2 + 8x + xy + 8y = (x^2 + 8x) + (xy + 8y)$
$$= x(x + 8) + y(x + 8)$$
$$= (x + 8)(x + y)$$

53. $m^2 + mn + 9m + 9n = (m^2 + mn) + (9m + 9n)$
$$= m(m + n) + 9(m + n)$$
$$= (m + n)(m + 9)$$

**55.** $x^2 - xy - 4x + 4y = (x^2 - xy) + (-4x + 4y)$
$$= x(x - y) - 4(x - y)$$
$$= (x - y)(x - 4)$$

**57.** $3x^2y + 6xy - 5x - 10 = (3x^2y + 6xy) + (-5x - 10)$
$$= 3xy(x + 2) - 5(x + 2)$$
$$= (x + 2)(3xy - 5)$$

**59.** (a) $(x + 2)(x + 3)$

(b) $x^2y^2(x + y)$

**60.** Step 1: $-x$ is replaced by $-5x + 4x$.

Step 2: The first two terms are grouped and the last two terms are grouped. The common factor is then removed from each group.

Step 3: The common factor of $(x - 5)$ is removed from the sum.

**61.** When we factor out $-5$ from the third and fourth terms, the remaining factor should be $x + 3$, not $x - 3$. This means that there is no common factor to remove.

**63.** $\left(x^{-3}\right)^{-4} = x^{(-3)(-4)} = x^{12}$

**65.** $\dfrac{\left(x^{-2}y^3\right)^{-1}}{\left(x^3\right)^{-2}} = \dfrac{\left(x^{-2}\right)^{-1}\left(y^3\right)^{-1}}{x^{-6}} = \dfrac{x^2y^{-3}}{x^{-6}}$

$$= x^{2-(-6)}y^{-3} = \dfrac{x^8}{y^3}$$

**67.** Let $x$ = number of liters of pure water to be added.
$$0.18(36) = 0.12(36 + x)$$
$$6.48 = 4.32 + 0.12x$$
$$2.16 = 0.12x$$
$$216 = 12x$$
$$18 = x$$
So 18 liters of pure water must be added.

## Exercises 8.2

**1.** $x^2 + 3x = x(x + 3)$

**3.** $x^2 + 3x + 2 = (x + 1)(x + 2)$

**5.** $x^2 - 3x + 2 = (x - 1)(x - 2)$

**7.** $x^2 + 3x - 2$ is not factorable.

**9.** $x^2 + x - 2 = (x - 1)(x + 2)$

**11.** $x^2 - x - 2 = (x + 1)(x - 2)$

**13.** $a^2 + 8a + 12 = (a + 2)(a + 6)$

**15.** $a^2 - a - 12 = (a - 4)(a + 3)$

**17.** $a^2 - a + 12$ is not factorable.

**19.** $a^2 - 12a = a(a - 12)$

**21.** $a - 12 + a^2 = a^2 + a - 12$
$$= (a + 4)(a - 3)$$

**23.** $y^2 + 11y + 28 = (y + 4)(y + 7)$

**25.** $x^2 - 5x - 36 = (x - 9)(x + 4)$

**27.** $a^2 + 6a - 40 = (a + 10)(a - 4)$

29. $x^2 - 17x + 30 = (x - 2)(x - 15)$

31. $x^2 - 9x = x(x - 9)$

33. $x^2 - 9 = (x - 3)(x + 3)$

35. $x^2 - 9x - 10 = (x - 10)(x + 1)$

37. $x^2 - 3xy + 2y^2 = (x - y)(x - 2y)$

39. $a^2 + 10a + 24 = (a + 4)(a + 6)$

41. $y^2 + 12y + 36 = (y + 6)(y + 6)$
    or $(y + 6)^2$

43. $y^2 - 36 = (y - 6)(y + 6)$

45. $x^2 - 7x - 18 = (x - 9)(x + 2)$

47. $r^2 - 3rs - 10s^2 = (r - 5s)(r + 2s)$

49. $c^2 - 6c + 5 = (c - 1)(c - 5)$

51. $4x^2 + 8x + 4 = 4(x^2 + 2x + 1)$
    $= 4(x + 1)(x + 1)$
    or $4(x + 1)^2$

53. $x^2 - 30 + x = x^2 + x - 30$
    $= (x + 6)(x - 5)$

55. $2x^2 - 50 = 2(x^2 - 25)$
    $= 2(x - 5)(x + 5)$

57. $x^2 - x - 20 = (x - 5)(x + 4)$

59. $x^2 - x + 20$ is not factorable.

61. $y^2 - 11y + 28 = (y - 4)(y - 7)$

63. $2y^2 + 2y - 84 = 2(y^2 + y - 42)$
    $= 2(y + 7)(y - 6)$

65. $49 - d^2 = (7 - d)(7 + d)$

67. $49 + d^2$ is not factorable.

69. $10x^2 - 40xy - 120y^2 = 10(x^2 - 4xy - 12y^2)$
    $= 10(x - 6y)(x + 2y)$

71. $a^2 + 13 + 14a = a^2 + 14a + 13$
    $= (a + 13)(a + 1)$

73. $6s^2 - 6s - 72 = 6(s^2 - s - 12)$
    $= 6(s - 4)(s + 3)$

75. $4x^2 - 64 = 4(x^2 - 16)$
    $= 4(x - 4)(x + 4)$

77. The factors of 10 are 1 and 10 and 2 and 5. Since the last term is positive, the signs in the parentheses must be the same.

   Possibilities:
   $(x + 1)(x + 10) = x^2 \underline{+ 11x} + 10$
   $(x - 1)(x - 10) = x^2 \underline{- 11x} + 10$
   $(x + 2)(x + 5) = x^2 \underline{+ 7x} + 10$
   $(x - 2)(x - 5) = x^2 \underline{- 7x} + 10$

   Therefore, the possible values of $k$ are $11, -11, 7$, and $-7$.

78. As in Exercise 77, the factors of 10 are 1 and 10 and 2 and 5. Since the last term is negative, the signs in the parentheses must be opposite.

   Possibilities:
   $(x + 1)(x - 10) = x^2 \underline{- 9x} + 10$
   $(x - 1)(x + 10) = x^2 \underline{+ 9x} + 10$
   $(x + 2)(x - 5) = x^2 \underline{- 3x} + 10$
   $(x - 2)(x + 5) = x^2 \underline{+ 3x} + 10$

   Therefore, the possible values of $b$ are $-9, 9, -3$, and $3$.

79.  There are infinitely many such integers.  We can always find such a "$c$" by choosing two integers that differ by 5 and forming the product.
$(x + \text{larger integer})(x - \text{smaller integer})$
For example, since 9 and 4 differ by 5, $(x + 9)(x - 4) = x^2 + 5x - 36$ will give the value $c = -36$. Clearly, we can find two integers that differ by 5 in infinitely many ways.  (Interestingly, if we ask for all <u>positive</u> integers $c$ with this property, there are only two answers:  $c = 4$ and $c = 6$.)

81.  $(5x^2 + y)^2 = (5x^2 + y)(5x^2 + y)$
$$= 25x^4 + 5x^2y + 5x^2y + y^2$$
$$= 25x^4 + 10x^2y + y^2$$

83.  $(2x - 7)(3x + 4) = 6x^2 + 8x - 21x - 28$
$$= 6x^2 - 13x - 28$$

85.  $28,700 = 2.87 \times 10^4$

87.  $\dfrac{(2,400)(0.003)}{(0.02)(0.004)} = \dfrac{(2.4 \times 10^3)(3 \times 10^{-3})}{(2 \times 10^{-2})(4 \times 10^{-3})}$
$$= \dfrac{(2.4)(3)}{(2)(4)} \times \dfrac{10^3 \cancel{10^{-3}}}{10^{-2} \cancel{10^{-3}}}$$
$$= 0.9 \times 10^5 = 9 \times 10^4$$
$$= 90,000$$

## Exercises 8.3

1.  $x^2 + 5x = x(x + 5)$

3.  $x^2 + 5x + 4 = (x + 1)(x + 4)$

5.  $x^2 + 5x - 4$ is not factorable.

7.  $3x^2 + 8x + 4 = (3x + 2)(x + 2)$

9.  $2x^2 + 11x + 12 = (2x + 3)(x + 4)$

11.  $2x^2 + 10x + 12 = 2(x^2 + 5x + 6)$
$$= 2(x + 2)(x + 3)$$

13.  $5x^2 - 27x + 10 = (5x - 2)(x - 5)$

15.  $5x^2 - 15x + 10 = 5(x^2 - 3x + 2)$
$$= 5(x - 1)(x - 2)$$

17.  $2y^2 - y - 6 = (2y + 3)(y - 2)$

19.  $5a^2 + 9a - 18 = (5a - 6)(a + 3)$

21.  $2t^2 + 7t + 6 = (2t + 3)(t + 2)$

23.  $2t^2 + 6t + 6 = 2(t^2 + 3t + 3)$

25.  $3w^2 - 6w - 30 = 3(w^2 - 2w - 10)$

27.  $3x^2 - 4x + 2$ is not factorable.

29.  $3x^2 - 14xy + 15y^2 = (3x - 5y)(x - 3y)$

31.  $6a^2 + 17a + 10 = (6a + 5)(a + 2)$

33.  $6a^2 + 17a - 10 = (3a + 10)(2a - 1)$

35.  $6a^2 - 18a - 24 = 6(a^2 - 3a - 4)$
$$= 6(a - 4)(a + 1)$$

37.  $x^2 - 36y^2 = (x - 6y)(x + 6y)$

39.  $4x^2 - 36y^2 = 4(x^2 - 9y^2)$
$$= 4(x - 3y)(x + 3y)$$

41.  $x^3 + 5x^2 - 24x = x(x^2 + 5x - 24)$
$$= x(x + 8)(x - 3)$$

43.  $x^2 + 5x^2 - 24x = 6x^2 - 24x$
$$= 6x(x - 4)$$

45. $4x^4 - 24x^3 + 32x^2 = 4x^2(x^2 - 6x + 8)$
$$= 4x^2(x - 2)(x - 4)$$

47. $6x^2y - 8xy^2 + 12xy = 2xy(3x - 4y + 6)$

49. $3x^2 - 7x - 48 = (3x - 16)(x + 3)$

51. $8x^2 - 32x = 8x(x - 4)$

53. $2x - x^2 + 15 = -x^2 + 2x + 15$
$$= -(x^2 - 2x - 15)$$
$$= -(x - 5)(x + 3)$$

55. $84xy - 16x^2y - 4x^3y = 4xy(21 - 4x - x^2)$
$$= -4xy(x^2 + 4x - 21)$$
$$= -4xy(x + 7)(x - 3)$$

57. $-x^2 + 25 = 25 - x^2 = (5 - x)(5 + x)$

59. The proposed factor $2x + 2$ has a common factor of 2, which would imply that the original trinomial $6x^2 - 5x - 4$ has a common factor of 2. This is not the case. We can eliminate eight possible factorizations of $6x^2 - 5x - 4$ in this way. In addition to $(3x - 2)(2x + 2)$, we can eliminate

$(3x - 1)(2x + 4)$    $(3x + 1)(2x - 4)$    $(3x + 2)(2x - 2)$    $(6x - 2)(x + 2)$
$(6x + 2)(x - 2)$    $(6x - 4)(x + 1)$    $(6x + 4)(x - 1)$

this leaves four possibilities:

$(6x - 1)(x + 4)$,    $(6x + 1)(x - 4)$,    $(3x + 4)(2x - 1)$,    and    $(3x - 4)(2x + 1)$,

the last of which is the correct one.

61. Let $f$ = number of franks sold.
Then $f + 20$ = number of knishes sold.
$$1.25f + 0.80(f + 20) = 169.75$$
$$1.25f + 0.80f + 16 = 169.75$$
$$2.05f + 16 = 169.75$$
$$2.05f = 153.75$$
$$f = 75$$
Then $f + 20 = 75 + 20 = 95$. So the vendor sold 75 franks and 95 knishes on the day in question.

## Exercises 8.4

1. $(x - 2)(x + 3) = 0$
$x - 2 = 0$ or $x + 3 = 0$
$x = 2$        $x = -3$

3. $(x - 2)(x + 3) = 6$
$x^2 + x - 6 = 6$
$x^2 + x - 12 = 0$
$(x + 4)(x - 3) = 0$
$x + 4 = 0$ or $x - 3 = 0$
$x = -4$        $x = 3$

CHECK $x = -4$:
$(x - 2)(x + 3) = 6$
$(-4 - 2)(-4 + 3) \stackrel{?}{=} 6$
$(-6)(-1) \stackrel{?}{=} 6$
$6 \stackrel{\checkmark}{=} 6$

CHECK $x = 3$:
$(x - 2)(x + 3) = 6$
$(3 - 2)(3 + 3) \stackrel{?}{=} 6$
$(1)(6) \stackrel{?}{=} 6$
$6 \stackrel{\checkmark}{=} 6$

9.  $x^2 - x - 6 = 0$
    $(x-3)(x+2) = 0$
    $x - 3 = 0$  or  $x + 2 = 0$
    $x = 3$ $\qquad$ $x = -2$

CHECK $x = 3$:
$$x^2 - x - 6 = 0$$
$$(3)^2 - (3) - 6 \stackrel{?}{=} 0$$
$$9 - 3 - 6 \stackrel{?}{=} 0$$
$$0 \stackrel{\checkmark}{=} 0$$

CHECK $x = -2$:
$$x^2 - x - 6 = 0$$
$$(-2)^2 - (-2) - 6 \stackrel{?}{=} 0$$
$$4 + 2 - 6 \stackrel{?}{=} 0$$
$$0 \stackrel{\checkmark}{=} 0$$

11. $x^2 - 3x = 10$
    $x^2 - 3x - 10 = 0$
    $(x-5)(x+2) = 0$
    $x - 5 = 0$  or  $x + 2 = 0$
    $x = 5$ $\qquad$ $x = -2$

CHECK $x = 5$:
$$x^2 - 3x = 10$$
$$(5)^2 - 3(5) \stackrel{?}{=} 10$$
$$25 - 15 \stackrel{?}{=} 10$$
$$10 \stackrel{\checkmark}{=} 10$$

CHECK $x = -2$:
$$x^2 - 3x = 10$$
$$(-2)^2 - 3(-2) \stackrel{?}{=} 10$$
$$4 + 6 \stackrel{?}{=} 10$$
$$10 \stackrel{\checkmark}{=} 10$$

15. $-m^2 = 8 - 9m$
    $-m^2 + 9x - 8 = 0$
    $m^2 - 9x + 8 = 0$
    $(m-1)(m-8) = 0$
    $m - 1 = 0$  or  $m - 8 = 0$
    $m = 1$ $\qquad$ $m = 8$

CHECK $m = 1$:
$$-m^2 = 8 - 9m$$
$$-(1)^2 \stackrel{?}{=} 8 - 9(1)$$
$$-1 \stackrel{?}{=} 8 - 9$$
$$-1 \stackrel{\checkmark}{=} -1$$

CHECK $m = 8$:
$$-m^2 = 8 - 9m$$
$$-(8)^2 \stackrel{?}{=} 8 - 9(8)$$
$$-64 \stackrel{?}{=} 8 - 72$$
$$-64 \stackrel{\checkmark}{=} -64$$

17. $p^2 + 3p = p(p+4)$
    $p^2 + 3p = p^2 + 4p$
    $3p = 4p$
    $p = 0$

CHECK $p = 0$:
$$p^2 + 3p = p(p+4)$$
$$(0)^2 + 3(0) \stackrel{?}{=} (0)(0+4)$$
$$0 + 0 \stackrel{?}{=} 0(4)$$
$$0 \stackrel{\checkmark}{=} 0$$

21. $5w^2 = 8w$
    $5w^2 - 8w = 0$
    $w(5w - 8) = 0$
    $w = 0$  or  $5w - 8 = 0$
    $\qquad\qquad$ $5w = 8$
    $\qquad\qquad$ $w = \dfrac{8}{5}$

CHECK $w = 0$:
$$5w^2 = 8w$$
$$5(0)^2 \stackrel{?}{=} 8(0)$$
$$5(0) \stackrel{?}{=} 8(0)$$
$$0 \stackrel{\checkmark}{=} 0$$

CHECK $w = \dfrac{8}{5}$:
$$5w^2 = 8w$$
$$5\left(\frac{8}{5}\right)^2 \stackrel{?}{=} 8\left(\frac{8}{5}\right)$$
$$\overset{1}{\cancel{5}}\left(\frac{64}{\underset{5}{25}}\right) \stackrel{?}{=} \frac{64}{5}$$
$$\frac{64}{5} \stackrel{\checkmark}{=} \frac{64}{5}$$

27.
$$2a(a+3) = 20$$
$$2a^2 + 6a = 20$$
$$2a^2 + 6a - 20 = 0$$
$$2(a^2 + 3a - 10) = 0$$
$$a^2 + 3a - 10 = 0$$
$$(a+5)(a-2) = 0$$
$$a + 5 = 0 \quad \text{or} \quad a - 2 = 0$$
$$a = -5 \qquad\quad a = 2$$

CHECK $a = -5$:
$$2a(a+3) = 20$$
$$2(-5)(-5+3) \overset{?}{=} 20$$
$$2(-5)(-2) \overset{?}{=} 20$$
$$20 \overset{\checkmark}{=} 20$$

CHECK $a = 2$:
$$2a(a+3) = 20$$
$$2(2)(2+3) \overset{?}{=} 20$$
$$2(2)(5) \overset{?}{=} 20$$
$$20 \overset{\checkmark}{=} 20$$

31.
$$a^2 - 4a - 2 = 2a^2 - 9a - 16$$
$$-4a - 2 = a^2 - 9a - 16$$
$$-2 = a^2 - 5a - 16$$
$$0 = a^2 - 5a - 14$$
$$0 = (a-7)(a+2)$$
$$0 = a - 7 \quad \text{or} \quad 0 = a + 2$$
$$7 = a \qquad\qquad -2 = a$$

CHECK $a = 7$:
$$a^2 - 4a - 2 = 2a^2 - 9a - 16$$
$$(7)^2 - 4(7) - 2 \overset{?}{=} 2(7)^2 - 9(7) - 16$$
$$49 - 28 - 2 \overset{?}{=} 98 - 63 - 16$$
$$19 \overset{\checkmark}{=} 19$$

CHECK $a = -2$:
$$a^2 - 4a - 2 = 2a^2 - 9a - 16$$
$$(-2)^2 - 4(-2) - 2 \overset{?}{=} 2(-2)^2 - 9(-2) - 16$$
$$4 + 8 - 2 \overset{?}{=} 8 + 18 - 16$$
$$10 \overset{\checkmark}{=} 10$$

35.
$$(x+3)^2 = 3x^2 - 10$$
$$x^2 + 6x + 9 = 3x^2 - 10$$
$$6x + 9 = 2x^2 - 10$$
$$9 = 2x^2 - 6x - 10$$
$$0 = 2x^2 - 6x - 19$$
The quadratic expression $2x^2 - 6x - 19$ cannot be factored.

39.
$$(x+2)^2 = 25$$
$$(x+2)(x+2) = 25$$
$$x^2 + 4x + 4 = 25$$
$$x^2 + 4x - 21 = 0$$
$$(x+7)(x-3) = 0$$
$$x + 7 = 0 \quad \text{or} \quad x - 3 = 0$$
$$x = -7 \qquad\quad x = 3$$

CHECK $x = -7$:
$$(x+2)^2 = 25$$
$$(-7+2)^2 \overset{?}{=} 25$$
$$(-5)^2 \overset{?}{=} 25$$
$$25 \overset{\checkmark}{=} 25$$

CHECK $x = 3$:
$$(x+2)^2 = 25$$
$$(3+2)^2 \overset{?}{=} 25$$
$$(5)^2 \overset{?}{=} 25$$
$$25 \overset{\checkmark}{=} 25$$

43.
$$(2x-4)(x+1) = (x-3)(x-2)$$
$$2x^2 - 2x - 4 = x^2 - 5x + 6$$
$$x^2 - 2x - 4 = -5x + 6$$
$$x^2 + 3x - 4 = 6$$
$$x^2 + 3x - 10 = 0$$
$$(x+5)(x-2) = 0$$
$$x + 5 = 0 \quad \text{or} \quad x - 2 = 0$$
$$x = -5 \qquad\quad x = 2$$

CHECK $x = -5$:

$$(2x - 4)(x + 1) = (x - 3)(x - 2)$$

$$[2(-5) - 4](-5 + 1) \overset{?}{=} (-5 - 3)(-5 - 2)$$

$$(-14)(-4) \overset{?}{=} (-8)(-7)$$

$$56 \overset{\checkmark}{=} 56$$

CHECK $x = 2$:

$$(2x - 4)(x + 1) = (x - 3)(x - 2)$$

$$[2(2) - 4](2 + 1) \overset{?}{=} (2 - 3)(2 - 2)$$

$$(0)(3) \overset{?}{=} (-1)(0)$$

$$0 \overset{\checkmark}{=} 0$$

45.  Let $L$ = length of the rectangle.
Then $3L - 2$ = width of the rectangle.

$$L(3L - 2) = 33$$
$$3L^2 - 2L = 33$$
$$3L^2 - 2L - 33 = 0$$
$$(3L - 11)(L + 3) = 0$$

$3L - 11 = 0$ or $L + 3 = 0$
$3L = 11 \qquad L = -3$

$$L = \frac{11}{3} \qquad \text{(reject)}$$

When $L = \frac{11}{3}$, $3L - 2 = \cancel{3}\left(\frac{11}{\cancel{3}}\right) - 2 = 11 - 2 = 9$. Thus, the dimensions of the rectangle

are $\frac{11}{3}$ or $3\frac{2}{3}$ ft. by 9 ft.

47.  Let $x$ = length of the side of the square painting.
Then $x + 2$ = length of the side of the framed painting.

$$(x + 2)^2 - x^2 = 28$$
$$x^2 + 4x + 4 - x^2 = 28$$
$$4x + 4 = 28$$
$$4x = 24$$
$$x = 6$$

The square painting is 6 in. by 6 in.

49.  Let $x$ = width of the path.

$$(2x + 5)(2x + 7) = 63$$
$$4x^2 + 24x + 35 = 63$$
$$4x^2 + 24x - 28 = 0$$
$$4(x^2 + 6x - 7) = 0$$
$$x^2 + 6x - 7 = 0$$
$$(x - 1)(x + 7) = 0$$

$x - 1 = 0$ or $x + 7 = 0$
$x = 1 \qquad x = -7$
$\qquad\qquad \text{(reject)}$

Thus, the width of the path is 1 ft.

51.
$$y^2 - 65y = 1200$$
$$y^2 - 65y - 1200 = 0$$
$$(y - 80)(y + 15) = 0$$

$y - 80 = 0$ or $y + 15 = 0$
$y = 80 \qquad y = -15$
$\qquad\qquad \text{(reject)}$

Thus, 80 sq. yd of carpeting must be sold.

53.   $240t - 16t^2 = 800$

$$240t = 16t^2 + 800$$
$$0 = 16t^2 - 240t + 800$$
$$0 = 16(t^2 - 15t + 50)$$
$$0 = t^2 - 15t + 50$$
$$0 = (t - 5)(t - 10)$$
$$0 = t - 5 \quad \text{or} \quad 0 = t - 10$$
$$5 = t \qquad\qquad 10 = t$$

The object will be 800 ft. high after 5 seconds (when it is rising) and again after 10 seconds (when it is falling).

55.   $$n^2 - n = 90$$
$$n^2 - n - 90 = 0$$
$$(n - 10)(n + 9) = 0$$
$$n - 10 = 0 \quad \text{or} \quad n + 9 = 0$$
$$n = 10 \qquad\qquad n = -9$$
$$\text{(reject)}$$

Thus, the league has 10 teams.

57.   (a)   We cannot set each of the factors equal to 7. The zero-product rule requires that the product of the factors be equal to 0.

      (b)   $x = 3$ is not a possible solution. The first factor on the left side of the equation can never be equal to zero. We can either ignore its presence or divide both sides of the equation by it. (This logic is valid for constant factors of a zero product, but <u>not</u> for variable factors.)

58.   Since the square of any real number must be non-negative, $x^2$ must be at least zero. Therefore, $x^2 + 4$ must be at least 4, which means that $x^2 + 4$ cannot ever equal 0. Thus, the equation $x^2 + 4 = 0$ cannot have any solution in the real number system.

## Exercises 8.5

1.   $\dfrac{3x + 12}{6} = \dfrac{\overset{}{\cancel{3}x}}{\cancel{6}} + \dfrac{\overset{2}{\cancel{12}}}{\cancel{6}} = \dfrac{x}{2} + 2$

      or $\dfrac{x + 4}{2}$

3.   $\dfrac{t^2 - 6t}{6t} = \dfrac{\overset{t}{\cancel{t^2}}}{\cancel{6t}} - \dfrac{\cancel{6t}}{\cancel{6t}} = \dfrac{t}{6} - 1$

      or $\dfrac{t - 6}{6}$

5.   $\dfrac{3x^2y - 9xy^2}{3xy} = \dfrac{\overset{x}{\cancel{3x^2y}}}{\cancel{3xy}} - \dfrac{\overset{3\ y}{\cancel{3xy^2}}}{\cancel{3xy}} = x - 3y$

7.   $\dfrac{3x^2y - 9xy^2}{6x^2y^2} = \dfrac{\overset{}{\cancel{3x^2y}}}{\cancel{6x^2y^2}} - \dfrac{\overset{3}{\cancel{9xy^2}}}{\cancel{6x^2y^2}}$

      $= \dfrac{1}{2y} - \dfrac{3}{2x}$

      or $\dfrac{x - 3y}{2xy}$

9.   $\dfrac{10a^2b^3c - 15ab^2c^2 - 20a^3b^2c^3}{5ab^2c} = \dfrac{\overset{2\ a\ b}{\cancel{10a^2b^3c}}}{\cancel{5ab^2c}} - \dfrac{\overset{3\quad c}{\cancel{15ab^2c^2}}}{\cancel{5ab^2c}} - \dfrac{\overset{4\ a^2\ c^2}{\cancel{20a^3b^2c^3}}}{\cancel{5ab^2c}}$

      $= 2ab - 3c - 4a^2c^2$

11.
$$
\begin{array}{r}
x - 5 \\
x+2 \overline{\smash{)}\ x^2 - 3x\ +\ 2} \\
\underline{-(x^2 + 2x)} \\
-5x\ +\ 2 \\
\underline{-(-5x\ -\ 10)} \\
12
\end{array}
$$

13.
$$
\begin{array}{r}
t + 2 \\
t-5 \overline{\smash{)}\ t^2 - 3t\ -\ 10} \\
\underline{-(t^2 - 5t)} \\
2t - 10 \\
\underline{-(2t\ -\ 10)} \\
0
\end{array}
$$

15.
$$
\begin{array}{r}
w + 1 \\
w+3 \overline{\smash{)}\ w^2 + 4w\ -\ 21} \\
\underline{-(w^2 + 3w)} \\
w - 21 \\
\underline{-(\ w\ +\ 3)} \\
-24
\end{array}
$$

17.
$$
\begin{array}{r}
2x - 1 \\
x-1 \overline{\smash{)}\ 2x^2 - 3x\ +7} \\
\underline{-(2x^2 - 2x)} \\
-x\ +7 \\
\underline{-(-x\ +1)} \\
6
\end{array}
$$

19.
$$
\begin{array}{r}
y^2 + 3y\ +7 \\
y-2 \overline{\smash{)}\ y^3 +\ y^2\ +\ y\ -\ 14} \\
\underline{-(y^3 - 2y^2)} \\
3y^2\ +\ y \\
\underline{-(3y^2\ -6y)} \\
7y\ -14 \\
\underline{-(7y\ -14)} \\
0
\end{array}
$$

21.
$$
\begin{array}{r}
2a^2 -\ a\ -\ 2 \\
a+1 \overline{\smash{)}\ 2a^3 +\ a^2\ -3a\ +2} \\
\underline{-(2a^3 + 2a^2)} \\
-a^2\ -3a \\
\underline{-(-a^2\ -\ a)} \\
-2a\ +2 \\
\underline{-(-2a\ -2)} \\
4
\end{array}
$$

23.
$$
\begin{array}{r}
x^2 - 4x\ +12 \\
x+3 \overline{\smash{)}\ x^3 -\ x^2\ +\ 0x\ +36} \\
\underline{-(x^3 + 3x^2)} \\
-4x^2\ +\ 0x \\
\underline{-(-4x^2\ -12x)} \\
12x\ +36 \\
\underline{-(12x\ +36)} \\
0
\end{array}
$$

25.
$$
\begin{array}{r}
x^3 + 2x^2\ +4x\ +8 \\
x-2 \overline{\smash{)}\ x^4 + 0x^3\ +0x^2\ +0x\ -16} \\
\underline{-(x^4 - 2x^3)} \\
2x^3\ +0x^2 \\
\underline{-(2x^3\ -4x^2)} \\
4x^2\ +0x \\
\underline{-(4x^2\ -8x)} \\
8x\ -16 \\
\underline{-(8x\ -16)} \\
0
\end{array}
$$

**27.**

$$
\begin{array}{r}
x^2 + 4x \phantom{xx} - 2 \phantom{0} \\
3x + 2 \overline{\smash{\big)}\ 3x^3 + 14x^2 + 2x - 4} \\
\underline{-(3x^3 + 2x^2)} \phantom{xxxxxxxxx} \\
12x^2 + 2x \phantom{xxx} \\
\underline{-(12x^2 + 8x)} \phantom{xx} \\
-6x - 4 \\
\underline{-(-6x - 4)} \\
0
\end{array}
$$

**29.**

$$
\begin{array}{r}
2t^2 + 5t \phantom{xx} - 4 \phantom{00} \\
2t - 5 \overline{\smash{\big)}\ 4t^3 + 0t^2 - 33t + 24} \\
\underline{-(4t^3 - 10t^2)} \phantom{xxxxxxxxx} \\
10t^2 - 33t \phantom{xxx} \\
\underline{-(10t^2 - 25t)} \phantom{xx} \\
-8t + 24 \\
\underline{-(-8t + 20)} \\
4
\end{array}
$$

**31.** To divide $5x^4 - 8x^3 + 7x^2 - 2x + 1$ by $x^2 + x - 1$, we would begin by asking what $x^2$ must be multiplied by in order to give a product of $5x^4$. When we determine this to be $5x^2$, we multiply $5x^2$ and $x^2 + x - 1$ and subtract the resulting product from $5x^4 - 8x^3 + 7x^2 - 2x + 1$. We continue this process, as shown below.

$$
\begin{array}{r}
5x^2 - 13x + 25 \phantom{xxxx} \\
x^2 + x - 1 \overline{\smash{\big)}\ 5x^4 - 8x^3 + 7x^2 - 2x + 1} \\
\underline{-(5x^4 + 5x^3 - 5x^2)} \phantom{xxxxxxxxxxxx} \\
-13x^3 + 12x^2 - 2x \phantom{xxxx} \\
\underline{-(-13x^3 - 13x^2 + 13x)} \phantom{xxx} \\
25x^2 - 15x + 1 \\
\underline{-(25x^2 + 25x - 25)} \\
-40x + 26
\end{array}
$$

The long division process ends when the degree of the remainder (in this case, 1) is less than the degree of the divisor (in this case, 2).

**33.** $-x^2 = -(-3)^2 = -9$

**35.** $(xy)^2 = (-3 \cdot 5)^2 = (-15)^2 = 225$

**37.** $x^{-2} = (-3)^{-2} = \dfrac{1}{(-3)^2} = \dfrac{1}{9}$

**39.** $(xy)^{-1} = (-3 \cdot 5)^{-1} = (-15)^{-1}$
$\phantom{(xy)^{-1}} = \dfrac{1}{-15} = -\dfrac{1}{15}$

## Chapter 8 Review Exercises

**1.** $x^2 + 7x + 12 = (x + 3)(x + 4)$

**3.** $x^2 + 7x = x(x + 7)$

**5.** $x^2 - 13x + 12 = (x - 1)(x - 12)$

**7.** $x^2 - 6xy - 27y^2 = (x + 3y)(x - 9y)$

**9.** $x^2 - 64 = (x - 8)(x + 8)$

**11.** $2x^2 + 9x + 10 = (2x + 5)(x + 2)$

**13.** $3x^2 - 6x - 24 = 3(x^2 - 2x - 8)$
$\phantom{3x^2 - 6x - 24} = 3(x - 4)(x + 2)$

**15.** $6a^2 + 36a + 48 = 6(a^2 + 6a + 8)$
$\phantom{6a^2 + 36a + 48} = 6(a + 2)(a + 4)$

**17.** $5x^3y - 80xy^3 = 5xy(x^2 - 16y^2)$
$\phantom{5x^3y - 80xy^3} = 5xy(x - 4y)(x + 4y)$

**19.** $x^2 + 9x = x(x + 9)$

**21.** $25t^2 - 1 = (5t-1)(5t+1)$

**23.** $30 - x^2 + x = -x^2 + x + 30$
$$= -(x^2 - x - 30)$$
$$= -(x-6)(x+5)$$

**25.** $12x - 3x^2 - 9 = -3x^2 + 12x - 9$
$$= -3(x^2 - 4x + 3)$$
$$= -3(x-1)(x-3)$$

**27.** $x(x-7) + 3(x-7) = (x-7)(x+3)$

**29.** $m^2 + 2mn + 9m + 18n = m(m+2n) + 9(m+2n)$
$$= (m+2n)(m+9)$$

**31.** $(x-5)(x+4) = 0$
$x - 5 = 0$ or $x + 4 = 0$
$x = 5 \qquad\qquad x = -4$

**33.** $(x-5)(x+4) = 36$
$x^2 + x - 20 = 36$
$x^2 + x - 56 = 0$
$(x-8)(x+7) = 0$
$x - 8 = 0$ or $x + 7 = 0$
$x = 8 \qquad\qquad x = -7$

CHECK $x = 8$:
$(x-5)(x+4) = 36$
$(8-5)(8+4) \overset{?}{=} 36$
$(3)(12) \overset{?}{=} 36$
$36 \overset{\checkmark}{=} 36$

CHECK $x = -7$:
$(x-5)(x+4) = 36$
$(-7-5)(-7+4) \overset{?}{=} 36$
$(-12)(-3) \overset{?}{=} 36$
$36 \overset{\checkmark}{=} 36$

**35.** $3x^2 + 8x = 35$
$3x^2 + 8x - 35 = 0$
$(3x-7)(x+5) = 0$
$3x - 7 = 0$ or $x + 5 = 0$
$3x = 7 \qquad\qquad x = -5$
$x = \dfrac{7}{3}$

CHECK $x = \dfrac{7}{3}$:
$3x^2 + 8x = 35$
$3\left(\dfrac{7}{3}\right)^2 + 8\left(\dfrac{7}{3}\right) \overset{?}{=} 35$
$\overset{1}{\cancel{3}}\left(\dfrac{49}{\underset{3}{\cancel{9}}}\right) + 8\left(\dfrac{7}{3}\right) \overset{?}{=} 35$
$\dfrac{49}{3} + \dfrac{56}{3} \overset{?}{=} 35$
$\dfrac{105}{3} \overset{?}{=} 35$
$35 \overset{\checkmark}{=} 35$

CHECK $x = -5$:
$3x^2 + 8x = 35$
$3(-5)^2 + 8(-5) \overset{?}{=} 35$
$3(25) + 8(-5) \overset{?}{=} 35$
$75 - 40 \overset{?}{=} 35$
$35 \overset{\checkmark}{=} 35$

**37.** $(2x-3)(x+2) = (x+6)(x-1)$
$2x^2 + x - 6 = x^2 + 5x - 6$
$\underline{-x^2 - 5x + 6 = -x^2 - 5x + 6}$
$x^2 - 4x \qquad\;\; = 0$
$x(x-4) = 0$
$x = 0$ or $x - 4 = 0$
$x = 0 \qquad\qquad x = 4$

CHECK $x = 0$:
$(2x - 3)(x + 2) = (x + 6)(x - 1)$
$[2(0) - 3](0 + 2) \overset{?}{=} (0 + 6)(0 - 1)$
$(0 - 3)(0 + 2) \overset{?}{=} (0 + 6)(0 - 1)$
$(-3)(2) \overset{?}{=} (6)(-1)$
$-6 \overset{\checkmark}{=} -6$

CHECK $x = 4$:
$(2x - 3)(x + 2) = (x + 6)(x - 1)$
$[2(4) - 3](4 + 2) \overset{?}{=} (4 + 6)(4 - 1)$
$(8 - 3)(4 + 2) \overset{?}{=} (4 + 6)(4 - 1)$
$(5)(6) \overset{?}{=} (10)(3)$
$30 \overset{\checkmark}{=} 30$

39.
$5x^2 = 80$
$x^2 = 16$
$x^2 - 16 = 0$
$(x - 4)(x + 4) = 0$
$x - 4 = 0$ or $x + 4 = 0$
$x = 4 \qquad x = -4$

CHECK $x = 4$:
$5x^2 = 80$
$5(4)^2 \overset{?}{=} 80$
$(5)(16) \overset{?}{=} 80$
$80 \overset{\checkmark}{=} 80$

CHECK $x = -4$:
$5x^2 = 80$
$5(-4)^2 \overset{?}{=} 80$
$(5)(16) \overset{?}{=} 80$
$80 \overset{\checkmark}{=} 80$

41. Let $w =$ width of the rectangle.
Then $2w - 3 =$ length of the rectangle.
$$w(2w - 3) = 14$$
$$2w^2 - 3w = 14$$
$$2w^2 - 3w - 14 = 0$$
$$(2w - 7)(w + 2) = 0$$
$$2w - 7 = 0 \quad \text{or} \quad w + 2 = 0$$
$$2w = 7 \qquad\qquad w = -2$$
$$w = \frac{7}{2} \qquad\qquad \text{(reject)}$$
When $w = \frac{7}{2}$, $2w - 3 = \cancel{2}\left(\dfrac{7}{\cancel{2}}\right) - 3 = 7 - 3 = 4$.

The dimensions of the rectangle are 4 cm. by $\dfrac{7}{2}$ or $3\dfrac{1}{2}$ cm.

43. Let $x =$ length of the side of the square garden.
$$(x + 4)^2 = 144$$
$$x^2 + 8x + 16 = 144$$
$$x^2 + 8x - 128 = 0$$
$$(x - 8)(x + 16) = 0$$
$$x - 8 = 0 \quad \text{or} \quad x + 16 = 0$$
$$x = 8 \qquad\qquad x = -2$$
$$\text{(reject)}$$
The dimensions of the garden are 8 ft. by 8 ft.

45.
$$n^2 - n = 56$$
$$n^2 - n - 56 = 0$$
$$(n - 8)(n + 7) = 0$$
$$n - 8 = 0 \quad \text{or} \quad n + 7 = 0$$
$$n = 8 \qquad\qquad n = -7$$
$$\text{(reject)}$$
There are 8 teams in the league.

47. $\dfrac{6r^2t - 4rt^3 + 10r^2t^4}{2rt^2} = \dfrac{\cancel{2rt}(3r - 2t^2 + 5rt^3)}{\underset{t}{\cancel{2rt^2}}}$

$= \dfrac{3r - 2t^2 + 5rt^2}{t}$

49.
$$
\begin{array}{r}
y^2 + y + 3 \\
y - 2 \overline{\smash{\big)}\ y^3 - y^2 + y - 1} \\
\underline{-(y^3 - 2y^2)} \\
y^2 + y \\
\underline{-(y^2 - 2y)} \\
3y - 1 \\
\underline{-(3y - 6)} \\
5
\end{array}
$$

51.
$$
\begin{array}{r}
2x^2 - x + 2 \\
3x - 5 \overline{\smash{\big)}\ 6x^3 - 13x^2 + 11x - 10} \\
\underline{-(6x^3 - 10x^2)} \\
-3x^2 + 11x \\
\underline{-(-3x^2 + 5x)} \\
6x - 10 \\
\underline{-(6x - 10)} \\
0
\end{array}
$$

53.
$$
\begin{array}{r}
16x^3 + 32x^2 + 64x + 128 \\
x - 2 \overline{\smash{\big)}\ 16x^4 + 0x^3 + 0x^2 + 0x - 64} \\
\underline{-(16x^4 - 32x^3)} \\
32x^3 + 0x^2 \\
\underline{-(32x^3 - 64x^2)} \\
64x^2 - 0x \\
\underline{-(64x^2 - 128x)} \\
128x - 64 \\
\underline{-(128x - 256)} \\
192
\end{array}
$$

## Chapter 8 Practice Test

1. $6x^3 + 12x^2 - 15x = 3x(2x^2 + 4x - 5)$

3. $x^2 + 9x + 8 = (x + 1)(x + 8)$

5. $4x^2 - 20x = 4x(x - 5)$

7. $6x^2 + 24x + 18 = 6(x^2 + 4x + 3)$
   $= 6(x + 1)(x + 3)$

9. $x^2 + 4x + 4 = (x + 2)(x + 2)$

11. $x^2y^2 - 9 = (xy - 3)(xy + 3)$

13. $\dfrac{12r^3t^2 - 18r^2t^2 + 20r^3t^4}{4r^3t^3} = \dfrac{\overset{3}{\cancel{12r^3t^2}}}{\underset{t}{\cancel{4r^3t^3}}} - \dfrac{\overset{9}{\cancel{18r^2t^2}}}{\underset{2rt}{\cancel{4r^3t^3}}} + \dfrac{\overset{5\ t}{\cancel{20r^3t^4}}}{\cancel{4r^3t^3}}$

$= \dfrac{3}{t} - \dfrac{9}{2rt} + 5t$

or $\dfrac{6r - 9 + 10rt^2}{2rt}$

15.
$$x^2 - 8x = 20$$
$$x^2 - 8x - 20 = 0$$
$$(x - 10)(x + 2) = 0$$
$$x - 10 = 0 \quad \text{or} \quad x + 2 = 0$$
$$x = 10 \qquad x = -2$$

**CHECK** $x = 10$:
$$x^2 - 8x = 20$$
$$(10)^2 - 8(10) \overset{?}{=} 20$$
$$100 - 80 \overset{?}{=} 20$$
$$20 \overset{\checkmark}{=} 20$$

**CHECK** $x = -2$:
$$x^2 - 8x = 20$$
$$(-2)^2 - 8(-2) \overset{?}{=} 20$$
$$4 + 16 \overset{?}{=} 20$$
$$20 \overset{\checkmark}{=} 20$$

17.
$$(x - 8)(x - 6) = 3$$
$$x^2 - 14x + 48 = 3$$
$$x^2 - 14x + 45 = 0$$
$$(x - 5)(x - 9) = 0$$
$$x - 5 = 0 \quad \text{or} \quad x - 9 = 0$$
$$x = 5 \qquad x = 9$$

**CHECK** $x = 5$:
$$(x - 8)(x - 6) = 3$$
$$(5 - 8)(5 - 6) \overset{?}{=} 3$$
$$(-3)(-1) \overset{?}{=} 3$$
$$3 \overset{\checkmark}{=} 3$$

**CHECK** $x = 9$:
$$(x - 8)(x - 6) = 3$$
$$(9 - 8)(9 - 6) \overset{?}{=} 3$$
$$(1)(3) \overset{?}{=} 3$$
$$3 \overset{\checkmark}{=} 3$$

19. Let $x =$ width of the path.
$$(2x + 9)^2 - 81 = 40$$
$$4x^2 + 36x + 81 - 81 = 40$$
$$4x^2 + 36x \qquad = 40$$
$$4x^2 + 36x - 40 = 0$$
$$x^2 + 9x - 10 = 0$$
$$(x - 1)(x + 10) = 0$$
$$x - 1 = 0 \quad \text{or} \quad x + 10 = 0$$
$$x = 1 \qquad x = -10$$
$$\text{(reject)}$$
The path is 1 ft. wide.

# Chapter 9
# More Rational Expressions

Exercises 9.1

1. $\dfrac{\overset{4}{\cancel{8x^3}}\overset{y^5}{\cancel{y^{10}}}}{\underset{5x^3}{\cancel{10x^6}\cancel{y^5}}} = \dfrac{4y^5}{5x^3}$

3. $\dfrac{5x-7x}{x^2-7x^2} = \dfrac{\overset{1}{\cancel{-2x}}}{\underset{3x}{\cancel{-6x^2}}} = \dfrac{1}{3x}$

5. $\dfrac{\overset{2\,(x+4)^4}{\cancel{6x^2(x+4)^5}}}{\underset{3x}{\cancel{9x^3(x+4)}}} = \dfrac{2(x+4)^4}{3x}$

7. $\dfrac{12a^2b+6c^3}{8a^2b+4c^3} = \dfrac{\overset{3}{\cancel{6}(2a^2b+c^3)}}{\underset{2}{\cancel{4}(2a^2b+c^3)}} = \dfrac{3}{2}$

9. $\dfrac{3x-6}{5x-10} = \dfrac{3\cancel{(x-2)}}{5\cancel{(x-2)}} = \dfrac{3}{5}$

11. $\dfrac{3x-6}{6x-12} = \dfrac{\overset{1}{\cancel{3}}\cancel{(x-2)}}{\underset{2}{\cancel{6}}\cancel{(x-2)}} = \dfrac{1}{2}$

13. $\dfrac{3x-6}{6x+12} = \dfrac{\cancel{3}(x-2)}{\underset{2}{\cancel{6}}(x+2)} = \dfrac{x-2}{2(x+2)}$

15. $\dfrac{5y}{10y+20} = \dfrac{\cancel{5}y}{\underset{2}{\cancel{10}}(y+2)} = \dfrac{y}{2(y+2)}$

17. $\dfrac{6x+18}{x^2-9} = \dfrac{6\cancel{(x+3)}}{(x-3)\cancel{(x+3)}}$
$\qquad = \dfrac{6}{x-3}$

19. $\dfrac{6x^2+18}{x^2-9} = \dfrac{6(x^2+3)}{x^2-9}$
(cannot be reduced)

21. $\dfrac{t^2+3t}{t^2+3t-10} = \dfrac{t(t+3)}{(t+5)(t-2)}$
(cannot be reduced)

23. $\dfrac{2x^2+x+x}{4x^3+4x^2} = \dfrac{2x^2+2x}{4x^3+4x^2}$
$\qquad = \dfrac{\overset{1}{\cancel{2x}}\cancel{(x+1)}}{\underset{2x}{\cancel{4x^2}}\cancel{(x+1)}} = \dfrac{1}{2x}$

25. $\dfrac{y^2-5y-6}{y^2-12y+36} = \dfrac{\cancel{(y-6)}(y+1)}{\cancel{(y-6)}(y-6)}$
$\qquad = \dfrac{y+1}{y-6}$

27. $\dfrac{s^2-2s-15}{s^2-6s+5} = \dfrac{\cancel{(s-5)}(s+3)}{\cancel{(s-5)}(s-1)}$
$\qquad = \dfrac{s+3}{s-1}$

29. $\dfrac{x^2(x+3)(x-4)}{x^4-4x^3} = \dfrac{x^2(x+3)\cancel{(x-4)}}{\underset{x}{\cancel{x^3}}\cancel{(x-4)}}$
$\qquad = \dfrac{x+3}{x}$

31. $\dfrac{3a^2+a-2}{a^2-a-2} = \dfrac{(3a-2)\cancel{(a+1)}}{(a-2)\cancel{(a+1)}}$
$\qquad = \dfrac{3a-2}{a-2}$

**33.** $\dfrac{4x^2 + 7x - 2}{x^2 + 4x + 4} = \dfrac{(4x - 1)(x + 2)}{(x + 2)(x + 2)}$

$\qquad\qquad = \dfrac{4x - 1}{x + 2}$

**35.** $\dfrac{x^2 - x + 3x - 8}{x^2 + 4x} = \dfrac{x^2 + 2x - 8}{x^2 + 4x}$

$\qquad\qquad = \dfrac{(x - 2)(x + 4)}{x(x + 4)}$

$\qquad\qquad = \dfrac{x - 2}{x}$

**37.** $\dfrac{6x^2 - 12x - 18}{3x^2 - 9x - 30} = \dfrac{6(x^2 - 2x - 3)}{3(x^2 - 3x - 10)}$

$\qquad\qquad = \dfrac{\overset{2}{6}(x - 3)(x + 1)}{\underset{}{3}(x - 5)(x + 2)}$

$\qquad\qquad = \dfrac{2(x - 3)(x + 1)}{(x - 5)(x + 2)}$

**39.** $\dfrac{6x^2 - 5x^2 - 4}{x^2 - 6x + 8} = \dfrac{x^2 - 4}{x^2 - 6x + 8}$

$\qquad\qquad = \dfrac{(x - 2)(x + 2)}{(x - 2)(x - 4)}$

$\qquad\qquad = \dfrac{x + 2}{x - 4}$

**41.** $\dfrac{x^2 - 7x + 10}{x^2 - 7x + 12} = \dfrac{(x - 2)(x - 5)}{(x - 3)(x - 4)}$

(cannot be reduced)

**43.** $\dfrac{y^3 - y^2 - 2y}{6y^2 - 24} = \dfrac{y(y^2 - y - 2)}{6(y^2 - 4)}$

$\qquad\qquad = \dfrac{y(y - 2)(y + 1)}{6(y - 2)(y + 2)}$

$\qquad\qquad = \dfrac{y(y + 1)}{6(y + 2)}$

**45.** $\dfrac{c^2 - 9c}{c^3 - 9c^2} = \dfrac{\overset{1}{c}(c - 9)}{\underset{c}{c^2}(c - 9)} = \dfrac{1}{c}$

**47.** The student was incorrect in each case, because he or she tried to cancel terms instead of factors. Neither of the fractions can be reduced because neither contains a common factor in its numerator and denominator. (in part (a), we could write

$\dfrac{x^2 + 5}{x} = \dfrac{\overset{x}{x^2}}{\underset{1}{x}} + \dfrac{5}{x} = x + \dfrac{5}{x}$ if we wished.)

**49.** $\dfrac{9x^2}{25y^6} \div \dfrac{xy}{15} = \dfrac{9\overset{x}{x^2}}{25y^6} \cdot \dfrac{\overset{3}{15}}{\underset{5}{xy}}$

$\qquad\qquad = \dfrac{27x}{5y^7}$

**51.** $\dfrac{5}{x} \div 10 = \dfrac{\overset{1}{5}}{x} \cdot \dfrac{1}{\underset{2}{10}} = \dfrac{1}{2x}$

**53.** $\dfrac{2a}{5} = \dfrac{a - 2}{3} \qquad \text{LCD} = 15$

$\overset{3}{15}\left(\dfrac{2a}{5}\right) = \overset{5}{15}\left(\dfrac{a - 2}{3}\right)$

$6a = 5a - 10$

$a = -10$

Exercises 9.2

1. $\dfrac{\overset{4}{\cancel{8}}x^2\overset{y^2}{\cancel{y}}}{\underset{3}{\cancel{9}}\cancel{y}z^{\underset{z^2}{3}}} \cdot \dfrac{\overset{4}{\cancel{12}}\overset{a}{\cancel{a^2}}\cancel{t}}{\underset{1}{\cancel{24}}\cancel{x^2}} = \dfrac{16ay^2}{3z^2}$

3. $\dfrac{3st^2}{5p} \div \dfrac{15t^2}{s^2} = \dfrac{\cancel{3}st^2}{5p} \cdot \dfrac{s^2}{\underset{5}{\cancel{15}}t^2}$

$= \dfrac{s^3}{25p}$

5. $\dfrac{12x^2y^3}{5z} \cdot 30xyz = \dfrac{12x^2y^3}{\underset{1}{\cancel{5}}\cancel{z}} \cdot \dfrac{\overset{6}{\cancel{30}}xy\cancel{z}}{1}$

$= 72x^3y^4$

7. $28a^2b^3z \div \dfrac{4a}{7b} = \dfrac{\overset{7}{\cancel{28}}\overset{a}{\cancel{a^2}}b^3z^4}{1} \cdot \dfrac{7b}{\underset{1}{\cancel{4}\cancel{a}}}$

$= 49ab^4z^4$

9. $\dfrac{x^2 + 4x}{x^2} \cdot \dfrac{x}{x^2 + 6x + 5} = \dfrac{\cancel{x}(x+4)}{\underset{\cancel{x}\,1}{x^2}} \cdot \dfrac{\overset{1}{\cancel{x}}}{(x+1)(x+5)}$

$= \dfrac{x+4}{(x+1)(x+5)}$

11. $\dfrac{x^2 + 4x}{x^2 + 4x + 4} \cdot \dfrac{x^2 - 4}{x^2 - 4x + 4} = \dfrac{x(x+4)}{(x+2)\cancel{(x+2)}} \cdot \dfrac{\cancel{(x-2)}\overset{1}{\cancel{(x+2)}}}{\cancel{(x-2)}(x-2)}$

$= \dfrac{x(x+4)}{(x+2)(x-2)}$

13. $\dfrac{r^2 - 4r - 5}{2r - 10} \div \dfrac{r^2 - 3r + 2}{4r^2} = \dfrac{r^2 - 4r - 5}{2r - 10} \cdot \dfrac{4r^2}{r^2 - 3r + 2}$

$= \dfrac{(r-5)(r+1)}{\underset{1}{2\cancel{(r-5)}}} \cdot \dfrac{\overset{2}{\cancel{4}}r^2}{(r-1)(r-2)}$

$= \dfrac{2r^2(r+1)}{(r-1)(r-2)}$

15. $\dfrac{m^2}{m^2 + 3m} \div \dfrac{3m^2}{m^2 + 6m} = \dfrac{m^2}{m^2 + 3m} \cdot \dfrac{m^2 + 6m}{3m^2}$

$= \dfrac{\overset{1}{\cancel{m^2}}}{\cancel{m}(m+3)} \cdot \dfrac{\cancel{m}(m+6)}{3\cancel{m^2}}$

$= \dfrac{m+6}{3(m+3)}$

17. $\dfrac{x^2 + 3x + 2}{x^2 + 2x} \cdot \dfrac{x}{x^2 + 2} = \dfrac{(x+1)(x+2)}{\cancel{x}(x+2)} \cdot \dfrac{\overset{1}{\cancel{x}}}{x^2 + 2}$

$$= \dfrac{x+1}{x^2+2}$$

19. $\dfrac{y^2 - 3y - 4}{y^2 - 2y - 8} \cdot \dfrac{y^2 + 4y + 4}{y^2 - 8y + 16} = \dfrac{(y-4)(y+1)}{(y-4)(y+2)} \cdot \dfrac{(y+2)(y+2)}{(y-4)(y-4)}$

$$= \dfrac{(y+1)(y+2)}{(y-4)(y-4)} \quad \text{or} \quad \dfrac{(y+1)(y+2)}{(y-4)^2}$$

21. $\dfrac{x^2 + 2x}{x^2 - x - 2} \cdot \dfrac{x-2}{x} = \dfrac{\cancel{x}(x+2)}{(x-2)(x+1)} \cdot \dfrac{\overset{1}{x-2}}{\cancel{x}}$

$$= \dfrac{x+2}{x+1}$$

23. $\dfrac{2x^2 + x - 15}{x^2 - 9} \cdot \dfrac{6x^2 + 7x + 1}{2x^2 - 3x - 5} = \dfrac{\overset{1}{(2x-5)}(x+3)}{(x-3)(x+3)} \cdot \dfrac{(6x+1)(x+1)}{(2x-5)(x+1)}$

$$= \dfrac{6x+1}{x-3}$$

25. $\dfrac{4a}{a+4} \cdot \dfrac{a+5}{5a} \div \dfrac{a^2 + 6a + 5}{a^2 + 5a + 4} = \left( \dfrac{4\cancel{a}}{a+4} \cdot \dfrac{a+5}{5\cancel{a}} \right) \div \dfrac{a^2 + 6a + 5}{a^2 + 5a + 4}$

$$= \dfrac{4(a+5)}{5(a+4)} \cdot \dfrac{a^2 + 5a + 4}{a^2 + 6a + 5}$$

$$= \dfrac{4(a+5)}{5(a+4)} \cdot \dfrac{\overset{1}{(a+1)(a+4)}}{(a+1)(a+5)}$$

$$= \dfrac{4}{5}$$

27. $\dfrac{x}{2} \div \dfrac{2}{x} \cdot \dfrac{x^2 - 16}{x^2 - 4x} = \left( \dfrac{x}{2} \div \dfrac{2}{x} \right) \cdot \dfrac{x^2 - 16}{x^2 - 4x}$

$$= \left( \dfrac{x}{2} \cdot \dfrac{x}{2} \right) \cdot \dfrac{x^2 - 16}{x^2 - 4x}$$

$$= \dfrac{\overset{x}{x^2}}{4} \cdot \dfrac{(x-4)(x+4)}{\cancel{x}(x-4)}$$

$$= \dfrac{x(x+4)}{4}$$

29. $\dfrac{t^2 + 2t}{t^2 + 2t + 1} \div \dfrac{2t^2 + 7t + 6}{t^2 + t} = \dfrac{t^2 + 2t}{t^2 + 2t + 1} \cdot \dfrac{t^2 + t}{2t^2 + 7t + 6}$

$$= \frac{t\,(t+2)}{(t+1)(t+1)} \cdot \frac{t\,(t+1)}{(2t+3)(t+2)}$$

$$= \frac{t^2}{(t+1)(2t+3)}$$

31. $\dfrac{2x^2 + 6x + 4x}{x^2 - 25} \cdot \dfrac{(x+5)^2}{4x - 2x} = \dfrac{2x^2 + 10x}{x^2 - 25} \cdot \dfrac{(x+5)^2}{2x}$

$$= \frac{\overset{1}{2x}(x+5)}{(x-5)(x+5)} \cdot \frac{(x+5)(x+5)}{\underset{1}{2x}}$$

$$= \frac{(x+5)(x+5)}{x - 5} \quad \text{or} \quad \frac{(x+5)^2}{x-5}$$

33. $\dfrac{x^3 y - xy^3}{8x^2 y + 4xy^2} \div \dfrac{(x-y)^2}{2x^2 + 3xy + y^2} = \dfrac{x^3 y - xy^3}{8x^2 y + 4xy^2} \cdot \dfrac{2x^2 + 3xy + y^2}{(x-y)^2}$

$$= \frac{xy(x^2 - y^2)}{4xy(2x+y)} \cdot \frac{(2x+y)(x+y)}{(x-y)^2}$$

$$= \frac{xy(x-y)(x+y)}{4xy(2x+y)} \cdot \frac{(2x+y)(x+y)}{(x-y)(x-y)}$$

$$= \frac{(x+y)(x+y)}{4(x-y)} \quad \text{or} \quad \frac{(x+y)^2}{4(x-y)}$$

35. $\left(\dfrac{x}{3} \cdot \dfrac{x}{4}\right) \cdot \dfrac{12}{x^2 - 3x} = \dfrac{x^2}{12} \cdot \dfrac{12}{x^2 - 3x}$

$$= \frac{\overset{x}{x^2}}{\underset{1}{12}} \cdot \frac{\overset{1}{12}}{x(x-3)}$$

$$= \frac{x}{x-3}$$

37. $\left(\dfrac{c}{2} + \dfrac{c}{5}\right) \div \dfrac{c^2 + 7c}{10} = \left(\dfrac{c(5)}{2(5)} + \dfrac{c(2)}{5(2)}\right) \div \dfrac{c^2 + 7c}{10}$

$$= \left(\frac{5c}{10} + \frac{2c}{10}\right) \div \frac{c^2 + 7c}{10}$$

$$= \frac{7c}{10} \div \frac{c^2 + 7c}{10}$$

$$= \frac{7c}{10} \cdot \frac{10}{c^2 + 7c}$$

$$= \frac{7c}{10} \cdot \frac{\overset{1}{10}}{c(c+7)} = \frac{7}{c+7}$$

39. $\left(\dfrac{w}{4} - \dfrac{9}{w}\right) \cdot \left(\dfrac{w+10}{2w} - \dfrac{5}{w}\right) = \left(\dfrac{w(w)}{4(w)} - \dfrac{9(4)}{w(4)}\right) \cdot \left(\dfrac{w+10}{2w} - \dfrac{5(2)}{w(2)}\right)$

$$= \left(\dfrac{w^2}{4w} - \dfrac{36}{4w}\right) \cdot \left(\dfrac{w+10}{2w} - \dfrac{10}{2w}\right)$$

$$= \left(\dfrac{w^2 - 36}{4w}\right) \cdot \left(\dfrac{w+10-10}{2w}\right)$$

$$= \dfrac{(w-6)(w+6)}{4w} \cdot \dfrac{\overset{1}{\cancel{w}}}{2\cancel{w}}$$

$$= \dfrac{(w-6)(w+6)}{8w}$$

41. Two cars leave from the same point and travel in opposite directions. The slower car, traveling at the rate of 20 kph, leaves two hours later than the faster one, which travels at 30 kph. How long has the slower car been traveling when the two cars are 310 km apart? (Let $t =$ time traveled by the slower car.)

42. Wilma invested a sum of money at an 8% interest rate and a second sum, $500 more than the first, at a 12% interest rate. How much money did she invest at each rate if her total interest on the two investments is $260? (Let $x =$ amount of money Wilma invested at 8%.)

43. $\dfrac{5}{6x} + \dfrac{3}{8x^2}$   LCD $= 24x^2$

$$\dfrac{5}{6x} = \dfrac{5(4x)}{6x(4x)} = \dfrac{20x}{24x^2}$$

$$\dfrac{3}{8x^2} = \dfrac{3(3)}{8x^2(3)} = \dfrac{9}{24x^2}$$

Then $\dfrac{5}{6x} + \dfrac{3}{8x^2} = \dfrac{20x}{24x^2} + \dfrac{9}{24x^2}$

$$= \dfrac{20x + 9}{24x^2}$$

45. $9 - \dfrac{6}{x} = \dfrac{9x}{x} - \dfrac{6}{x} = \dfrac{9x-6}{x}$

47. Let $x =$ cost of an orchestra seat.
Then $x - 4 =$ cost of a balcony seat.
$$250x + 150(x-4) = 3400$$
$$250x + 150x - 600 = 3400$$
$$400x - 600 = 3400$$
$$400x = 4000$$
$$x = 10$$
They should charge $10 for an orchestra seat.

Exercises 9.3

1. $\dfrac{5x-2}{4x+8}+\dfrac{3x+2}{4x+8}=\dfrac{5x-2+3x+2}{4x+8}$

$$=\dfrac{8x}{4x+8}=\dfrac{\overset{2}{\cancel{8}}x}{\cancel{4}(x+2)}$$

$$=\dfrac{2x}{x+2}$$

3. $\dfrac{5x-2}{4x+8}-\dfrac{3x+2}{4x+8}=\dfrac{5x-2-(3x+2)}{4x+8}$

$$=\dfrac{5x-2-3x-2}{4x+8}$$

$$=\dfrac{2x-4}{4x+8}=\dfrac{\cancel{2}(x-2)}{\underset{2}{\cancel{4}}(x+2)}$$

$$=\dfrac{x-2}{2(x+2)}$$

7. $\dfrac{x+7}{x+2}-\dfrac{x+3}{x+2}+\dfrac{x-2}{x+2}=\dfrac{(x+7)-(x+3)+(x-2)}{x+2}$

$$=\dfrac{x+7-x-3+x-2}{x+2}$$

$$=\dfrac{\overset{1}{\cancel{x+2}}}{\underset{1}{\cancel{x+2}}}=\dfrac{1}{1}=1$$

11. $\dfrac{5}{2x}+\dfrac{4}{x+2}=\dfrac{5(x+2)}{2x(x+2)}+\dfrac{4(2x)}{(x+2)(2x)}$

$$=\dfrac{5x+10}{2x(x+2)}+\dfrac{8x}{2x(x+2)}$$

$$=\dfrac{5x+10+8x}{2x(x+2)}=\dfrac{13x+10}{2x(x+2)}$$

13. $\dfrac{5}{\cancel{2}x}\cdot\dfrac{\overset{2}{\cancel{4}}}{x+2}=\dfrac{10}{x(x+2)}$

17. $\dfrac{7}{a+7}-\dfrac{5}{a+5}=\dfrac{7(a+5)}{(a+7)(a+5)}-\dfrac{5(a+7)}{(a+5)(a+7)}$

$$=\dfrac{7a+35}{(a+7)(a+5)}-\dfrac{5a+35}{(a+7)(a+5)}$$

$$=\dfrac{7a+35-(5a+35)}{(a+7)(a+5)}$$

$$=\dfrac{2a}{(a+7)(a+5)}$$

19. $\dfrac{4}{3x^2} - \dfrac{2}{x^2 + 3x} = \dfrac{4}{3x^2} - \dfrac{2}{x(x+3)}$

$$= \dfrac{4(x+3)}{3x^2(x+3)} - \dfrac{2(3x)}{x(x+3)(3x)}$$

$$= \dfrac{4x + 12}{3x^2(x+3)} - \dfrac{6x}{3x^2(x+3)}$$

$$= \dfrac{4x + 12 - 6x}{3x^2(x+3)}$$

$$= \dfrac{-2x + 12}{3x^2(x+3)} = \dfrac{2(-x+6)}{3x^2(x+3)}$$

23. $\dfrac{4}{x^2 + 4x} - \dfrac{2}{x^2 - 4x} = \dfrac{4}{x(x+4)} - \dfrac{2}{x(x-4)}$

$$= \dfrac{4(x-4)}{x(x+4)(x-4)} - \dfrac{2(x+4)}{x(x-4)(x+4)}$$

$$= \dfrac{4x - 16}{x(x+4)(x-4)} - \dfrac{2x + 8}{x(x+4)(x-4)}$$

$$= \dfrac{4x - 16 - (2x + 8)}{x(x+4)(x-4)}$$

$$= \dfrac{4x - 16 - 2x - 8}{x(x+4)(x-4)}$$

$$= \dfrac{2x - 24}{x(x+4)(x-4)} = \dfrac{2(x-12)}{x(x+4)(x-4)}$$

25. $2 - \dfrac{x}{x-1} = \dfrac{2}{1} - \dfrac{x}{x-1}$

$$= \dfrac{2(x-1)}{1(x-1)} - \dfrac{x}{x-1}$$

$$= \dfrac{2x - 2}{x-1} - \dfrac{x}{x-1}$$

$$= \dfrac{2x - 2 - x}{x-1} = \dfrac{x-2}{x-1}$$

**29.** $\dfrac{3}{x^2 - 16} - \dfrac{3}{2x^2 + 8x} = \dfrac{3}{(x-4)(x+4)} - \dfrac{3}{2x(x+4)}$

$$= \dfrac{3(2x)}{(x-4)(x+4)(2x)} - \dfrac{3(x-4)}{2x(x+4)(x-4)}$$

$$= \dfrac{6x}{2x(x+4)(x-4)} - \dfrac{3x-12}{2x(x+4)(x-4)}$$

$$= \dfrac{6x - (3x-12)}{2x(x+4)(x-4)}$$

$$= \dfrac{6x - 3x + 12}{2x(x+4)(x-4)}$$

$$= \dfrac{3x + 12}{2x(x+4)(x-4)}$$

$$= \dfrac{3(x+4)}{2x(x+4)(x-4)} = \dfrac{3}{2x(x-4)}$$

**33.** $\dfrac{x}{x^2 + 6x + 9} + \dfrac{1}{x^2 + 4x + 3} = \dfrac{x}{(x+3)(x+3)} + \dfrac{1}{(x+1)(x+3)}$

$$= \dfrac{x(x+1)}{(x+3)(x+3)(x+1)} + \dfrac{1(x+3)}{(x+1)(x+3)(x+3)}$$

$$= \dfrac{x^2 + x}{(x+1)(x+3)(x+3)} + \dfrac{x+3}{(x+1)(x+3)(x+3)}$$

$$= \dfrac{x^2 + x + x + 3}{(x+1)(x+3)(x+3)}$$

$$= \dfrac{x^2 + 2x + 3}{(x+1)(x+3)(x+3)} \quad \text{or} \quad \dfrac{x^2 + 2x + 3}{(x+1)(x+3)^2}$$

**37.** $5 - \dfrac{1}{3x - 6} + \dfrac{3}{x^2 - 2x} = \dfrac{5}{1} - \dfrac{1}{3(x-2)} + \dfrac{3}{x(x-2)}$

$$= \dfrac{5(3x)(x-2)}{1(3x)(x-2)} - \dfrac{1(x)}{3(x-2)(x)} + \dfrac{3(3)}{x(x-2)(3)}$$

$$= \dfrac{15x^2 - 30x}{3x(x-2)} - \dfrac{x}{3x(x-2)} + \dfrac{9}{3x(x-2)}$$

$$= \dfrac{15x^2 - 30x - x + 9}{3x(x-2)}$$

$$= \dfrac{15x^2 - 31x + 9}{3x(x-2)}$$

**39.**
$$\frac{5}{x^2+x+6} - \frac{3}{x^2+3x} + \frac{2}{x^2-2x} = \frac{5}{(x+3)(x-2)} - \frac{3}{x(x+3)} + \frac{2}{x(x-2)}$$

$$= \frac{5(x)}{(x+3)(x-2)(x)} - \frac{3(x-2)}{x(x+3)(x-2)} + \frac{2(x+3)}{x(x-2)(x+3)}$$

$$= \frac{5x}{x(x+3)(x-2)} - \frac{3x-6}{x(x+3)(x-2)} + \frac{2x+6}{x(x+3)(x-2)}$$

$$= \frac{5x-(3x-6)+2x+6}{x(x+3)(x-2)}$$

$$= \frac{5x-3x+6+2x+6}{x(x+3)(x-2)}$$

$$= \frac{4x+12}{x(x+3)(x-2)} = \frac{4\,\cancel{(x+3)}}{x\,\cancel{(x+3)}(x-2)}$$

$$= \frac{4}{x(x-2)}$$

**45.**
$$\frac{1-\dfrac{1}{a}}{\dfrac{1}{a}-\dfrac{1}{a^2}} = \frac{a^2\left(1-\dfrac{1}{a}\right)}{a^2\left(\dfrac{1}{a}-\dfrac{1}{a^2}\right)}$$

$$= \frac{a^2\cdot 1 - \dfrac{\overset{a}{\cancel{a^2}}}{1}\cdot\dfrac{1}{\cancel{a}}}{\dfrac{\overset{a}{\cancel{a^2}}}{1}\cdot\dfrac{1}{\cancel{a}} - \dfrac{\overset{1}{\cancel{a^2}}}{1}\cdot\dfrac{1}{\cancel{a^2}}}$$

$$= \frac{a^2-a}{a-1} = \frac{a\cancel{(a-1)}}{\cancel{a-1}} = \frac{a}{1} = a$$

**47.**
$$\frac{\dfrac{a}{2}-\dfrac{b}{4}}{\dfrac{4}{b^2}-\dfrac{1}{a^2}} = \frac{4a^2b^2\left(\dfrac{a}{2}-\dfrac{b}{4}\right)}{4a^2b^2\left(\dfrac{4}{b^2}-\dfrac{1}{a^2}\right)}$$

$$= \frac{\dfrac{\overset{2}{\cancel{4}}a^2b^2}{1}\cdot\dfrac{a}{\cancel{2}} - \dfrac{\cancel{4}a^2b^2}{1}\cdot\dfrac{b}{\cancel{4}}}{\dfrac{4a^2\cancel{b^2}}{1}\cdot\dfrac{4}{\cancel{b^2}} - \dfrac{4\cancel{a^2}b^2}{1}\cdot\dfrac{1}{\cancel{a^2}}}$$

$$= \frac{2a^3b^2-a^2b^3}{16a^2-4b^2} = \frac{a^2b^2(2a-b)}{4(4a^2-b^2)}$$

$$= \frac{a^2b^2\cancel{(2a-b)}}{4\cancel{(2a-b)}(2a+b)} = \frac{a^2b^2}{4(2a+b)}$$

49. $$\dfrac{X-a}{\dfrac{s}{n}} = \dfrac{78-70}{\dfrac{6.2}{20}}$$

$$= \dfrac{8}{\dfrac{6.2}{20}} = 8 \cdot \dfrac{20}{6.2}$$

$$= \dfrac{160}{6.2} = 25.8$$

51. In the solution to Example 12, replace the 150-mile distance between cities A and B by $x$. Then the time going from A to B is $\dfrac{x}{60}$ and the time returning from B to A is $\dfrac{x}{40}$. Therefore:

Average rate for entire trip $= \dfrac{\text{Total distance}}{\text{Total time}} = \dfrac{x+x}{\dfrac{x}{60} + \dfrac{x}{40}}$

$$= \dfrac{2x}{\dfrac{2x}{120} + \dfrac{3x}{120}}$$

$$= \dfrac{2x}{\dfrac{5x}{120}}$$

$$= 2\cancel{x} \cdot \dfrac{\overset{24}{\cancel{120}}}{\cancel{5x}}$$

$$= 48$$

Thus, the average rate of speed for the entire trip is 48 mph. independent of the distance between the cities.

52. (a) $$\left(x + \dfrac{2}{x}\right)\left(x - \dfrac{3}{x}\right) = x^2 - \dfrac{\cancel{x}}{1}\left(\dfrac{3}{\cancel{x}}\right) + \dfrac{\cancel{x}}{1}\left(\dfrac{2}{\cancel{x}}\right) - \left(\dfrac{2}{x}\right)\left(\dfrac{3}{x}\right)$$

$$= x^2 - 3 + 2 - \dfrac{6}{x^2}$$

$$= x^2 - 1 + \dfrac{6}{x^2}$$

(b) $$x + \dfrac{2}{x} = \dfrac{x}{1} + \dfrac{2}{x} \qquad\qquad x - \dfrac{3}{x} = \dfrac{x}{1} - \dfrac{3}{x}$$

$$= \dfrac{x(x)}{1(x)} + \dfrac{2}{x} \qquad\qquad = \dfrac{x(x)}{1(x)} - \dfrac{3}{x}$$

$$= \dfrac{x^2}{x} + \dfrac{2}{x} = \dfrac{x^2+2}{x} \qquad\qquad = \dfrac{x^2}{x} - \dfrac{3}{x} = \dfrac{x^2-3}{x}$$

Then $\left(x + \dfrac{2}{x}\right)\left(x - \dfrac{3}{x}\right) = \dfrac{x^2+2}{x} \cdot \dfrac{x^2-3}{x} = \dfrac{(x^2+2)(x^2-3)}{x^2}$

The second method is easier. If the instructions ask for the answer in the form of a single fraction, we get such an answer directly in part (b). In part (a), there would be further work to do.

53. $\dfrac{x-1}{2} = \dfrac{x}{3}$   LCD $= 6$

$$\overset{3}{\cancel{6}}\left(\dfrac{x-1}{\cancel{2}}\right) = \overset{2}{\cancel{6}}\left(\dfrac{x}{\cancel{3}}\right)$$

$$3(x-1) = 2x$$
$$3x - 3 = 2x$$
$$-3 = -x$$
$$3 = x$$

55. $\dfrac{a}{5} - \dfrac{a}{6} = \dfrac{1}{2}$   LCD $= 30$

$$30\left(\dfrac{a}{5} - \dfrac{a}{6}\right) = 30\left(\dfrac{1}{2}\right)$$

$$\overset{6}{\cancel{30}}\left(\dfrac{a}{\cancel{5}}\right) - \overset{5}{\cancel{30}}\left(\dfrac{a}{\cancel{6}}\right) = \overset{15}{\cancel{30}}\left(\dfrac{1}{\cancel{2}}\right)$$

$$6a - 5a = 15$$
$$a = 15$$

## Exercises 9.4

1. $$\dfrac{x}{2} + \dfrac{x}{3} = 10$$

$$6\left(\dfrac{x}{2} + \dfrac{x}{3}\right) = 6 \cdot 10$$

$$\dfrac{\overset{3}{\cancel{6}}}{1} \cdot \dfrac{x}{\underset{1}{\cancel{2}}} + \dfrac{\overset{2}{\cancel{6}}}{1} \cdot \dfrac{x}{\underset{1}{\cancel{3}}} = 60$$

$$3x + 2x = 60$$
$$5x = 60$$
$$x = 12$$

**CHECK** $x = 12$:

$$\dfrac{(12)}{2} + \dfrac{(12)}{3} \overset{?}{=} 10$$

$$6 + 4 \overset{?}{=} 10$$

$$10 \overset{\checkmark}{=} 10$$

5. $$\dfrac{a+1}{3} + \dfrac{a+3}{4} = 4$$

$$12\left(\dfrac{a+1}{3} + \dfrac{a+3}{4}\right) = 12 \cdot 4$$

$$\dfrac{\overset{4}{\cancel{12}}}{1} \cdot \dfrac{a+1}{\underset{1}{\cancel{3}}} + \dfrac{\overset{3}{\cancel{12}}}{1} \cdot \dfrac{a+3}{\underset{1}{\cancel{4}}} = 48$$

$$4(a+1) + 3(a+3) = 48$$
$$4a + 4 + 3a + 9 = 48$$
$$7a + 13 = 48$$
$$7a = 35$$
$$a = 5$$

**CHECK** $a = 5$:

$$\dfrac{(5)+1}{3} + \dfrac{(5)+3}{4} \overset{?}{=} 4$$

$$\dfrac{6}{3} + \dfrac{8}{4} \overset{?}{=} 4$$

$$2 + 2 \overset{?}{=} 4$$

$$4 \overset{\checkmark}{=} 4$$

9. $\dfrac{x}{2} + \dfrac{x}{3} + \dfrac{x}{4}$   LCD $= 12$

$$= \dfrac{x(6)}{2(6)} + \dfrac{x(4)}{3(4)} + \dfrac{x(3)}{4(3)}$$

$$= \dfrac{6x}{12} + \dfrac{4x}{12} + \dfrac{3x}{12}$$

$$= \dfrac{6x + 4x + 3x}{12} = \dfrac{13x}{12}$$

**13.**
$$\frac{x+1}{2} - \frac{3}{x} = \frac{x}{2}$$
$$2x\left(\frac{x+1}{2} - \frac{3}{x}\right) = 2x\left(\frac{x}{2}\right)$$
$$\frac{\cancel{2}x}{1} \cdot \frac{x+1}{\cancel{2}} - \frac{2\cancel{x}}{1} \cdot \frac{3}{\cancel{x}} = \frac{\cancel{2}x}{1} \cdot \frac{x}{\cancel{2}}$$
$$x(x+1) - 6 = x^2$$
$$x^2 + x - 6 = x^2$$
$$x - 6 = 0$$
$$x = 6$$

CHECK $x = 6$:
$$\frac{(6)+1}{2} - \frac{3}{(6)} \overset{?}{=} \frac{(6)}{2}$$
$$\frac{7}{2} - \frac{3}{6} \overset{?}{=} \frac{6}{2}$$
$$\frac{7}{2} - \frac{1}{2} \overset{?}{=} \frac{6}{2}$$
$$\frac{6}{2} \overset{\checkmark}{=} \frac{6}{2}$$

**17.**
$$\frac{x}{5} + \frac{x-1}{4} + \frac{x-3}{2} = 3$$
$$20\left(\frac{x}{5} + \frac{x-1}{4} + \frac{x-3}{2}\right) = 20 \cdot 3$$
$$\frac{\overset{4}{\cancel{20}}}{1} \cdot \frac{x}{\cancel{5}} + \frac{\overset{5}{\cancel{20}}}{1} \cdot \frac{x-1}{\cancel{4}} + \frac{\overset{10}{\cancel{20}}}{1} \cdot \frac{x-3}{\cancel{2}} = 60$$
$$4x + 5(x-1) + 10(x-3) = 60$$
$$4x + 5x - 5 + 10x - 30 = 60$$
$$19x - 35 = 60$$
$$19x = 95$$
$$x = 5$$

CHECK $x = 5$:
$$\frac{(5)}{5} + \frac{(5)-1}{4} + \frac{(5)-3}{2} \overset{?}{=} 3$$
$$\frac{5}{5} + \frac{4}{4} + \frac{2}{2} \overset{?}{=} 3$$
$$1 + 1 + 1 \overset{?}{=} 3$$
$$3 \overset{\checkmark}{=} 3$$

**23.**
$$\frac{2r+1}{3} - \frac{r+1}{5} = \frac{r+8}{6}$$
$$30\left(\frac{2r+1}{3} - \frac{r+1}{5}\right) = 30\left(\frac{r+8}{6}\right)$$
$$\frac{\overset{10}{\cancel{30}}}{1} \cdot \frac{2r+1}{\cancel{3}} - \frac{\overset{6}{\cancel{30}}}{1} \cdot \frac{r+1}{\cancel{5}} = \frac{\overset{5}{\cancel{30}}}{1} \cdot \frac{r+8}{\cancel{6}}$$
$$10(2r+1) - 6(r+1) = 5(r+8)$$
$$20r + 10 - 6r - 6 = 5r + 40$$
$$14r + 4 = 5r + 40$$
$$9r + 4 = 40$$
$$9r = 36$$
$$r = 4$$

CHECK $r = 4$:
$$\frac{2(4)+1}{3} - \frac{(4)+1}{5} \overset{?}{=} \frac{(4)+8}{6}$$
$$\frac{9}{3} - \frac{5}{5} \overset{?}{=} \frac{12}{6}$$
$$3 - 1 \overset{?}{=} 2$$
$$2 \overset{\checkmark}{=} 2$$

25.
$$\frac{3}{x} - \frac{2}{3} = \frac{2}{x}$$

$$3x\left(\frac{3}{x} - \frac{2}{3}\right) = 3x\left(\frac{2}{x}\right)$$

$$\frac{3\cancel{x}}{1} \cdot \frac{3}{\cancel{x}} - \frac{\cancel{3}x}{1} \cdot \frac{2}{\cancel{3}} = \frac{3\cancel{x}}{1} \cdot \frac{2}{\cancel{x}}$$

$$9 - 2x = 6$$

$$-2x = -3$$

$$x = \frac{3}{2}$$

CHECK $x = \dfrac{3}{2}$:

$$\frac{3}{\left(\dfrac{3}{2}\right)} - \frac{2}{3} \overset{?}{=} \frac{2}{\left(\dfrac{3}{2}\right)}$$

$$\frac{3}{1} \cdot \frac{2}{3} - \frac{2}{3} \overset{?}{=} \frac{2}{1} \cdot \frac{2}{3}$$

$$2 - \frac{2}{3} \overset{?}{=} \frac{4}{3}$$

$$\frac{4}{3} \overset{\vee}{=} \frac{4}{3}$$

29. $\dfrac{5}{2x} - \dfrac{8}{x+2}$   LCD $= 2x(x+2)$

$$= \frac{5(x+2)}{2x(x+2)} - \frac{8(2x)}{2x(x+2)}$$

$$= \frac{5x+10}{2x(x+2)} - \frac{16x}{2x(x+2)}$$

$$= \frac{5x+10-16x}{2x(x+2)}$$

$$= \frac{-11x+10}{2x(x+2)}$$

31.
$$\frac{4}{x-1} - \frac{5}{8} = \frac{3}{2x-2}$$

$$\frac{4}{x-1} - \frac{5}{8} = \frac{3}{2(x-1)}$$

$$8(x-1)\left(\frac{4}{x-1} - \frac{5}{8}\right) = 8(x-1)\left(\frac{3}{2(x-1)}\right)$$

$$\frac{8(x-1)}{1} \cdot \frac{4}{x-1} - \frac{8(x-1)}{1} \cdot \frac{5}{8} = \frac{\overset{4}{8}(x-1)}{1} \cdot \frac{3}{2(x-1)}$$

$$32 - 5(x-1) = 12$$

$$32 - 5x + 5 = 12$$

$$-5x + 37 = 12$$

$$-5x = -25$$

$$x = 5$$

CHECK $x = 5$:

$$\frac{4}{(5)-1} - \frac{5}{8} \overset{?}{=} \frac{3}{2(5)-2}$$

$$\frac{4}{4} - \frac{5}{8} \overset{?}{=} \frac{3}{8}$$

$$\frac{8}{8} - \frac{5}{8} \overset{?}{=} \frac{3}{8}$$

$$\frac{3}{8} \overset{\vee}{=} \frac{3}{8}$$

**33.**

$$\frac{8}{x-2} + 3 = \frac{x+6}{x-2}$$

$$(x-2)\left(\frac{8}{x-2} + 3\right) = (x-2)\left(\frac{x+6}{x-2}\right)$$

$$\frac{\overset{1}{\cancel{x-2}}}{1} \cdot \frac{8}{\cancel{x-2}} - \frac{x-2}{1} \cdot \frac{3}{1} = \frac{\overset{1}{\cancel{x-2}}}{1} \cdot \frac{x+6}{\cancel{x-2}}$$

$$8 + 3(x-2) = x+6$$

$$8 + 3x - 6 = x+6$$

$$3x + 2 = x+6$$

$$2x + 2 = 6$$

$$2x = 4$$

$$x = 2$$

**CHECK** $x = 2$:

$$\frac{8}{(2)-2} + 3 \overset{?}{=} \frac{(2)+6}{(2)-2}$$

$$\frac{8}{0} + 3 \overset{?}{=} \frac{8}{0}$$

$\frac{8}{0}$ is not defined, so $x = 2$ is not a solution. Since $x = 2$ does not satisfy the original equation, no solution can be found.

**35.** $\dfrac{4}{x+2} - \dfrac{3}{x+6}$    LCD $= (x+2)(x+6)$

$$= \frac{4(x+6)}{(x+2)(x+6)} - \frac{3(x+2)}{(x+2)(x+6)} = \frac{4x+24}{(x+2)(x+6)} - \frac{3x+6}{(x+2)(x+6)}$$

$$= \frac{4x+24-(3x+6)}{(x+2)(x+6)} = \frac{4x+24-3x-6}{(x+2)(x+6)}$$

$$= \frac{x+18}{(x+2)(x+6)}$$

**39.**

$$\frac{5}{x^2-x} - \frac{1}{2x-2} = \frac{1}{x}$$

$$\frac{5}{x(x-1)} - \frac{1}{2(x-1)} = \frac{1}{x}$$

$$2x(x-1)\left(\frac{5}{x(x-1)} - \frac{1}{2(x-1)}\right) = 2x(x-1)\left(\frac{1}{x}\right)$$

$$\frac{2\cancel{x}(\cancel{x-1})}{1} \cdot \frac{5}{\cancel{x}(\cancel{x-1})} - \frac{\overset{}{\cancel{2}}x(\cancel{x-1})}{1} \cdot \frac{1}{\cancel{2}(\cancel{x-1})} = \frac{2\cancel{x}(x-1)}{1} \cdot \frac{1}{\cancel{x}}$$

$$10 - x = 2(x-1)$$

$$10 - x = 2x - 2$$

$$10 = 3x - 2$$

$$12 = 3x$$

$$4 = x$$

CHECK $x = 4$:

$$\frac{5}{(4)^2 - (4)} - \frac{1}{2(4) - 2} \stackrel{?}{=} \frac{1}{(4)}$$

$$\frac{5}{16 - 4} - \frac{1}{8 - 2} \stackrel{?}{=} \frac{1}{4}$$

$$\frac{5}{12} - \frac{1}{6} \stackrel{?}{=} \frac{1}{4}$$

$$\frac{5}{12} - \frac{2}{12} \stackrel{?}{=} \frac{1}{4}$$

$$\frac{3}{12} \stackrel{?}{=} \frac{1}{4}$$

$$\frac{1}{4} \stackrel{\checkmark}{=} \frac{1}{4}$$

45. $\dfrac{2}{x^2 - x} + \dfrac{8}{x^2 - 1}$    LCD $= x(x - 1)(x + 1)$

$$= \frac{2}{x(x - 1)} + \frac{8}{(x - 1)(x + 1)}$$

$$= \frac{2(x + 1)}{x(x - 1)(x + 1)} + \frac{8(x)}{x(x - 1)(x + 1)} = \frac{2x + 2}{x(x - 1)(x + 1)} + \frac{8x}{x(x - 1)(x + 1)}$$

$$= \frac{2x + 2 + 8x}{x(x - 1)(x + 1)} = \frac{10x + 2}{x(x - 1)(x + 1)}$$

49.

$$\frac{5}{x^2 - 2x - 3} = \frac{4}{x^2 - 3x - 4}$$

$$\frac{5}{(x - 3)(x + 1)} = \frac{4}{(x - 4)(x + 1)}$$

$$\frac{(x - 3)(x - 4)(x + 1)}{1}\left(\frac{5}{(x - 3)(x + 1)}\right) = \frac{(x - 3)(x - 4)(x + 1)}{1}\left(\frac{4}{(x - 4)(x + 1)}\right)$$

$$5(x - 4) = 4(x - 3)$$

$$5x - 20 = 4x - 12$$

$$x - 20 = -12$$

$$x = 8$$

CHECK $x = 8$:

$$\frac{5}{(8)^2 - 2(8) - 3} \stackrel{?}{=} \frac{4}{(8)^2 - 3(8) - 4}$$

$$\frac{5}{64 - 16 - 3} \stackrel{?}{=} \frac{4}{64 - 24 - 4}$$

$$\frac{5}{45} \stackrel{?}{=} \frac{4}{36}$$

$$\frac{1}{9} \stackrel{\checkmark}{=} \frac{1}{9}$$

**51.**

$$\frac{9}{x^2 - 3x + 2} - \frac{2}{x - 1} = \frac{1}{x - 2}$$

$$\frac{9}{(x - 1)(x - 2)} - \frac{2}{x - 1} = \frac{1}{x - 2}$$

$$(x - 1)(x - 2)\left(\frac{9}{(x - 1)(x - 2)} - \frac{2}{x - 1}\right) = (x - 1)(x - 2)\left(\frac{1}{x - 2}\right)$$

$$\frac{\cancel{(x - 1)}^1 \cancel{(x - 2)}}{1} \cdot \frac{9}{\cancel{(x - 1)(x - 2)}_1} - \frac{\cancel{(x - 1)}(x - 2)}{1} \cdot \frac{2}{\cancel{x - 1}_1} = \frac{(x - 1)\cancel{(x - 2)}}{1} \cdot \frac{1}{\cancel{x - 2}_1}$$

$$9 - 2(x - 2) = x - 1$$

$$9 - 2x + 4 = x - 1$$

$$13 - 2x = x - 1$$

$$14 = 3x$$

$$\frac{14}{3} = x$$

**53.** Let $w =$ width of the rectangle.

Then $2w + 3 =$ length of the rectangle.

$$\frac{w}{2w + 3} = \frac{2}{5}$$

$$5(2w + 3)\left(\frac{w}{2w + 3}\right) = \cancel{5}(2w + 3)\left(\frac{2}{\cancel{5}}\right)$$

$$5w = 2(2w + 3)$$

$$5w = 4w + 6$$

$$w = 6$$

Then $2w + 3 = 2(6) + 3 = 15$.

The dimensions of the rectangle are 6 by 15.

**55.** The line passing through the points $(1, c)$ and $(-5, 3)$ has slope $= \dfrac{c - 3}{1 - (-5)} = \dfrac{c - 3}{6}$.

The line passing through the points $(4, 3)$ and $(7, -2)$ has slope $= \dfrac{-2 - 3}{7 - 4} = \dfrac{-5}{3}$.

Since parallel lines have equal slopes, $\dfrac{c - 3}{6} = \dfrac{-5}{3}$

$$\cancel{6}\left(\frac{c - 3}{\cancel{6}}\right) = \overset{2}{\cancel{6}}\left(\frac{-5}{\cancel{3}}\right)$$

$$c - 3 = -10$$

$$c = -7.$$

**57.** 99.996%

59.
$$n = \frac{100x}{0.6x + 5}$$
$$160 = \frac{100x}{0.6x + 5}$$
$$160(0.6x + 5) = \frac{100x}{0.6x + 5}(0.6x + 5)$$
$$160(0.6x) + 160(5) = 100x$$
$$96x + 800 = 100x$$
$$800 = 4x$$
$$200 = x$$

It would take 200 acres to support 160 deer.

61. Using the given formula, we construct the following table:

| P | C |
|-----|-------|
| 97 | 970 |
| 98 | 1470 |
| 99 | 2970 |
| 99.5 | 5970 |
| 99.9 | 29970 |

As the percentage of pollutants removed approaches 100%, the cost of the procedure skyrockets.

62. All of the given values satisfy the equation.
$$\frac{x + 3}{2} - \frac{2x + 5}{4} = \frac{2(x + 3)}{2(2)} - \frac{2x + 5}{4}$$
$$= \frac{2x + 6}{4} - \frac{2x + 5}{4} = \frac{2x + 6 - (2x + 5)}{4}$$
$$= \frac{2x + 6 - 2x - 5}{4} = \frac{1}{4}$$

This calculation shows that the given equation is always true, no matter what value of $x$ is chosen. Such an equation is called an identity.

63. The length of a rectangle is five less than three times its width. Find the dimensions of the rectangle if its perimeter is 62 cm. (Let $x$ = width of the rectangle.)

65. Let $x$ = the number of students who take longer than four years to graduate. Then $2600 + x$ = the number of graduates altogether.
$$\frac{2600}{x} = \frac{13}{3}$$
$$\frac{3x}{1}\left(\frac{2600}{x}\right) = \frac{3x}{1}\left(\frac{13}{3}\right)$$
$$7800 = 13x$$
$$600 = x$$

Then $2600 + x = 2600 + 600 = 3200$. So there were 3200 graduates during the last five years.

## Exercises 9.5

3. $3x + 6y = 18$

$$\begin{array}{ccc} -3x & & -3x \\ \hline & 6y = 18 - 3x \end{array}$$

$$\frac{\cancel{6}y}{\cancel{6}} = \frac{18 - 3x}{6}$$

$$y = \frac{18 - 3x}{6} = \frac{\cancel{3}(6 - x)}{\underset{2}{\cancel{6}}}$$

$$= \frac{6 - x}{2}$$

7. $3(m + 2p) = 4(p - m)$

$$3m + 6p = 4p - 4m$$

$$\begin{array}{ccc} +4m & & +4m \\ \hline 7m + 6p = & 4p \end{array}$$

$$\begin{array}{ccc} & -6p & -6p \\ \hline 7m & = -2p \end{array}$$

$$\frac{\cancel{7}m}{\cancel{7}} = \frac{-2p}{7}$$

$$m = \frac{-2p}{7} = -\frac{2p}{7}$$

11. $2r + 3(r - 5) = r - 5s + 1$

$$2r + 3r - 15 = r - 5s + 1$$

$$5r - 15 = r - 5s + 1$$

$$\begin{array}{ccc} -r & & -r \\ \hline 4r & = & -5s + 1 \end{array}$$

$$\begin{array}{ccc} +15 & & +15 \\ \hline 4r & = & -5s + 16 \end{array}$$

$$\frac{\cancel{4}r}{\cancel{4}} = \frac{-5s + 16}{4}$$

$$r = \frac{-5s + 16}{4}$$

15.

$$\frac{x}{2} + \frac{y}{3} = \frac{x}{3} + \frac{y}{4} - \frac{1}{6}$$

$$12\left(\frac{x}{2} + \frac{y}{3}\right) = 12\left(\frac{x}{3} + \frac{y}{4} - \frac{1}{6}\right)$$

$$\frac{\overset{6}{\cancel{12}}}{1} \cdot \frac{x}{\underset{1}{\cancel{2}}} + \frac{\overset{4}{\cancel{12}}}{1} \cdot \frac{y}{\underset{1}{\cancel{3}}} = \frac{\overset{4}{\cancel{12}}}{1} \cdot \frac{x}{\underset{1}{\cancel{3}}} + \frac{\overset{3}{\cancel{12}}}{1} \cdot \frac{y}{\underset{1}{\cancel{4}}} - \frac{\overset{2}{\cancel{12}}}{1} \cdot \frac{1}{\underset{1}{\cancel{6}}}$$

$$6x + 4y = 4x + 3y - 2$$

$$\begin{array}{ccc} -4x & & -4x \\ \hline 2x + 4y = & 3y - 2 \end{array}$$

$$\begin{array}{ccc} -4y = & -4y \\ \hline 2x & = & -y - 2 \end{array}$$

$$\frac{\cancel{2}x}{\cancel{2}} = \frac{-y - 2}{2}$$

$$x = \frac{-y - 2}{2}$$

21.
$$2pn - \frac{y}{3} = 3n + ay + p$$

$$3\left(2pn - \frac{y}{3}\right) = 3(3n + ay + p)$$

$$6pn - \frac{\overset{1}{\cancel{3}}}{1} \cdot \frac{y}{\underset{1}{\cancel{3}}} = 9n + 3ay + 3p$$

$$6pn - y = 9n + 3ay + 3p$$

$$\underline{+y \qquad\qquad\qquad +y}$$

$$6pn = 9n + 3ay + 3p + y$$

$$\underline{-9n \qquad = -9n}$$

$$6pn - 9n = 3ay + 3p + y$$

$$\underline{-3p \qquad\qquad -3p}$$

$$6pn - 9n - 3p = 3ay + y$$

$$6pn - 9n - 3p = (3a + 1)y$$

$$\frac{6pn - 9n - 3p}{3a + 1} = \frac{(3a + 1)y}{3a + 1}$$

$$\frac{6pn - 9n - 3p}{3a + 1} = y$$

23.
$$y = \frac{u + 1}{u - 1}$$

$$(u - 1)y = \frac{\cancel{(u - 1)}}{1} \frac{(u + 1)}{\cancel{(u - 1)}}$$

$$(u - 1)y = u + 1$$

$$uy - y = u + 1$$

$$\underline{-u \qquad -u}$$

$$uy - y - u = 1$$

$$\underline{+y \qquad\qquad +y}$$

$$uy - u = 1 + y$$

$$u(y - 1) = 1 + y$$

$$\frac{u\cancel{(y - 1)}}{\cancel{y - 1}} = \frac{1 + y}{y - 1}$$

$$u = \frac{1 + y}{y - 1}$$

29.
$$A = \frac{1}{2}h(b_1 + b_2)$$

$$2A = 2\left[\frac{1}{2}h(b_1 + b_2)\right]$$

$$2A = h(b_1 + b_2)$$

$$\frac{2A}{b_1 + b_2} = h$$

33.
$$C = \frac{5}{9}(F - 32)$$

$$9C = 9\left[\frac{5}{9}(F - 32)\right]$$

$$9C = 5(F - 32)$$

$$\frac{9C}{5} = F - 32$$

$$\frac{9C}{5} + 32 = F$$

37. 
$$\frac{x-\mu}{s} < 2$$

$$\cancel{s}\left(\frac{x-\mu}{\cancel{s}}\right) < s(2) \quad \text{(since } s \text{ is positive)}$$

$$x - \mu < 2s$$

$$x < 2s + \mu$$

39.
$$\frac{1}{f} = \frac{1}{f_1} + \frac{1}{f_2}$$

$$ff_1f_2\left(\frac{1}{f}\right) = ff_1f_2\left(\frac{1}{f_1} + \frac{1}{f_2}\right)$$

$$\frac{ff_1f_2}{1} \cdot \frac{1}{\cancel{f}} = \frac{f\cancel{f_1}f_2}{1} \cdot \frac{1}{\cancel{f_1}} + \frac{ff_1\cancel{f_2}}{1} \cdot \frac{1}{\cancel{f_2}}$$

$$f_1f_2 = ff_2 + ff_1$$

$$f_1f_2 = f(f_1 + f_2)$$

$$\frac{f_1f_2}{f_1 + f_2} = f$$

41. The ratio of men to women on the faculty of a local college is 9 to 10. If there are 600 women on the faculty, how many men are on the faculty? (Let $x =$ the number of men.)

43. $x^3x^4(x^3)^4 = x^3x^4x^{12}$

$\qquad = x^{3+4+12}$

$\qquad = x^{19}$

45. $2^{-3} + 3^{-2} = \dfrac{1}{2^3} + \dfrac{1}{3^2} = \dfrac{1}{8} + \dfrac{1}{9}$

$\qquad = \dfrac{9 \cdot 1}{9 \cdot 8} + \dfrac{1 \cdot 8}{9 \cdot 8}$

$\qquad = \dfrac{9}{72} + \dfrac{8}{72} = \dfrac{17}{72}$

## Exercises 9.6

3. Let $x =$ the number of people who preferred Brand X.
Then $200 - x =$ the number of people who did not.

$$\frac{x}{200 - x} = \frac{13}{12}$$

$$\frac{12(200-x)}{1} \cdot \frac{x}{200-x} = \frac{12(200-x)}{1} \cdot \frac{13}{12}$$

$$12x = 13(200 - x)$$

$$12x = 2600 - 13x$$

$$25x = 2600$$

$$x = 104$$

Then $200 - x = 200 - 104 = 96$. Thus, 104 people preferred Brand X.

CHECK: $\dfrac{104}{96} = \dfrac{\cancel{8}(13)}{\cancel{8}(12)} = \dfrac{13}{12}$ and $104 + 96 = 200$.

5. Let $x$ = the number of hits Joe must get in his next 50 at-bats.

$$\frac{120 + x}{400 + 50} = 0.400 = \frac{400}{1000} = \frac{2}{5}$$

$$\frac{120 + x}{450} = \frac{2}{5}$$

$$\frac{450}{1} \cdot \frac{120 + x}{450} = \frac{\overset{90}{\cancel{450}}}{1} \cdot \frac{2}{\cancel{5}}$$

$$120 + x = 180$$

$$x = 60$$

This is impossible, since the largest possible value of $x$ is 50. Therefore, Joe cannot raise his average to .400 in his next 50 at-bats.

9. Let $x$ = the number of hours that the electrician needs to complete the job working alone.
Then $2x$ = the number of hours that the apprentice needs to complete the job working alone.

$$6\left(\frac{1}{x}\right) + 6\left(\frac{1}{2x}\right) = 1$$

$$\frac{6}{1} \cdot \frac{1}{x} + \frac{\overset{3}{\cancel{6}}}{1} \cdot \frac{1}{\cancel{2}x} = 1$$

$$\frac{6}{x} + \frac{3}{x} = 1$$

$$\frac{9}{x} = 1$$

$$\frac{\cancel{x}}{1} \cdot \frac{9}{\cancel{x}} = x \cdot 1$$

$$9 = x$$

Thus, the electrician can complete the job in 9 hours, working alone. The apprentice can complete the job in $2x = 2(9) = 18$ hours, working alone.

CHECK: In one hour, the electrician completes $\frac{1}{9}$ of the job, so in 6 hours, he completes $\frac{6}{9} = \frac{2}{3}$ of the job. In one hour, the apprentice completes $\frac{1}{18}$ of the job, so in 6 hours, he completes $\frac{6}{18} = \frac{1}{3}$ of the job. $\frac{2}{3} + \frac{1}{3} = 1$ (the entire job).

11. Let $r$ = rate of the train (in kph).
Then $r + 100$ = rate of the plane (in kph).

$$\frac{300}{r} = \frac{500}{r + 100}$$

$$\frac{r(r + 100)}{1} \cdot \frac{300}{\cancel{r}} = \frac{r\cancel{(r + 100)}}{1} \cdot \frac{500}{\cancel{r + 100}}$$

$$300(r + 100) = 500r$$

$$300r + 30000 = 500r$$

$$30000 = 200r$$

$$150 = r$$

Thus, the train travels at the rate of 150 kph and the plane travels at the rate of 250 kph.

CHECK: At the rate of 150 kph, the train covers 300 kilometers in $\dfrac{300}{150} = 2$ hours. At the rate of 250 kph, the plane covers 500 kilometers in $\dfrac{500}{250} = 2$ hours, the same amount of time.

15. Let $t$ = number of hours that Ronnie walks.
Then $3 - t$ = number of hours that Ronnie jogs.
$$6t = 14(3 - t)$$
$$6t = 42 - 14t$$
$$20t = 42$$
$$t = \frac{42}{20} = \frac{21}{10} = 2.1$$
Then the distance to Ronnie's friend's house is $6(2.1) = 12.6$ kilometers.
CHECK: Ronnie walks $6(2.1) = 12.6$ kilometers to his friend's house and jogs $14(3 - 2.1) = 14(0.9) = 12.6$ kilometers back, the same distance.

17. Let $x$ = smaller number.
Then $3x$ = larger number.
$$\frac{1}{x} + \frac{1}{3x} = \frac{5}{3}$$
$$3x\left(\frac{1}{x} + \frac{1}{3x}\right) = 3x\left(\frac{5}{3}\right)$$
$$\frac{3\not{x}}{1} \cdot \frac{1}{\not{x}} + \frac{3\not{x}}{1} \cdot \frac{1}{3\not{x}} = \frac{\not{3}x}{1} \cdot \frac{5}{\not{3}}$$
$$3 + 1 = 5x$$
$$4 = 5x$$
$$\frac{4}{5} = x$$
Then $3x = 3\left(\dfrac{4}{5}\right) = \dfrac{12}{5}$. Thus, the numbers are $\dfrac{4}{5}$ and $\dfrac{12}{5}$.

CHECK: $\dfrac{12}{5}$ is three times $\dfrac{4}{5}$, and $\dfrac{1}{\left(\dfrac{12}{5}\right)} + \dfrac{1}{\left(\dfrac{4}{5}\right)} = \dfrac{5}{12} + \dfrac{5}{4}$
$$= \frac{5}{12} + \frac{15}{12} = \frac{20}{12} = \frac{5}{3}.$$

19. Let $x$ = number to be added.
$$\frac{3 + x}{5 + x} = \frac{5}{6}$$
$$\frac{6(5 + x)}{1} \cdot \frac{3 + x}{5 + x} = \frac{\not{6}(5 + x)}{1} \cdot \frac{5}{\not{6}}$$
$$6(3 + x) = 5(5 + x)$$
$$18 + 6x = 25 + 5x$$
$$18 + x = 25$$
$$x = 7$$
Thus, the number to be added to both the numerator and denominator is 7.

CHECK: $\dfrac{3+(7)}{5+(7)} = \dfrac{10}{12} = \dfrac{5}{6}.$

21. Let $x =$ numerator of the fraction.
    Then $x + 2 =$ denominator of the fraction.

$$\frac{x+2}{x} = \frac{x}{x+2}$$

$$\frac{\not x(x+2)}{1} \cdot \frac{x+2}{\not x} = \frac{x(\not{x+2})}{1} \cdot \frac{x}{\not{x+2}}$$

$$(x+2)(x+2) = x^2$$

$$x^2 + 4x + 4 = x^2$$

$$4x + 4 = 0$$

$$4x = -4$$

$$x = -1$$

Then $x + 2 = (-1) + 2 = 1$. Thus, the fraction is $\dfrac{-1}{1}$.

CHECK: 1 is two more than $-1$, and the reciprocal of $\dfrac{-1}{1}$ is $\dfrac{1}{-1}$, which is equal to the original fraction.

25.

$$\frac{1}{R_T} = \frac{1}{R_1} + \frac{1}{R_2}$$

$$\frac{1}{R_T} = \frac{1}{20} + \frac{1}{30}$$

$$60R_T\left(\frac{1}{R_T}\right) = 60R_T\left(\frac{1}{20} + \frac{1}{30}\right)$$

$$\frac{60\not R_T}{1} \cdot \frac{1}{\not R_T} = \frac{\overset{3}{\not{60}}R_T}{1} \cdot \frac{1}{\not{20}} + \frac{\overset{2}{\not{60}}R_T}{1} \cdot \frac{1}{\not{30}}$$

$$60 = 3R_T + 2R_T$$

$$60 = 5R_T$$

$$12 = R_T$$

The total resistance is 12 ohms.

27.

$$\frac{1}{f} = \frac{1}{d_s} + \frac{1}{d_i}$$

$$\frac{1}{f} = \frac{1}{6} + \frac{1}{3}$$

$$6f\left(\frac{1}{f}\right) = 6f\left(\frac{1}{6} + \frac{1}{3}\right)$$

$$\frac{6f}{1} \cdot \frac{1}{f} = \frac{\not{6}f}{1} \cdot \frac{1}{\not{6}} + \frac{\overset{2}{\not{6}}f}{1} \cdot \frac{1}{\not{3}}$$

$$6 = f + 2f$$

$$6 = 3f$$

$$2 = f$$

The focal length of the lens is 2 cm.

29. Let $h$ = number to hours Taisha needs to finish painting the room.

$$2\left(\frac{1}{5}\right) + h\left(\frac{1}{8}\right) = 1$$

$$\frac{2}{5} + \frac{h}{8} = 1$$

$$40\left(\frac{2}{5} + \frac{h}{8}\right) = 40 \cdot 1$$

$$\frac{\overset{8}{\cancel{40}}}{1} \cdot \frac{2}{\cancel{5}} + \frac{\overset{5}{\cancel{40}}}{1} \cdot \frac{h}{\cancel{8}} = 40$$

$$16 + 5h = 40$$

$$5h = 24$$

$$h = \frac{24}{5} \text{ or } 4\frac{4}{5}$$

Thus, it takes $2 + 4\frac{4}{5} = 6\frac{4}{5}$ hours to paint the room.

31. If someone can do a job in $h$ hours, then he or she can complete $\frac{1}{h}$ of the job per hour. This is the rate at which the person works. When this rate is multiplied by the number of hours that the person works, we get the fraction part of the entire job that he or she accomplishes.

## Chapter 9 Review Exercises

1. $\dfrac{x^2}{x^2 + 2}$ cannot be reduced.

3. $\dfrac{x^2 + 3x - 4}{x^2 - 16} = \dfrac{(x + 4)(x - 1)}{(x - 4)(x + 4)}$

$$= \dfrac{x - 1}{x - 4}$$

5. $\dfrac{a^2 + 8a + 16}{a^2 + 6a + 8} = \dfrac{(a + 4)(a + 4)}{(a + 2)(a + 4)}$

$$= \dfrac{a + 4}{a + 2}$$

7. $\dfrac{3z^2 - 12}{3z^2 + 9z + 6} = \dfrac{3(z^2 - 4)}{3(z^2 + 3z + 2)}$

$$= \dfrac{3(z - 2)(z + 2)}{3(z + 1)(z + 2)}$$

$$= \dfrac{z - 2}{z + 1}$$

9. $\dfrac{x}{x + 2} + \dfrac{x}{2} = \dfrac{x(2)}{(x + 2)(2)} + \dfrac{x(x + 2)}{2(x + 2)}$

$$= \dfrac{2x}{2(x + 2)} + \dfrac{x^2 + 2x}{2(x + 2)}$$

$$= \dfrac{2x + x^2 + 2x}{2(x + 2)} = \dfrac{x^2 + 4x}{2(x + 2)}$$

$$= \dfrac{x(x + 4)}{2(x + 2)}$$

11. $\dfrac{3}{2x+4} + \dfrac{6}{x^2+2x} = \dfrac{3}{2(x+2)} + \dfrac{6}{x(x+2)}$

$$= \dfrac{3(x)}{2(x+2)(x)} + \dfrac{6(2)}{x(x+2)(2)}$$

$$= \dfrac{3x}{2x(x+2)} + \dfrac{12}{2x(x+2)}$$

$$= \dfrac{3x+12}{2x(x+2)} = \dfrac{3(x+4)}{2x(x+2)}$$

13. $\dfrac{x^2-5x-6}{2x-12} \div \dfrac{x^2+2x+1}{8x^2} = \dfrac{x^2-5x-6}{2x-12} \cdot \dfrac{8x^2}{x^2+2x+1}$

$$= \dfrac{\cancel{(x-6)}(x+1)}{\cancel{2}(x-6)} \cdot \dfrac{\overset{4}{\cancel{8}}x^2}{(x+1)(x+1)}$$

$$= \dfrac{4x^2}{x+1}$$

15. $\dfrac{5}{z^2+z-6} - \dfrac{3}{z^2+3z} = \dfrac{5}{(z+3)(z-2)} - \dfrac{3}{z(z+3)}$

$$= \dfrac{5(z)}{(z+3)(z-2)(z)} - \dfrac{3(z-2)}{z(z+3)(z-2)}$$

$$= \dfrac{5z}{z(z+3)(z-2)} - \dfrac{3z-6}{z(z+3)(z-2)}$$

$$= \dfrac{5z-(3z-6)}{z(z+3)(z-2)} = \dfrac{5z-3z+6}{z(z+3)(z-2)}$$

$$= \dfrac{2z+6}{z(z+3)(z-2)} = \dfrac{2\cancel{(z+3)}}{z\cancel{(z+3)}(z-2)}$$

$$= \dfrac{2}{z(z-2)}$$

17. $2 + \dfrac{3}{x+2} - \dfrac{1}{x} = \dfrac{2}{1} + \dfrac{3}{x+2} - \dfrac{1}{x}$

$$= \dfrac{2[x(x+2)]}{1[x(x+2)]} + \dfrac{3(x)}{(x+2)(x)} - \dfrac{1(x+2)}{x(x+2)}$$

$$= \dfrac{2x^2+4x}{x(x+2)} + \dfrac{3x}{x(x+2)} - \dfrac{x+2}{x(x+2)}$$

$$= \dfrac{2x^2+4x+3x-(x+2)}{x(x+2)}$$

$$= \dfrac{2x^2+4x+3x-x-2}{x(x+2)}$$

$$= \dfrac{2x^2+6x-2}{x(x+2)} = \dfrac{2(x^2+3x-1)}{x(x+2)}$$

19.
$$\frac{5}{x} - \frac{2}{3x} = \frac{13}{6}$$

$$6x\left(\frac{5}{x} - \frac{2}{3x}\right) = 6x\left(\frac{13}{6}\right)$$

$$\frac{6\cancel{x}}{1} \cdot \frac{5}{\cancel{x}} - \frac{\cancel{6}\overset{2}{\cancel{x}}}{1} \cdot \frac{2}{\cancel{3}\cancel{x}} = \frac{\cancel{6}x}{1} \cdot \frac{13}{\cancel{6}}$$

$$30 - 4 = 13x$$

$$26 = 13x$$

$$2 = x$$

21.
$$\frac{y+2}{y} + \frac{4}{y+2} = \frac{y}{y+2}$$

$$y(y+2)\left(\frac{y+2}{y} + \frac{4}{y+2}\right) = y(y+2)\left(\frac{y}{y+2}\right)$$

$$\frac{\cancel{y}(y+2)}{1} \cdot \frac{y+2}{\cancel{y}} + \frac{y\cancel{(y+2)}}{1} \cdot \frac{4}{\cancel{y+2}} = \frac{y\cancel{(y+2)}}{1} \cdot \frac{y}{\cancel{y+2}}$$

$$(y+2)(y+2) + 4y = y^2$$

$$y^2 + 4y + 4 + 4y = y^2$$

$$y^2 + 8y + 4 = y^2$$

$$8y + 4 = 0$$

$$8y = -4$$

$$y = \frac{-4}{8} = -\frac{1}{2}$$

23.
$$\frac{x+2}{x-3} + \frac{4}{3} = \frac{5}{x-3}$$

$$3(x-3)\left(\frac{x+2}{x-3} + \frac{4}{3}\right) = 3(x-3)\left(\frac{5}{x-3}\right)$$

$$\frac{3\cancel{(x-3)}}{1} \cdot \frac{x+2}{\cancel{x-3}} - \frac{\cancel{3}(x-3)}{1} \cdot \frac{4}{\cancel{3}} = \frac{3\cancel{(x-3)}}{1} \cdot \frac{5}{\cancel{x-3}}$$

$$3(x+2) + 4(x-3) = 15$$

$$3x + 6 + 4x - 12 = 15$$

$$7x - 6 = 15$$

$$7x = 21$$

$$x = 3$$

This value does not check in the original equation, since it produces fractions with denominators of 0. Thus, the equation has no solution.

25.
$$\begin{aligned} 3x - 4y + 7 &= 8x - 7y + 3 \\ -8x &\quad -8x \\ \hline -5x - 4y + 7 &= -7y + 3 \\ +4y &\quad +4y \\ \hline -5x \quad + 7 &= -3y + 3 \\ -7 &\quad -7 \\ \hline -5x &= -3y - 4 \\ \frac{\cancel{-5}x}{\cancel{-5}} &= \frac{-3y - 4}{-5} \\ x &= \frac{-3y - 4}{-5} \text{ or } \frac{3y + 4}{5} \end{aligned}$$

27. Let $x$ = number of black marbles in the bag.
Then $x + 80$ = number of red marbles in the bag.

$$\frac{x + 80}{x} = \frac{7}{5}$$

$$\frac{5\cancel{x}}{1} \cdot \frac{x + 80}{\cancel{x}} = \frac{\cancel{5}x}{1} \cdot \frac{7}{\cancel{5}}$$

$$5(x + 80) = 7x$$

$$5x + 400 = 7x$$

$$400 = 2x$$

$$200 = x$$

Then $x + 80 = 200 + 80 = 280$. Thus, there are 200 black marbles and 280 red marbles in the bag, or 480 marbles altogether.

CHECK: 280 is 80 more than 200, and $\dfrac{280}{200} = \dfrac{7}{5}$.

29. Let $x$ = number of hours John needs to do the job alone.

$$4\left(\frac{1}{x}\right) + 4\left(\frac{1}{6}\right) = 1$$

$$\frac{4}{x} + \frac{4}{6} = 1$$

$$6x\left(\frac{4}{x} + \frac{4}{6}\right) = 6x \cdot 1$$

$$\frac{6\cancel{x}}{1} \cdot \frac{4}{\cancel{x}} + \frac{\cancel{6}x}{1} \cdot \frac{4}{\cancel{6}} = 6x$$

$$24 + 4x = 6x$$

$$24 = 2x$$

$$12 = x$$

Thus, John needs 12 hours to do the job alone.

CHECK: In 4 hours, Susan does $4\left(\dfrac{1}{6}\right) = \dfrac{4}{6} = \dfrac{2}{3}$ of the job; in 4 hours, John does $4\left(\dfrac{1}{12}\right) = \dfrac{4}{12} = \dfrac{1}{3}$ of the job. $\dfrac{2}{3} + \dfrac{1}{3} = 1$ (the entire job).

31. Let $w$ = width of the rectangle.
Then $2w - 5$ = length of the rectangle.

$$\frac{2w - 5}{w} = \frac{5}{3}$$

$$3w\left(\frac{2w - 5}{w}\right) = 3w\left(\frac{5}{3}\right)$$

$$3(2w - 5) = 5w$$

$$6w - 15 = 5w$$

$$-15 = -w$$

$$15 = w$$

Then $2w - 5 = 2(15) - 5 = 30 - 5 = 25$. So the dimensions are 15 by 25.

CHECK: $\dfrac{25}{15} = \dfrac{5}{3}$ and $25 = 2(15) - 5$.

## Chapter 9 Practice Test

1. $\dfrac{4x^2}{x^2 - 4x} = \dfrac{4\overset{x}{\cancel{x}^2}}{\cancel{x}(x - 4)} = \dfrac{4x}{x - 4}$

3. $\dfrac{3}{x + 3} + \dfrac{2}{x + 2} = \dfrac{3(x + 2)}{(x + 3)(x + 2)} + \dfrac{2(x + 3)}{(x + 2)(x + 3)}$

$$= \dfrac{3x + 6}{(x + 3)(x + 2)} + \dfrac{2x + 6}{(x + 3)(x + 2)}$$

$$= \dfrac{3x + 6 + 2x + 6}{(x + 3)(x + 2)} = \dfrac{5x + 12}{(x + 3)(x + 2)}$$

5. $\dfrac{5}{2x} - \dfrac{10}{x^2 + 4x} = \dfrac{5}{2x} - \dfrac{10}{x(x + 4)}$

$$= \dfrac{5(x + 4)}{2x(x + 4)} - \dfrac{10(2)}{x(x + 4)(2)}$$

$$= \dfrac{5x + 20}{2x(x + 4)} - \dfrac{20}{2x(x + 4)}$$

$$= \dfrac{5x + 20 - 20}{2x(x + 4)} = \dfrac{5\cancel{x}}{2\cancel{x}(x + 4)}$$

$$= \dfrac{5}{2(x + 4)}$$

7. $\dfrac{2x - 5}{x^2 - 3x} - \dfrac{3x + 7}{x^2 - 3x} + \dfrac{6x - 3}{x^2 - 3x} = \dfrac{2x - 5 - (3x + 7) + 6x - 3}{x^2 - 3x}$

$$= \dfrac{2x - 5 - 3x - 7 + 6x - 3}{x^2 - 3x}$$

$$= \dfrac{5x - 15}{x^2 - 3x} = \dfrac{5(\cancel{x - 3})}{x(\cancel{x - 3})} = \dfrac{5}{x}$$

9.
$$at + b = \frac{3t}{2} + 7$$

$$2(at + b) = 2\left(\frac{3t}{2} + 7\right)$$

$$2at \qquad + 2b = \quad 3t + 14$$

$$\underline{-3t \qquad\qquad -3t}$$

$$2at - 3t + 2b = \qquad 14$$

$$\underline{\qquad -2b \qquad\qquad -2b}$$

$$2at - 3t \quad = \quad 14 - 2b$$

$$(2a - 3)t \quad = \quad 14 - 2b$$

$$t \quad = \quad \frac{14 - 2b}{2a - 3}$$

11. Let $t$ = number of hours needed to go from A to B.
Then $14 - t$ = number of hours needed to go from B to A.

$$45t = 60(14 - t)$$
$$45t = 840 - 60t$$
$$105t = 840$$
$$t = 8$$

Thus, the distance between the towns is $45(8) = 360$ kilometers.

CHECK: Time going $= \dfrac{360}{45} = 8$. Time returning $= \dfrac{360}{60} = 6$ hours. $8 + 6 \overset{\checkmark}{=} 14$ hours.

13. Let $x$ = number of hours Haleema works.
Then $9 - x$ = number of hours Andrea works.

$$x\left(\frac{1}{8}\right) + (9 - x)\left(\frac{1}{12}\right) = 1$$

$$\frac{x}{8} + \frac{9 - x}{12} = 1$$

$$24\left(\frac{x}{8} + \frac{9 - x}{12}\right) = 24 \cdot 1$$

$$\overset{3}{\frac{24}{1}} \cdot \frac{x}{\overset{}{8}} + \overset{2}{\frac{24}{1}} \cdot \frac{9 - x}{\overset{}{12}} = 24$$

$$3x + 2(9 - x) = 24$$

$$3x + 18 - 2x = 24$$

$$x + 18 = 24$$

$$x = 6$$

Then $9 - x = 9 - (6) = 3$. So Haleema works on the puzzle for 6 hours and Andrea works on the puzzle for 3 hours.

CHECK: In 6 hours, Haleema completes $6\left(\dfrac{1}{8}\right) = \dfrac{6}{8} = \dfrac{3}{4}$ of the puzzle; in 3 hours, Andrea completes $3\left(\dfrac{1}{12}\right) = \dfrac{3}{12} = \dfrac{1}{4}$ of the puzzle. $\dfrac{3}{4} + \dfrac{1}{4} \overset{\checkmark}{=} 1$.

# Chapters 7 - 9
# Cumulative Review

1. $(x+8)(x-5) = x^2 - 5x + 8x - 40$
$$= x^2 + 3x - 40$$

3. $(a+b+c)(a+b-c) = a(a+b-c) + b(a+b-c) + c(a+b-c)$
$$= a^2 + ab - ac + ab + b^2 - bc + ac + bc - c^2$$
$$= a^2 + 2ab + b^2 - c^2$$

5. $(5a - 3c)(4a + 3c) = 20a^2 + 15ac - 12ac - 9c^2$
$$= 20a^2 + 3ac - 9c^2$$

7. $(x+3)(x-12) + (x+6)^2 = (x+3)(x-12) + (x+6)(x+6)$
$$= x^2 - 12x + 3x - 36 + x^2 + 6x + 6x + 36$$
$$= 2x^2 + 3x$$

9. $(a-3)(2a+3)(2a-3) = (a-3)(4a^2 - 6a + 6a - 9)$
$$= (a-3)(4a^2 - 9)$$
$$= 4a^3 - 9a - 12a^2 + 27$$
$$= 4a^3 - 12a^2 - 9a + 27$$

11. $(x^2 - xy + 3y^2) + (5x^2 - 8y^2) + (y^2 - 6x^2) = x^2 + 5x^2 - 6x^2 - xy + 3y^2 - 8y^2 + y^2$
$$= -xy - 4y^2$$

13. (a) degree of $5x^4 - 3x^2 + 6x - 1$ is 4.

    (b) coefficient of second degree term is $-3$.

15. $\dfrac{x^2 + 8x}{2x} = \dfrac{\overset{x}{\cancel{x^2}}}{2\cancel{x}} + \dfrac{\overset{4}{\cancel{8x}}}{2\cancel{x}}$

$$= \frac{x}{2} + 4 \text{ or } \frac{x+8}{2}$$

17.
$$
\begin{array}{r}
y \phantom{^2} - 1 \phantom{)} \\
y-2 \overline{)\ y^2 - 3y + 4} \\
\underline{-(y^2 - 2y)} \phantom{xx} \\
-y + 4 \\
\underline{-(-y + 2)} \\
2
\end{array}
$$

19.

$$
\begin{array}{r}
6x^2 + 8x + 9 \\
3x - 4 \overline{)\ 18x^3 + 0x^2 - 5x - 28} \\
\underline{-(18x^3 - 24x^2)} \\
24x^2 - 5x \\
\underline{-(24x^2 - 32x)} \\
27x - 28 \\
\underline{-(27x - 36)} \\
8
\end{array}
$$

21. $\quad 5^0 + 2^{-3} + 2^{-4} = 1 + \dfrac{1}{2^3} + \dfrac{1}{2^4}$

$\qquad\qquad\qquad\quad = 1 + \dfrac{1}{8} + \dfrac{1}{16} = \dfrac{16}{16} + \dfrac{2}{16} + \dfrac{1}{16}$

$\qquad\qquad\qquad\quad = \dfrac{16 + 2 + 1}{16} = \dfrac{19}{16}$

23. $\quad \dfrac{(x^2)^3}{x^2 x^3} = \dfrac{x^6}{x^5} = x^{6-5} = x$

25. $\quad \dfrac{(2x^3)^4}{4(x^5)^3} = \dfrac{2^4(x^3)^4}{4(x^5)^3} = \dfrac{16x^{12}}{4x^{15}}$

$\qquad\qquad\quad = \dfrac{16}{4} \cdot \dfrac{x^{12}}{x^{15}} = \dfrac{4}{1} \cdot x^{-3}$

$\qquad\qquad\quad = \dfrac{4}{1} \cdot \dfrac{1}{x^3} = \dfrac{4}{x^3}$

27. $\quad \dfrac{(3a^{-3}t^2)^{-3}}{(a^{-1}t^{-2})^2} = \dfrac{3^{-3}(a^{-3})^{-3}(t^2)^{-3}}{(a^{-1})^2(t^{-2})^2}$

$\qquad\qquad\quad = \dfrac{3^{-3}a^9 t^{-6}}{a^{-2}t^{-4}} = \dfrac{1}{3^3} \cdot \dfrac{a^9}{a^{-2}} \cdot \dfrac{t^{-6}}{t^{-4}}$

$\qquad\qquad\quad = \dfrac{1}{27}a^{11}t^{-2} = \dfrac{1}{27} \cdot \dfrac{a^{11}}{1} \cdot \dfrac{1}{t^2}$

$\qquad\qquad\quad = \dfrac{a^{11}}{27t^2}$

29. $\quad 0.000439 = 4.39 \times 10^{-4}$

31. $\quad \dfrac{(4 \times 10^{-3})(5 \times 10^4)}{2 \times 10^{-3}} = \dfrac{(\overset{2}{\cancel{4}})(5)}{\underset{1}{\cancel{2}}} \times \dfrac{10^{-3}10^4}{10^{-3}}$

$\qquad\qquad\qquad\quad = 10 \times \dfrac{10^1}{10^{-3}} = 10 \times 10^4$

$\qquad\qquad\qquad\quad = 1 \times 10^5 \ \text{ or } \ 10^5$

33. $\quad x^2 + 6x + 5 = (x+1)(x+5)$
35. $\quad x^2 - 5x + 6 = (x-2)(x-3)$

37. $\quad 6x^3 y - 12xy^2 - 9x^2 y = 3xy(2x^2 - 4y - 3x)$

39. $\quad u^2 - 49 = (u-7)(u+7)$
41. $\quad 2r^2 + r - 15 = (2r-5)(r+3)$

43. $\quad 5t^2 + 10t + 15 = 5(t^2 + 2t + 3)$
45. $\quad 6x^2 - 17xy + 12y^2 = (3x-4y)(2x-3y)$

47. $\quad x^2 + 16x = x(x+16)$

49. $x^2 + ax + xy + ay = (x^2 + ax) + (xy + ay)$

$\qquad = x(x + a) + y(x + a)$

$\qquad = (x + a)(x + y)$

51. $x^2 - 4x - ax + 4a = (x^2 - 4x) + (-ax + 4a)$

$\qquad = x(x - 4) - a(x - 4)$

$\qquad = (x - 4)(x - a)$

53. $\dfrac{x^2 - 4x}{x^2 - 16} = \dfrac{x(x - 4)}{(x + 4)(x - 4)} = \dfrac{x}{x + 4}$

55. $\dfrac{3}{4x} + \dfrac{5}{x + 4} = \dfrac{3(x + 4)}{4x(x + 4)} + \dfrac{5(4x)}{(x + 4)(4x)}$

$\qquad = \dfrac{3x + 12}{4x(x + 4)} + \dfrac{20x}{4x(x + 4)}$

$\qquad = \dfrac{3x + 12 + 20x}{4x(x + 4)} = \dfrac{23x + 12}{4x(x + 4)}$

57. $\dfrac{x^2 - 5x}{10x} \cdot \dfrac{x^2}{x^2 - 25} = \dfrac{x(x - 5)}{10\cancel{x}} \cdot \dfrac{\overset{x}{\cancel{x^2}}}{(x - 5)(x + 5)} = \dfrac{x^2}{10(x + 5)}$

59. $\dfrac{6}{x^2 + 2x} - \dfrac{4}{x^2 - 2x} = \dfrac{6}{x(x + 2)} - \dfrac{4}{x(x - 2)}$

$\qquad = \dfrac{6(x - 2)}{x(x + 2)(x - 2)} - \dfrac{4(x + 2)}{x(x - 2)(x + 2)}$

$\qquad = \dfrac{6x - 12}{x(x + 2)(x - 2)} - \dfrac{4x + 8}{x(x + 2)(x - 2)}$

$\qquad = \dfrac{6x - 12 - (4x + 8)}{x(x + 2)(x - 2)} = \dfrac{6x - 12 - 4x - 8}{x(x + 2)(x - 2)}$

$\qquad = \dfrac{2x - 20}{x(x + 2)(x - 2)} = \dfrac{2(x - 10)}{x(x + 2)(x - 2)}$

61. $\dfrac{\dfrac{a}{2} - \dfrac{8}{a}}{\dfrac{a^2 - 8a + 16}{4}} = \dfrac{4a\left(\dfrac{a}{2} - \dfrac{8}{a}\right) \cdot}{4a\left(\dfrac{a^2 - 8a + 16}{4}\right)} = \dfrac{\dfrac{\overset{2}{\cancel{4a}}}{1} \cdot \dfrac{a}{\cancel{2}} - \dfrac{4\cancel{a}}{1} \cdot \dfrac{8}{\cancel{a}}}{\dfrac{\cancel{4a}}{1} \cdot \dfrac{a^2 - 8a + 16}{\cancel{4}}}$

$\qquad = \dfrac{2a^2 - 32}{a(a^2 - 8a + 16)} = \dfrac{2(a^2 - 16)}{a(a^2 - 8a + 16)}$

$\qquad = \dfrac{2(a - 4)(a + 4)}{a(a - 4)(a - 4)} = \dfrac{2(a + 4)}{a(a - 4)}$

**63.** $\dfrac{2z+9}{4z+12} - \dfrac{5z+8}{4z+12} + \dfrac{3z+1}{4z+12} = \dfrac{2z+9-(5z+8)+3z+1}{4z+12}$

$$= \dfrac{2z+9-5z-8+3z+1}{4z+12}$$

$$= \dfrac{2}{4z+12} = \dfrac{\overset{1}{\cancel{2}}}{\underset{2}{\cancel{4}}(z+3)} = \dfrac{1}{2(z+3)}$$

**65.**

$$\dfrac{4}{3a+6} - \dfrac{3}{2} = \dfrac{5}{6a+12}$$

$$\dfrac{4}{3(a+2)} - \dfrac{3}{2} = \dfrac{5}{6(a+2)}$$

$$6(a+2)\left(\dfrac{4}{3(a+2)} - \dfrac{3}{2}\right) = 6(a+2)\left(\dfrac{5}{6(a+2)}\right)$$

$$\dfrac{\overset{2}{\cancel{6(a+2)}}}{1} \cdot \dfrac{4}{\cancel{3(a+2)}} - \dfrac{\overset{3}{\cancel{6}}(a+2)}{1} \cdot \dfrac{3}{\cancel{2}} = \dfrac{\cancel{6(a+2)}}{1} \cdot \dfrac{5}{\cancel{6(a+2)}}$$

$$8 - 9(a+2) = 5$$

$$8 - 9a - 18 = 5$$

$$-9a - 10 = 5$$

$$-9a = 15$$

$$a = \dfrac{15}{-9} = -\dfrac{5}{3}$$

**67.**

$$\dfrac{3y}{5} - a = 4y + 3a - 6$$

$$5\left(\dfrac{3y}{5} - a\right) = 5(4y + 3a - 6)$$

$$3y - 5a = 20y + 15a - 30$$

$$\begin{array}{rl} \underline{-20y} & \quad \underline{-20y} \\ -17y - 5a = & 15a - 30 \\ \underline{\phantom{-17y} +5a} & \quad \underline{+5a} \\ -17y \quad = & 20a - 30 \\ y \quad = & \dfrac{20a - 30}{-17} = \dfrac{30 - 20a}{17} \end{array}$$

**69.**

$$\frac{x+8}{8} - \frac{x+6}{6} = \frac{x+3}{3} - \frac{x+4}{4}$$

$$24\left(\frac{x+8}{8} - \frac{x+6}{6}\right) = 24\left(\frac{x+3}{3} - \frac{x+4}{4}\right)$$

$$\frac{\overset{3}{\cancel{24}}}{1} \cdot \frac{x+8}{\cancel{8}} - \frac{\overset{4}{\cancel{24}}}{1} \cdot \frac{x+6}{\cancel{6}} = \frac{\overset{8}{\cancel{24}}}{1} \cdot \frac{x+3}{\cancel{3}} - \frac{\overset{6}{\cancel{24}}}{1} \cdot \frac{x+4}{\cancel{4}}$$

$$3(x+8) - 4(x+6) = 8(x+3) - 6(x+4)$$

$$3x + 24 - 4x - 24 = 8x + 24 - 6x - 24$$

$$-x = 2x$$

$$0 = 3x$$

$$0 = x$$

**71.**

$$\frac{3c+1}{3c-2} = \frac{3c}{3c-1}$$

$$\frac{(3c-2)(3c-1)}{1}\left(\frac{3c+1}{3c-2}\right) = \frac{(3c-2)(3c-1)}{1}\left(\frac{3c}{3c-1}\right)$$

$$(3c-1)(3c+1) = (3c-2)(3c)$$

$$9c^2 + 3c - 3c - 1 = 9c^2 - 6c$$

$$9c^2 - 1 = 9c^2 - 6c$$

$$\underline{-9c^2 \qquad\qquad -9c^2}$$

$$-1 = -6c$$

$$\frac{1}{6} = c$$

**73.**

$$3x^2 - 5x - 7 = x^2 - 6x + 8$$

$$\underline{-x^2 \qquad\qquad -x^2}$$

$$2x^2 - 5x - 7 = -6x + 8$$

$$\underline{+6x \qquad\qquad +6x}$$

$$2x^2 + x - 7 = 8$$

$$\underline{-8 \qquad\qquad -8}$$

$$2x^2 + x - 15 = 0$$

$$(2x - 5)(x + 3) = 0$$

$$2x - 5 = 0 \quad \text{or} \quad x + 3 = 0$$

$$2x = 5 \quad \text{or} \quad x = -3$$

$$x = \frac{5}{2}$$

75.     $(x+4)(x-6) = (x-2)(x+5)$
$$x^2 - 6x + 4x - 24 = x^2 + 5x - 2x - 10$$
$$x^2 - 2x - 24 = \quad x^2 + 3x - 10$$
$$\underline{-x^2 \qquad\qquad -x^2}$$
$$-2x - 24 = \qquad 3x - 10$$
$$\underline{+2x \qquad\qquad +2x}$$
$$-24 = \qquad 5x - 10$$
$$\underline{+10 \qquad\qquad +10}$$
$$-14 = \qquad 5x$$
$$-\frac{14}{5} = \qquad x$$

77.    Let $x =$ number of hours Pat needs working alone.
$$5\left(\frac{1}{8}\right) + 5\left(\frac{1}{x}\right) = 1$$
$$\frac{5}{8} + \frac{5}{x} = 1$$
$$8x\left(\frac{5}{8} + \frac{5}{x}\right) = 8x(1)$$
$$\cancel{8}x\left(\frac{5}{\cancel{8}}\right) + 8\cancel{x}\left(\frac{5}{\cancel{x}}\right) = 8x$$
$$5x + 40 = 8x$$
$$40 = 3x$$
$$\frac{40}{3} = x$$

So Pat needs $\dfrac{40}{3}$ or $13\dfrac{1}{3}$ hours to overhaul the engine working alone.

# Chapters 7 - 9
# Cumulative Practice Test

1.  $(x - 2y)(x^2 - 3xy - y^2) = x(x^2 - 3xy - y^2) - 2y(x^2 - 3xy - y^2)$
$$= x^3 - 3x^2y - xy^2 - 2x^2y + 6xy^2 + 2y^3$$
$$= x^3 - 5x^2y + 5xy^2 + 2y^3$$

3.  $2a(3a - 5) + (a - 6)(a - 4) = 6a^2 - 10a + a^2 - 4a - 6a + 24$
$$= 7a^2 - 20a + 24$$

5.  $\dfrac{(2x^3)^{-4}}{(x^{-3})^3} = \dfrac{2^{-4}(x^3)^{-4}}{(x^{-3})^3} = \dfrac{x^{-12}}{2^4 \cdot x^{-9}}$
$$= \dfrac{x^{-3}}{16} = \dfrac{1}{16x^3}$$

7.

$$\begin{array}{r}
4x^2 + 5x\ \ + 15 \\
x-2{\overline{\smash{\big)}\,4x^3 - 3x^2\ +\ 5x\ - 20}} \\
\underline{-(4x^3 - 8x^2)\phantom{xxxxxxxxxxxx}} \\
5x^2\ +\ 5x\phantom{xxxxx} \\
\underline{-(5x^2\ - 10x)\phantom{xxx}} \\
15x\ - 20 \\
\underline{-(15x\ - 30)} \\
10
\end{array}$$

9.  (a)   $0.000916 = 9.16 \times 10^{-4}$

(b)   $916{,}000 = 9.16 \times 10^5$

11.  $x^3 - x^2 - x + 1 - (x^3 - 3x + x^2 - 5x) = x^3 - x^2 - x + 1 - (x^3 + x^2 - 8x)$

$$= x^3 - x^2 - x + 1 - x^3 - x^2 + 8x$$

$$= -2x^2 + 7x + 1$$

13.  $6a^2b^5 - 3ab^3 = 3ab^3(2ab^2 - 1)$         15.   $6x^2 - 36x + 72 = 6(x^2 - 6x + 12)$

17.  $a(a+5) - 7(a+5) = (a+5)(a-7)$

19.  $2u^4 - 32 = 2(u^4 - 16)$

$$= 2(u^2 + 4)(u^2 - 4)$$

$$= 2(u^2 + 4)(u+2)(u-2)$$

21.  $\dfrac{t^2 - t - 6}{t^2 + t - 6} = \dfrac{(t-3)(t+2)}{(t+3)(t-2)}$

**(cannot be reduced)**

23.  $\dfrac{w^2 - 3w - 10}{4w^2 + 8w} \cdot \dfrac{w^2}{w^2 - 10w + 25} = \dfrac{(w-5)(w+2)}{4w(w+2)} \cdot \dfrac{\overset{w}{w^2}}{(w-5)(w-5)}$

$$= \dfrac{w}{4(w-5)}$$

25.  $\dfrac{u^2 - 9u}{u^2} \div (u^2 - 81) = \dfrac{u^2 - 9u}{u^2} \cdot \dfrac{1}{u^2 - 81}$

$$= \dfrac{\overset{1}{u(u-9)}}{\underset{u}{u^2}} \cdot \dfrac{1}{(u+9)(u-9)}$$

$$= \dfrac{1}{u(u+9)}$$

27.
$$\frac{9}{4t-12} - \frac{2}{3} = \frac{11}{12t-36}$$

$$\frac{9}{4(t-3)} - \frac{2}{3} = \frac{11}{12(t-3)}$$

$$12(t-3)\left(\frac{9}{4(t-3)} - \frac{2}{3}\right) = 12(t-3)\left(\frac{11}{12(t-3)}\right)$$

$$\frac{\overset{3}{\cancel{12}(t-3)}}{1} \cdot \frac{9}{\underset{1}{\cancel{4(t-3)}}} - \frac{\overset{4}{\cancel{12}(t-3)}}{1} \cdot \frac{2}{\underset{1}{\cancel{3}}} = \frac{\overset{1}{\cancel{12}(t-3)}}{1} \cdot \frac{11}{\underset{1}{\cancel{12(t-3)}}}$$

$$27 - 8(t-3) = 11$$
$$27 - 8t + 24 = 11$$
$$-8t + 51 = 11$$
$$-8t = -40$$
$$t = 5$$

29.
$$\frac{10}{x+4} + \frac{3}{5} = \frac{6-x}{x+4}$$

$$5(x+4)\left(\frac{10}{x+4} + \frac{3}{5}\right) = 5(x+4)\left(\frac{6-x}{x+4}\right)$$

$$\frac{5(x+4)}{1} \cdot \frac{10}{\underset{1}{(x+4)}} + \frac{\overset{}{\cancel{5}(x+4)}}{1} \cdot \frac{3}{\underset{1}{\cancel{5}}} = \frac{5(x+4)}{1} \cdot \frac{6-x}{\underset{1}{(x+4)}}$$

$$50 + 3(x+4) = 5(6-x)$$
$$50 + 3x + 12 = 30 - 5x$$
$$3x + 62 = 30 - 5x$$
$$8x + 62 = 30$$
$$8x = -32$$
$$x = -4$$

When this value of $x$ is checked in the original equation, it produces fractions with zero denominators. Therefore, the equation has no solution.

31.   Let $x =$ Roni's speed (in kph).

Then $x + 5 =$ Lamar's speed (in kph).
$$\frac{140}{x} = \frac{160}{x+5}$$

$$\frac{\cancel{x}(x+5)}{1} \cdot \frac{140}{\cancel{x}} = \frac{x(x+5)}{1} \cdot \frac{160}{x+5}$$

$$140(x+5) = 160x$$
$$140x + 700 = 160x$$
$$700 = 20x$$
$$35 = x$$

Since Roni drives 140 km at a speed of 35 kph, her driving time is 4 hours.

# Chapter 10
# Radical Expressions

Exercises 10.1

1. $\sqrt{4} = 2$ because $2 \cdot 2 = 4$

3. $-\sqrt{4} = -2$

5. $\sqrt{-4}$ is undefined

11. $\sqrt{64} = 8$ because $8 \cdot 8 = 64$

15. $\sqrt{169} = 13$ because $13 \cdot 13 = 169$

19. $\sqrt{289} = 17$ because $17 \cdot 17 = 289$

23. $\sqrt{3}\sqrt{3} = 3$

27. $\left(\sqrt{11}\right)^2 = 11$

31. $\sqrt{x}\sqrt{x} = x$

35. $\left(\sqrt{7}\right)^5 = \sqrt{7} \cdot \sqrt{7} \cdot \sqrt{7} \cdot \sqrt{7} \cdot \sqrt{7}$

$$= \quad 7 \quad \cdot \quad 7 \quad \cdot \sqrt{7}$$
$$= 49\sqrt{7}$$

37. $\sqrt{25 - 9} = \sqrt{16} = 4$

39. $\sqrt{25} - \sqrt{9} = 5 - 3 = 2$

41. $\left(\sqrt{25} - \sqrt{9}\right)^2 = (5 - 3)^2 = 2^2 = 4$

43. $\left(\sqrt{25 - 9}\right)^2 = \left(\sqrt{16}\right)^2 = 16$

45. $20.62$

47. $25.24$

49. $8.58$

51. $1.45$

53. $0.19$

55. $\sqrt{17} = 4.12$     correct to 1 place
$\phantom{\sqrt{17}} = 4.123$     correct to 2 places
$\phantom{\sqrt{17}} = 4.1231$     correct to 3 places

57. $\sqrt{110} = 10.49$     correct to 1 place
$\phantom{\sqrt{110}} = 10.488$     correct to 2 places
$\phantom{\sqrt{110}} = 10.4881$     correct to 3 places

59. Since $64 < 73 < 81$, $\sqrt{64} < \sqrt{73} < \sqrt{81}$. That is, $8 < \sqrt{73} < 9$.
$\sqrt{73} = 8.54$, rounded to two decimal places.

61. Since $196 < 217 < 225$, $\sqrt{196} < \sqrt{217} < \sqrt{225}$. That is, $14 < \sqrt{217} < 15$.
$\sqrt{217} = 14.73$, rounded to two decimal places.

63. Since $20^2 = 400$, $30^2 = 900$, and 648 is between 400 and 900, $\sqrt{648}$ must be between 20 and 30.
Consider the following table:

$$0^2 = \underline{0} \quad 1^2 = \underline{1} \quad 2^2 = \underline{4} \quad 3^2 = \underline{9} \quad 4^2 = 1\underline{6}$$
$$5^2 = 2\underline{5} \quad 6^2 = 3\underline{6} \quad 7^2 = 4\underline{9} \quad 8^2 = 6\underline{4} \quad 9^2 = 8\underline{1}$$

This implies that any perfect square must end in one of the digits underlined: 0, 1, 4, 9, 6, or 5. Therefore, any number that ends in either 2, 3, 7, or 8 cannot be a perfect square. So 648 cannot be a perfect square.

64. Using the same argument as in (63), $\sqrt{841}$ must be between 20 and 30. Since 841 ends in the digit 1, there are only two possibilities, if 841 is a perfect square: $(21)^2 = 841$ or $(29)^2 = 841$. Since 841 is closer to 900 than it is to 400, we try $(29)^2$ first, since 29 is closer to 30 than it is to 20. Since $29 \cdot 29 = 841$, we conclude that 841 *is* a perfect square, and that $\sqrt{841} = 29$.

65. It is incorrect to claim that $\sqrt{1+1} = \sqrt{1} + \sqrt{1}$. In fact, if $a$ and $b$ are positive, it is <u>never</u> true that $\sqrt{a+b} = \sqrt{a} + \sqrt{b}$. That is, the square root of the sum of two positive numbers is never equal to the sum of the square roots of those numbers.

66. For any non-negative number $a$, $\sqrt{a}\sqrt{a} = a$ tells us that $\sqrt{a}$ is the non-negative quantity whose square is equal to $a$.

67.
$$\frac{x^2 - 6x}{x^2} \cdot \frac{2x}{x^2 - 12x + 36} = \frac{\cancel{x}(\cancel{x-6})}{\overset{\cancel{x}}{\underset{1}{x^2}}} \cdot \frac{2\cancel{x}}{(\cancel{x-6})(x-6)}$$

$$= \frac{2}{x-6}$$

69.
$$\frac{3}{2x} + \frac{5}{x+2} = \frac{3(x+2)}{2x(x+2)} + \frac{5(2x)}{(x+2)(2x)}$$

$$= \frac{3(x+2) + 5(2x)}{2x(x+2)} = \frac{3x + 6 + 10x}{2x(x+2)}$$

$$= \frac{13x + 6}{2x(x+2)}$$

71. Let $x =$ number of regular bulbs.

Then $40 - x =$ number of long-life bulbs.

$$0.75x + 0.89(40 - x) = 31.68$$
$$100[0.75x + 0.89(40 - x)] = 100(31.68)$$
$$75x + 89(40 - x) = 3168$$
$$75x + 3560 - 89x = 3168$$
$$3560 - 14x = 3168$$
$$-14x = -392$$
$$x = 28$$

Then $40 - x = 40 - 28 = 12$, so 28 regular bulbs and 12 long-life bulbs are sold.

CHECK: $28(0.75) + 12(0.89) = 21.00 + 10.68 = 31.68$ and $28 + 12 \overset{\checkmark}{=} 40$.

Exercises 10.2

3. $\sqrt{18} = \sqrt{9 \cdot 2} = \sqrt{9}\sqrt{2} = 3\sqrt{2}$

7. $\sqrt{50} = \sqrt{25 \cdot 2} = \sqrt{25}\sqrt{2} = 5\sqrt{2}$

11. $\sqrt{x^6} = x^3$, because $x^3 x^3 = x^6$

13. $\sqrt{x^7} = \sqrt{x^6 \cdot x} = \sqrt{x^6}\sqrt{x} = x^3\sqrt{x}$

15. $\sqrt{16x^{16}} = \sqrt{16}\sqrt{x^{16}} = 4x^8$

19. $\sqrt{40x^8} = \sqrt{4x^8}\sqrt{10} = \sqrt{4}\sqrt{x^8}\sqrt{10}$
$= 2x^4\sqrt{10}$

23. $\sqrt{12x^5} = \sqrt{4x^4}\sqrt{3x} = \sqrt{4}\sqrt{x^4}\sqrt{3x}$
$= 2x^2\sqrt{3x}$

27. $\sqrt{x^6 + y^8}$ cannot be simplified

31. $\sqrt{28x^9y^6} = \sqrt{4x^8y^6}\sqrt{7x} = \sqrt{4}\sqrt{x^8}\sqrt{y^6}\sqrt{7x}$
$= 2x^4y^3\sqrt{7x}$

37. $\sqrt{48x^6y^8z^9} = \sqrt{16x^6y^8z^8}\sqrt{3z}$
$= \sqrt{16}\sqrt{x^6}\sqrt{y^8}\sqrt{z^8}\sqrt{3z}$
$= 4x^3y^4z^4\sqrt{3z}$

39. $\sqrt{\dfrac{4}{9}} = \dfrac{\sqrt{4}}{\sqrt{9}} = \dfrac{2}{3}$

41. $\sqrt{\dfrac{7}{25}} = \dfrac{\sqrt{7}}{\sqrt{25}} = \dfrac{\sqrt{7}}{5}$

43. $\sqrt{\dfrac{5}{6}} = \dfrac{\sqrt{5}}{\sqrt{6}} = \dfrac{\sqrt{5}\cdot\sqrt{6}}{\sqrt{6}\cdot\sqrt{6}} = \dfrac{\sqrt{30}}{6}$

47. $\dfrac{1}{\sqrt{2}} = \dfrac{1\cdot\sqrt{2}}{\sqrt{2}\cdot\sqrt{2}} = \dfrac{\sqrt{2}}{2}$

49. $\dfrac{18}{\sqrt{10}} = \dfrac{18\cdot\sqrt{10}}{\sqrt{10}\cdot\sqrt{10}} = \dfrac{\overset{9}{\cancel{18}}\sqrt{10}}{\underset{5}{\cancel{10}}} = \dfrac{9\sqrt{10}}{5}$

53. $\dfrac{15}{2\sqrt{7}} = \dfrac{15\cdot\sqrt{7}}{2\sqrt{7}\cdot\sqrt{7}} = \dfrac{15\sqrt{7}}{2\cdot7} = \dfrac{15\sqrt{7}}{14}$

57. $\dfrac{8x}{\sqrt{2x}} = \dfrac{8x\sqrt{2x}}{\sqrt{2x}\sqrt{2x}} = \dfrac{\overset{4}{\cancel{8x}}\sqrt{2x}}{\cancel{2x}} = 4\sqrt{2x}$

61. $\dfrac{x^2}{\sqrt{xy}} = \dfrac{x^2\sqrt{xy}}{\sqrt{xy}\sqrt{xy}} = \dfrac{\overset{x}{\cancel{x^2}}\sqrt{xy}}{\cancel{xy}} = \dfrac{x\sqrt{xy}}{y}$

63. $\dfrac{\sqrt{8}}{\sqrt{6}} = \dfrac{\sqrt{8}\cdot\sqrt{6}}{\sqrt{6}\cdot\sqrt{6}} = \dfrac{\sqrt{48}}{6} = \dfrac{\sqrt{16}\sqrt{3}}{6}$
$= \dfrac{\overset{2}{\cancel{4}}\sqrt{3}}{\underset{3}{\cancel{6}}} = \dfrac{2\sqrt{3}}{3}$

65. $r = \sqrt{\dfrac{A}{P}} - 1$; $A = 10,000$ and $P = 8,000$
$r = \sqrt{\dfrac{10,000}{8,000}} - 1 = \sqrt{\dfrac{5}{4}} - 1$
$= 1.1180 - 1 = 0.1180$ or $11.80\%$

67. $s = 8.3\sqrt{L}$; $L = 50$
$s = 8.3\sqrt{50} = 8.3(7.1)$
$= 59$ mph (rounded to the nearest mph)

**69.** $t = \sqrt{\dfrac{h}{16}}; \ h = 1,250$

$t = \sqrt{\dfrac{1250}{16}}$

$= 8.8$ seconds (rounded to the nearest tenth)

**71.** $d = \sqrt{8,000m}; \ m = 6$

$d = \sqrt{8,000 \cdot 6} = \sqrt{48,000}$

$= 219$ miles (rounded to the nearest mile)

**73.** $I = \sqrt{\dfrac{W}{R}}; \ W = 200, R = 70$

$I = \sqrt{\dfrac{200}{70}} = \sqrt{\dfrac{20}{7}}$

$= 1.7$ amps (rounded to the nearest tenth)

**75.**

| $x$ | $y$ | $\sqrt{xy}$ | $\sqrt{x}\sqrt{y}$ |
|---|---|---|---|
| 0 | 1 | $\sqrt{0(1)} = \sqrt{0} = 0$ | $\sqrt{0}\sqrt{1} = 0(1) = 0$ |
| 4 | 9 | $\sqrt{4(9)} = \sqrt{36} = 6$ | $\sqrt{4}\sqrt{9} = 2(3) = 6$ |
| 9 | 16 | $\sqrt{9(16)} = \sqrt{144} = 12$ | $\sqrt{9}\sqrt{16} = 3(4) = 12$ |
| 4 | 5 | $\sqrt{4(5)} = \sqrt{4}\sqrt{5} = 2\sqrt{5}$ | $\sqrt{4}\sqrt{5} = 2\sqrt{5}$ |
| 9 | 7 | $\sqrt{9(7)} = \sqrt{9}\sqrt{7} = 3\sqrt{7}$ | $\sqrt{9}\sqrt{7} = 3\sqrt{7}$ |

It appears from the table that in general $\sqrt{xy} = \sqrt{x}\sqrt{y}$.

| $x$ | $y$ | $\sqrt{x+y}$ | $\sqrt{x} + \sqrt{y}$ |
|---|---|---|---|
| 0 | 1 | $\sqrt{0+1} = \sqrt{1} = 1$ | $\sqrt{0} + \sqrt{1} = 0 + 1 = 1$ |
| 4 | 9 | $\sqrt{4+9} = \sqrt{13}$ | $\sqrt{4} + \sqrt{9} = 2 + 3 = 5$ |
| 9 | 16 | $\sqrt{9+16} = \sqrt{25} = 5$ | $\sqrt{9} + \sqrt{16} = 3 + 4 = 7$ |
| 4 | 5 | $\sqrt{4+5} = \sqrt{9} = 3$ | $\sqrt{4} + \sqrt{5} = 2 + \sqrt{5}$ |
| 9 | 7 | $\sqrt{9+7} = \sqrt{16} = 4$ | $\sqrt{9} + \sqrt{7} = 3 + \sqrt{7}$ |

However, it is **not** true that in general $\sqrt{x+y} = \sqrt{x} + \sqrt{y}$.

**76.** When we square a real number other than 0 or 1, we get an answer that is different from the original number. So it is incorrect to say that $\dfrac{2}{\sqrt{5}} = \dfrac{2^2}{(\sqrt{5})^2}$. When we rationalize the denominator properly, we multiply $\dfrac{2}{\sqrt{5}}$ by $\dfrac{\sqrt{5}}{\sqrt{5}}$. This means that we multiply by 1, which does **not** change the value of the original number.

**77.** $\dfrac{1}{\sqrt{5}} = \dfrac{1 \cdot \sqrt{5}}{\sqrt{5} \cdot \sqrt{5}} = \dfrac{\sqrt{5}}{5}$

$\dfrac{1}{\sqrt{5}} = \dfrac{1}{2.2360} = 0.4472,$ correct to 3 decimal places

$\dfrac{\sqrt{5}}{5} = \dfrac{2.2360}{5} = 0.4472,$ correct to 3 decimal places

It is easier to compute $\dfrac{\sqrt{5}}{5}$ than to compute $\dfrac{1}{\sqrt{5}}$, since $\dfrac{1}{\sqrt{5}}$ involves division by a decimal quantity, whereas $\dfrac{\sqrt{5}}{5}$ does not.

78. $\dfrac{\sqrt{3}}{\sqrt{7}} = \dfrac{1.7321}{2.6458} = 0.6547$, correct to 3 decimal places

$\dfrac{\sqrt{3}}{\sqrt{7}} = \dfrac{\sqrt{3} \cdot \sqrt{7}}{\sqrt{7} \cdot \sqrt{7}} = \dfrac{\sqrt{21}}{7} = \dfrac{4.5828}{7} = 0.6547$, correct to 3 decimal places

79. $4x + 3y = 8$

$$
\begin{array}{rl}
\dfrac{-3y \qquad -3y}{4x \quad = 8 - 3y} & \\
\dfrac{\cancel{4}x}{\cancel{4}} & = \dfrac{8 - 3y}{4} \\
x & = \dfrac{8 - 3y}{4}
\end{array}
$$

81. $4(a + 3b) - 2(a - 5b) = 6 - a + \quad b$

$4a + 12b - 2a + 10b = 6 - a + \quad b$

$\qquad\qquad 2a + 22b = 6 - a + \quad b$

$$
\dfrac{+a \qquad\qquad\qquad +a}{3a + 22b = 6 \qquad + \quad b}
$$

$$
\dfrac{-22b \qquad\qquad -22b}{3a \qquad = 6 \qquad -21b}
$$

$\dfrac{\cancel{3}a}{\cancel{3}} = \dfrac{6 - 21b}{3}$

$a = \dfrac{6 - 21b}{3}$

$a = \dfrac{\cancel{3}(2 - 7b)}{\cancel{3}}$

$a = 2 - 7b$

83. Let $t =$ number of hours the slower train travels until the trains meet.

Then $t - 2 =$ number of hours the faster train travels until the trains meet.

$110t + 140(t - 2) = 595$

$110t + 140t - 280 = 595$

$\qquad 250t - 280 = 595$

$\qquad\quad 250t \qquad = 875$

$\qquad\qquad t \qquad = \dfrac{875}{250} = \dfrac{7}{2}$ or $3\dfrac{1}{2}$

Since the slower train left at 11:00 A.M., the trains will meet $3\dfrac{1}{2}$ hours later, or at 2:30 P.M.

CHECK: $110\left(\dfrac{7}{2}\right) = 385$ and $140\left(\dfrac{3}{2}\right) = 210$. $385 + 210 \overset{\checkmark}{=} 595$.

## Exercises 10.3

3. $\sqrt{5} + 2\sqrt{5} + 3\sqrt{5} = (1 + 2 + 3)\sqrt{5}$
$$= 6\sqrt{5}$$

7. $4\sqrt{6} - \sqrt{6} = (4 - 1)\sqrt{6} = 3\sqrt{6}$

11. $3\sqrt{5} + 5\sqrt{3}$ cannot be simplified.

15. $\sqrt{5}\sqrt{5} = 5$

19. $\sqrt{5} + \sqrt{5} = 2\sqrt{5}$

21. $\sqrt{5} + 3\sqrt{7} - 4\sqrt{5} - 5\sqrt{7} = \sqrt{5} - 4\sqrt{5} + 3\sqrt{7} - 5\sqrt{7}$
$$= -3\sqrt{5} - 2\sqrt{7}$$

27. $2(\sqrt{5} - \sqrt{3}) + 3(\sqrt{3} - \sqrt{5}) = 2\sqrt{5} - 2\sqrt{3} + 3\sqrt{3} - 3\sqrt{5}$
$$= 2\sqrt{5} - 3\sqrt{5} - 2\sqrt{3} + 3\sqrt{3}$$
$$= -\sqrt{5} + \sqrt{3}$$

31. $6(\sqrt{m} - \sqrt{n}) - (3\sqrt{m} + 6\sqrt{n}) = 6\sqrt{m} - 6\sqrt{n} - 3\sqrt{m} - 6\sqrt{n}$
$$= 6\sqrt{m} - 3\sqrt{m} - 6\sqrt{n} - 6\sqrt{n}$$
$$= 3\sqrt{m} - 12\sqrt{n}$$

33. $\sqrt{8} + \sqrt{18} = \sqrt{4}\sqrt{2} + \sqrt{9}\sqrt{2}$
$$= 2\sqrt{2} + 3\sqrt{2} = 5\sqrt{2}$$

35. $\sqrt{25} + \sqrt{24} = 5 + \sqrt{4}\sqrt{6} = 5 + 2\sqrt{6}$

39. $4\sqrt{12} - \sqrt{75} = 4\sqrt{4}\sqrt{3} - \sqrt{25}\sqrt{3}$
$$= 4(2\sqrt{3}) - 5\sqrt{3}$$
$$= 8\sqrt{3} - 5\sqrt{3} = 3\sqrt{3}$$

43. $3\sqrt{72} - 5\sqrt{32} = 3(\sqrt{36}\sqrt{2}) - 5(\sqrt{16}\sqrt{2})$
$$= 3(6\sqrt{2}) - 5(4\sqrt{2})$$
$$= 18\sqrt{2} - 20\sqrt{2} = -2\sqrt{2}$$

45. $5\sqrt{36} + 4\sqrt{30} = 5 \cdot 6 + 4\sqrt{30}$
$$= 30 + 4\sqrt{30}$$

49. $\sqrt{12w} + \sqrt{27w} = \sqrt{4}\sqrt{3w} + \sqrt{9}\sqrt{3w}$
$$= 2\sqrt{3w} + 3\sqrt{3w}$$
$$= 5\sqrt{3w}$$

53. $\sqrt{20y^3} - \sqrt{45y^3} = \sqrt{4y^2}\sqrt{5y} - \sqrt{9y^2}\sqrt{5y}$
$$= 2y\sqrt{5y} - 3y\sqrt{5y} = -y\sqrt{5y}$$

55. $x\sqrt{28xy^3} + y\sqrt{63x^3y} = x\sqrt{4y^2}\sqrt{7xy} + y\sqrt{9x^2}\sqrt{7xy}$
$$= x(2y)\sqrt{7xy} + y(3x)\sqrt{7xy}$$
$$= 2xy\sqrt{7xy} + 3xy\sqrt{7xy} = 5xy\sqrt{7xy}$$

57. $\dfrac{\sqrt{32x^3y^2}}{2xy} - \sqrt{8x} = \dfrac{\sqrt{16x^2y^2}\sqrt{2x}}{2xy} - \sqrt{4}\sqrt{2x}$
$$= \dfrac{\overset{2}{\cancel{4xy}}\sqrt{2x}}{\underset{1}{\cancel{2xy}}} - 2\sqrt{2x}$$
$$= 2\sqrt{2x} - 2\sqrt{2x} = 0$$

61. $\sqrt{27} + \dfrac{4}{\sqrt{3}} = \sqrt{9}\sqrt{3} + \dfrac{4 \cdot \sqrt{3}}{\sqrt{3} \cdot \sqrt{3}}$
$$= 3\sqrt{3} + \dfrac{4}{3}\sqrt{3}$$
$$= \left(3 + \dfrac{4}{3}\right)\sqrt{3} = \dfrac{13}{3}\sqrt{3}$$

65. $\sqrt{\dfrac{2}{7}} + \sqrt{\dfrac{7}{2}} = \dfrac{\sqrt{2}}{\sqrt{7}} + \dfrac{\sqrt{7}}{\sqrt{2}}$
$$= \dfrac{\sqrt{2} \cdot \sqrt{7}}{\sqrt{7} \cdot \sqrt{7}} + \dfrac{\sqrt{7} \cdot \sqrt{2}}{\sqrt{2} \cdot \sqrt{2}}$$
$$= \dfrac{\sqrt{14}}{7} + \dfrac{\sqrt{14}}{2}$$
$$= \dfrac{1}{7}\sqrt{14} + \dfrac{1}{2}\sqrt{14}$$
$$= \left(\dfrac{1}{7} + \dfrac{1}{2}\right)\sqrt{14} = \dfrac{9}{14}\sqrt{14}$$

67. $\sqrt{80} = 8.9442719$
$\sqrt{80} = \sqrt{16 \cdot 5} = \sqrt{16}\sqrt{5} = 4\sqrt{5} = 4(2.360679) = 8.9442716$
The two results are the same, to 6 places. These results should be equal, and only appear to differ because of rounding off.

68. $\sqrt{150} = 12.247448$
$\sqrt{150} = \sqrt{25 \cdot 6} = \sqrt{25}\sqrt{6} = 5\sqrt{6} = 5(2.4494897) = 12.247448$

69. $4x - 3y = 12$

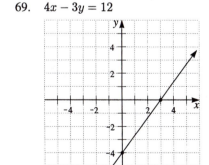

71. $x = 2$

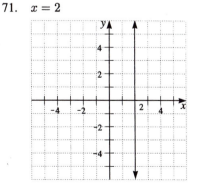

73.  Let $x$ = number of hours that the associate worked.

Then $14 - x$ = number of hours that the law clerk worked.

$$120x + 40(14 - x) = 1200$$
$$120x + 560 - 40x = 1200$$
$$80x + 560 = 1200$$
$$80x = 640$$
$$x = 8$$

Then $14 - x = 14 - 8 = 6$, so the associate worked 8 hours and the law clerk worked 6 hours.

## Exercises 10.4

3.  $\sqrt{3}\sqrt{5}\sqrt{13} = \sqrt{3 \cdot 5 \cdot 13} = \sqrt{195}$

7.  $\sqrt{6}\sqrt{24} = \sqrt{6}\left(\sqrt{4}\sqrt{6}\right) = \sqrt{6}\left(2\sqrt{6}\right)$
$$= 2\left(\sqrt{6}\sqrt{6}\right) = 2 \cdot 6 = 12$$

11.  $\sqrt{3}\left(\sqrt{5} + \sqrt{6}\right) = \sqrt{3}\sqrt{5} + \sqrt{3}\sqrt{6}$
$$= \sqrt{15} + \sqrt{18}$$
$$= \sqrt{15} + \sqrt{9}\sqrt{2}$$
$$= \sqrt{15} + 3\sqrt{2}$$

15.  $\sqrt{3}\left(2\sqrt{3} - 3\sqrt{2}\right) = \sqrt{3}\left(2\sqrt{3}\right) - \sqrt{3}\left(3\sqrt{2}\right)$
$$= 2\left(\sqrt{3}\sqrt{3}\right) - 3\left(\sqrt{3}\sqrt{2}\right)$$
$$= 2 \cdot 3 - 3\sqrt{6} = 6 - 3\sqrt{6}$$

19.  $3\sqrt{2}\left(\sqrt{2} - 4\right) + \sqrt{2}\left(5 - \sqrt{2}\right) = 3\sqrt{2}\left(\sqrt{2}\right) - 3\sqrt{2}(4) + \sqrt{2}(5) - \sqrt{2}\sqrt{2}$
$$= 3\left(\sqrt{2}\sqrt{2}\right) - 3 \cdot 4\sqrt{2} + 5\sqrt{2} - 2$$
$$= 3 \cdot 2 - 12\sqrt{2} + 5\sqrt{2} - 2$$
$$= 6 - 12\sqrt{2} + 5\sqrt{2} - 2$$
$$= 4 - 7\sqrt{2}$$

21.  $4\sqrt{x}\left(\sqrt{x} - \sqrt{2}\right) - \sqrt{x}\left(3\sqrt{x} - 2\sqrt{2}\right) = 4\sqrt{x}\sqrt{x} - 4\sqrt{x}\sqrt{2} - 3\sqrt{x}\sqrt{x} + 2\sqrt{x}\sqrt{2}$
$$= 4x - 4\sqrt{2x} - 3x + 2\sqrt{2x}$$
$$= x - 2\sqrt{2x}$$

**25.** $\left(\sqrt{x}+\sqrt{3}\right)^2 = \left(\sqrt{x}+\sqrt{3}\right)\left(\sqrt{x}+\sqrt{3}\right)$

$$= \sqrt{x}\sqrt{x}+\sqrt{x}\sqrt{3}+\sqrt{3}\sqrt{x}+\sqrt{3}\sqrt{3}$$
$$= x+\sqrt{3x}+\sqrt{3x}+3$$
$$= x+2\sqrt{3x}+3$$

**27.** $\left(\sqrt{x}+\sqrt{3}\right)\left(\sqrt{x}-\sqrt{3}\right) = \sqrt{x}\sqrt{x}-\sqrt{x}\sqrt{3}+\sqrt{3}\sqrt{x}-\sqrt{3}\sqrt{3}$

$$= x-\sqrt{3x}+\sqrt{3x}-3 = x-3$$

**29.** $\left(3\sqrt{2}-2\sqrt{5}\right)^2 = \left(3\sqrt{2}-2\sqrt{5}\right)\left(3\sqrt{2}-2\sqrt{5}\right)$

$$= \left(3\sqrt{2}\right)\left(3\sqrt{2}\right)-\left(3\sqrt{2}\right)\left(2\sqrt{5}\right)-\left(2\sqrt{5}\right)\left(3\sqrt{2}\right)+\left(2\sqrt{5}\right)\left(2\sqrt{5}\right)$$
$$= 3\cdot3\sqrt{2}\sqrt{2}-3\cdot2\sqrt{2}\sqrt{5}-2\cdot3\sqrt{5}\sqrt{2}+2\cdot2\sqrt{5}\sqrt{5}$$
$$= 9\cdot2-6\sqrt{10}-6\sqrt{10}+4\cdot5$$
$$= 18-6\sqrt{10}-6\sqrt{10}+20$$
$$= 38-12\sqrt{10}$$

**35.** $\left(\sqrt{28}-\sqrt{24}\right)\left(\sqrt{7}-\sqrt{6}\right) = \left(\sqrt{4}\sqrt{7}-\sqrt{4}\sqrt{6}\right)\left(\sqrt{7}-\sqrt{6}\right)$

$$= \left(2\sqrt{7}-2\sqrt{6}\right)\left(\sqrt{7}-\sqrt{6}\right)$$
$$= 2\sqrt{7}\sqrt{7}-2\sqrt{7}\sqrt{6}-2\sqrt{6}\sqrt{7}+2\sqrt{6}\sqrt{6}$$
$$= 2\cdot7-2\sqrt{42}-2\sqrt{42}+2\cdot6$$
$$= 14-2\sqrt{42}-2\sqrt{42}+12$$
$$= 26-4\sqrt{42}$$

**39.** $\left(\sqrt{x}+2\right)^2 - \left(\sqrt{x+2}\right)^2 = \left(\sqrt{x}+2\right)\left(\sqrt{x}+2\right)-(x+2)$

$$= \sqrt{x}\sqrt{x}+2\sqrt{x}+2\sqrt{x}+4-(x+2)$$
$$= x+2\sqrt{x}+2\sqrt{x}+4-x-2$$
$$= 4\sqrt{x}+2$$

**43.** $\dfrac{\sqrt{54}}{\sqrt{3}} = \sqrt{\dfrac{54}{3}} = \sqrt{18} = \sqrt{9}\sqrt{2} = 3\sqrt{2}$

47. $\dfrac{\sqrt{a^2 b^5}}{\sqrt{ab^8}} = \sqrt{\dfrac{\overset{a}{\cancel{a^2}} b^5}{\underset{b^3}{\cancel{ab^8}}}} = \sqrt{\dfrac{a}{b^3}} = \dfrac{\sqrt{a}}{\sqrt{b^3}}$

$\qquad = \dfrac{\sqrt{a}}{\sqrt{b^2}\sqrt{b}} = \dfrac{\sqrt{a}}{b\sqrt{b}}$

$\qquad = \dfrac{\sqrt{a}\sqrt{b}}{b\sqrt{b}\sqrt{b}} = \dfrac{\sqrt{ab}}{b \cdot b}$

$\qquad = \dfrac{\sqrt{ab}}{b^2}$

49. $\dfrac{10}{4 - \sqrt{11}} = \dfrac{10\left(4 + \sqrt{11}\right)}{\left(4 - \sqrt{11}\right)\left(4 + \sqrt{11}\right)}$

$\qquad = \dfrac{10\left(4 + \sqrt{11}\right)}{16 - 11} = \dfrac{\overset{2}{\cancel{10}}\left(4 + \sqrt{11}\right)}{\cancel{5}}$

$\qquad = 2\left(4 + \sqrt{11}\right) \text{ or } 8 + 2\sqrt{11}$

53. $\dfrac{\sqrt{3}}{2 + \sqrt{3}} = \dfrac{\sqrt{3}\left(2 - \sqrt{3}\right)}{\left(2 + \sqrt{3}\right)\left(2 - \sqrt{3}\right)}$

$\qquad = \dfrac{2\sqrt{3} - \sqrt{3}\sqrt{3}}{4 - 3} = \dfrac{2\sqrt{3} - 3}{1}$

$\qquad = 2\sqrt{3} - 3$

55. $\dfrac{\sqrt{5} + \sqrt{3}}{\sqrt{5} - \sqrt{3}} = \dfrac{\left(\sqrt{5} + \sqrt{3}\right)\left(\sqrt{5} + \sqrt{3}\right)}{\left(\sqrt{5} - \sqrt{3}\right)\left(\sqrt{5} + \sqrt{3}\right)}$

$\qquad = \dfrac{\sqrt{5}\sqrt{5} + \sqrt{5}\sqrt{3} + \sqrt{3}\sqrt{5} + \sqrt{3}\sqrt{3}}{5 - 3}$

$\qquad = \dfrac{5 + \sqrt{15} + \sqrt{15} + 3}{5 - 3} = \dfrac{8 + 2\sqrt{15}}{2}$

$\qquad = \dfrac{\cancel{2}\left(4 + \sqrt{15}\right)}{\cancel{2}} = 4 + \sqrt{15}$

**59.** $\dfrac{6}{3-\sqrt{7}} - \dfrac{21}{\sqrt{7}} = \dfrac{6\left(3+\sqrt{7}\right)}{\left(3-\sqrt{7}\right)\left(3+\sqrt{7}\right)} - \dfrac{21\sqrt{7}}{\sqrt{7}\sqrt{7}}$

$$= \dfrac{18+6\sqrt{7}}{9-7} - \dfrac{21\sqrt{7}}{7}$$

$$= \dfrac{\overset{3}{\cancel{6}}\left(3+\sqrt{7}\right)}{\cancel{2}} - \dfrac{\overset{3}{\cancel{21}}\sqrt{7}}{\cancel{7}}$$

$$= 9 + 3\sqrt{7} - 3\sqrt{7} = 9$$

**61.** $\left(3+\sqrt{5}\right)^2 + \dfrac{8}{3-\sqrt{5}} = \left(3+\sqrt{5}\right)\left(3+\sqrt{5}\right) + \dfrac{8\left(3+\sqrt{5}\right)}{\left(3-\sqrt{5}\right)\left(3+\sqrt{5}\right)}$

$$= 9 + 3\sqrt{5} + 3\sqrt{5} + 5 + \dfrac{24+8\sqrt{5}}{9-5}$$

$$= 14 + 6\sqrt{5} + \dfrac{\overset{2}{\cancel{8}}\left(3+\sqrt{5}\right)}{\cancel{4}}$$

$$= 14 + 6\sqrt{5} + 6 + 2\sqrt{5}$$

$$= 20 + 8\sqrt{5}$$

**65.** $\dfrac{12-\sqrt{20}}{10} = \dfrac{12-\sqrt{4}\sqrt{5}}{10}$

$$= \dfrac{12-2\sqrt{5}}{10} = \dfrac{\cancel{2}\left(6-\sqrt{5}\right)}{\underset{5}{\cancel{10}}}$$

$$= \dfrac{6-\sqrt{5}}{5}$$

**67.** $\left(2+\sqrt{10}\right)^2 - 4\left(2+\sqrt{10}\right) - 6 = \left(2+\sqrt{10}\right)\left(2+\sqrt{10}\right) - 4\left(2+\sqrt{10}\right) - 6$

$$= 4 + 2\sqrt{10} + 2\sqrt{10} + 10 - 8 - 4\sqrt{10} - 6$$

$$\overset{\cdot}{=} 0$$

**69.** (a) The "2" in the numerator is under the square root and is thus $\sqrt{2}$. This cannot be canceled with the "2" in the denominator.

(b) The cancellation is not valid since 2 is not a common factor of the numerator. Remember that terms cannot be canceled.

**71.** $x^2 - 6x - 7 = (x-7)(x+1)$ 
**73.** $x^2 + 10x - 16$ is not factorable.

**75.** $w^3 - 8w^2 + 16w = w(w^2 - 8w + 16)$
$$= w(w-4)(w-4)$$
**77.** $6u^2 - 5u - 14 = (6u+7)(u-2)$

61. $\left(3+\sqrt{5}\right)^2 + \dfrac{8}{3-\sqrt{5}} = \left(3+\sqrt{5}\right)\left(3+\sqrt{5}\right) + \dfrac{8\left(3+\sqrt{5}\right)}{\left(3-\sqrt{5}\right)\left(3+\sqrt{5}\right)}$

$$= 9 + 3\sqrt{5} + 3\sqrt{5} + 5 + \dfrac{24+8\sqrt{5}}{9-5}$$

$$= 14 + 6\sqrt{5} + \dfrac{\overset{2}{\cancel{8}}\left(3+\sqrt{5}\right)}{\cancel{4}}$$

$$= 14 + 6\sqrt{5} + 6 + 2\sqrt{5}$$

$$= 20 + 8\sqrt{5}$$

65. $\dfrac{12-\sqrt{20}}{10} = \dfrac{12-\sqrt{4}\sqrt{5}}{10}$

$$= \dfrac{12-2\sqrt{5}}{10} = \dfrac{\cancel{2}\left(6-\sqrt{5}\right)}{\underset{5}{\cancel{10}}}$$

$$= \dfrac{6-\sqrt{5}}{5}$$

67. $\left(2+\sqrt{10}\right)^2 - 4\left(2+\sqrt{10}\right) - 6 = \left(2+\sqrt{10}\right)\left(2+\sqrt{10}\right) - 4\left(2+\sqrt{10}\right) - 6$

$$= 4 + 2\sqrt{10} + 2\sqrt{10} + 10 - 8 - 4\sqrt{10} - 6$$

$$= 0$$

69. (a)  The "2" in the numerator is under the square root and is thus $\sqrt{2}$. This cannot be canceled with the "2" in the denominator.

   (b)  The cancellation is not valid since 2 is not a common factor of the numerator. Remember that terms cannot be canceled.

71.  $x^2 - 6x - 7 = (x-7)(x+1)$         73.  $x^2 + 10x - 16$ is not factorable.

75.  $w^3 - 8w^2 + 16w = w(w^2 - 8w + 16)$         77.  $6u^2 - 5u - 14 = (6u+7)(u-2)$
$$= w(w-4)(w-4)$$

79.  Let $x$ = price of a ticket sold at the door (in dollars).

   Then $x - 1.50$ = price of a ticket sold in advance (in dollars).

$$200(x-1.50) + 75x = 1075$$
$$200x - 300 + 75x = 1075$$
$$275x - 300 = 1075$$
$$275x = 1375$$
$$x = 5$$

   So a ticket sold at the door cost $5.

## Exercises 10.5

1. $\quad \sqrt{x} = 5 \qquad$ **CHECK** $x = 25$:

$\quad \left(\sqrt{x}\right)^2 = (5)^2 \qquad \sqrt{x} = 5$

$\qquad x = 25 \qquad \sqrt{25} \overset{?}{=} 5$

$\qquad\qquad\qquad 5 \overset{\vee}{=} 5$

5. $\quad \sqrt{u} = 6 \qquad$ **CHECK** $u = 36$:

$\quad \left(\sqrt{u}\right)^2 = (6)^2 \qquad \sqrt{u} = 6$

$\qquad u = 36 \qquad \sqrt{36} \overset{?}{=} 6$

$\qquad\qquad\qquad 6 \overset{\vee}{=} 6$

9. $\quad \sqrt{x+3} = 10 \qquad$ **CHECK** $x = 97$:

$\quad \left(\sqrt{x+3}\right)^2 = (10)^2 \qquad \sqrt{x+3} = 10$

$\qquad x + 3 = 100 \qquad \sqrt{97+3} \overset{?}{=} 10$

$\qquad x = 97 \qquad\quad \sqrt{100} \overset{?}{=} 10$

$\qquad\qquad\qquad\qquad 10 \overset{\vee}{=} 10$

11. $\quad \sqrt{x} + 3 = 10 \qquad$ **CHECK** $x = 49$:

$\qquad\quad -3 \quad -3 \qquad \sqrt{x} + 3 = 10$

$\quad \sqrt{x} \quad\;\; = 7 \qquad \sqrt{49} + 3 \overset{?}{=} 10$

$\quad \left(\sqrt{x}\right)^2 = (7)^2 \qquad 7 + 3 \overset{?}{=} 10$

$\qquad x \quad\;\; = 49 \qquad\quad 10 \overset{\vee}{=} 10$

17. $\quad \sqrt{5-4x} = 7 \qquad$ **CHECK** $x = -11$:

$\quad \left(\sqrt{5-4x}\right)^2 = (7)^2 \qquad \sqrt{5-4x} = 7$

$\qquad 5 - 4x = 49 \qquad \sqrt{5-4(-11)} \overset{?}{=} 7$

$\qquad -4x = 44 \qquad\quad \sqrt{5+44} \overset{?}{=} 7$

$\qquad x = -11 \qquad\qquad \sqrt{49} \overset{?}{=} 7$

$\qquad\qquad\qquad\qquad\qquad 7 \overset{\vee}{=} 7$

21. $\quad \sqrt{1+x} + 8 = \quad 4 \qquad$ **CHECK** $x = 15$:

$\qquad\qquad -8 \quad -8 \qquad \sqrt{1+x} + 8 = 4$

$\quad \sqrt{1+x} \quad\;\; = -4 \qquad \sqrt{1+15} + 8 \overset{?}{=} 4$

$\quad \left(\sqrt{1+x}\right)^2 = (-4)^2 \qquad \sqrt{16} + 8 \overset{?}{=} 4$

$\qquad 1 + x \quad\;\; = 16 \qquad\quad 4 + 8 \overset{?}{=} 4$

$\qquad x \quad\;\; = 15 \qquad\qquad 12 \neq 4$

So the given equation has no solution.

27. $\quad 10 - 2\sqrt{y} = \quad 4 \qquad$ **CHECK** $y = 9$:

$\quad -10 \qquad\quad -10 \qquad 10 - 2\sqrt{y} = 4$

$\quad\quad -2\sqrt{y} = \; -6 \qquad 10 - 2\sqrt{9} \overset{?}{=} 4$

$\quad\quad\quad \sqrt{y} = \quad 3 \qquad 10 - 2(3) \overset{?}{=} 4$

$\quad \left(\sqrt{y}\right)^2 = (3)^2 \qquad 10 - 6 \overset{?}{=} 4$

$\quad\quad\quad y = \quad 9 \qquad\qquad 4 \overset{\vee}{=} 4$

29. $\quad 10 - \sqrt{2y} = \quad 2 \qquad$ **CHECK** $y = 32$:

$\quad -10 \qquad\quad -10 \qquad 10 - \sqrt{2y} = 2$

$\quad\quad -\sqrt{2y} = \; -8 \qquad 10 - \sqrt{2(32)} \overset{?}{=} 2$

$\quad\quad\quad \sqrt{2y} = \quad 8 \qquad 10 - \sqrt{64} \overset{?}{=} 2$

$\quad \left(\sqrt{2y}\right)^2 = (8)^2 \qquad 10 - 8 \overset{?}{=} 2$

$\quad\quad\quad 2y = \quad 64 \qquad\qquad 2 \overset{\vee}{=} 2$

$\quad\quad\quad y = \quad 32$

33. $\quad 6\sqrt{c} - 3 = \quad 3\sqrt{c} \qquad$ **CHECK** $c = 1$:

$\quad -6\sqrt{c} \qquad\quad -6\sqrt{c} \qquad 6\sqrt{c} - 3 = 3\sqrt{c}$

$\quad\quad\quad -3 = -3\sqrt{c} \qquad 6\sqrt{1} - 3 \overset{?}{=} 3\sqrt{1}$

$\quad\quad\quad\; 1 = \quad \sqrt{c} \qquad 6(1) - 3 \overset{?}{=} 3(1)$

$\quad\quad (1)^2 = \left(\sqrt{c}\right)^2 \qquad\quad 6 - 3 \overset{?}{=} 3$

$\quad\quad\quad\; 1 = \quad c \qquad\qquad\quad 3 \overset{\vee}{=} 3$

49. $\sqrt{5x-1} = \sqrt{3x+9}$      CHECK $x = 5$:

$$\left(\sqrt{5x-1}\right)^2 = \left(\sqrt{3x+9}\right)^2 \qquad \sqrt{5x-1} = \sqrt{3x+9}$$

$$5x - 1 = 3x + 9 \qquad\qquad \sqrt{5(5)-1} \overset{?}{=} \sqrt{3(5)+9}$$

$$\underline{\quad -3x \qquad\quad -3x \quad} \qquad\quad \sqrt{25-1} \overset{?}{=} \sqrt{15+9}$$

$$2x - 1 = \qquad 9 \qquad\qquad\qquad \sqrt{24} \overset{\checkmark}{=} \sqrt{24}$$

$$2x \quad = \qquad 10$$

$$x \quad = \qquad 5$$

51. $I = \sqrt{\dfrac{W}{R}}$; $I = 12$, $R = 9$

$$12 = \sqrt{\frac{W}{9}}$$

$$(12)^2 = \left(\sqrt{\frac{W}{9}}\right)^2$$

$$144 = \frac{W}{9}$$

$$1296 = W$$

53. $t = \sqrt{\dfrac{h}{16}}$; $t = 3.4$

$$3.4 = \sqrt{\frac{h}{16}}$$

$$(3.4)^2 = \left(\sqrt{\frac{h}{16}}\right)^2$$

$$11.56 = \frac{h}{16}$$

$$184.96 = h$$

The height of the building is 185 ft (rounded to the nearest foot).

55. $d = \sqrt{8000a}$; $d = 150$

$$150 = \sqrt{8000a}$$

$$(150)^2 = \left(\sqrt{8000a}\right)^2$$

$$22500 = 8000a$$

$$2.8125 = a$$

The altitude of the plane is 2.8 miles (rounded to the nearest tenth).

57. $r = \sqrt{\dfrac{A}{P}} - 1$; $A = 5000$, $r = 0.06$

$$0.06 = \sqrt{\frac{5000}{P}} - 1$$

$$1.06 = \sqrt{\frac{5000}{P}}$$

$$(1.06)^2 = \left(\sqrt{\frac{5000}{P}}\right)^2$$

$$1.1236 = \frac{5000}{P}$$

$$1.1236P = 5000$$

$$P = 4449.98$$

The original investment must be $4,450 (rounded to the nearest dollar).

59. The equations $x = 5$ and $x^2 = 25$ are not equivalent, since their solution sets are not the same. The solution set of the second equation includes $-5$, which is not in the solution set of the first equation. In general, squaring both sides of an equation does not yield an equivalent equation, since the "squared" equation might have a solution that fails to satisfy the original equation.

60. The equation $\sqrt{x} + 3 = 2$ implies that $\sqrt{x} = -1$. We know this is impossible because by definition, the $\sqrt{\phantom{x}}$ symbol stands for the non-negative square root. Therefore, the given equation has no solution.

61. $\dfrac{8x^3 + 12x - 4x^2}{2x} = \dfrac{\overset{4\ x^2}{\cancel{8}\cancel{x^3}}}{\cancel{2}\cancel{x}} + \dfrac{\overset{6}{\cancel{12}\cancel{x}}}{\cancel{2}\cancel{x}} - \dfrac{\overset{2\ x}{\cancel{4}\cancel{x^2}}}{\cancel{2}\cancel{x}} = 4x^2 + 6 - 2x$

63.
$$
\begin{array}{r}
3x\ -2\phantom{)} \\
x-2\overline{)\ 3x^2 - 8x\ +4} \\
\underline{-(3x^2 - 6x)\phantom{+4}} \\
-2x\ +4 \\
\underline{-(-2x\ +4)} \\
0
\end{array}
$$

## Chapter 10 Review Exercises

1. $\sqrt{49} = 7$

3. $\sqrt{-16}$ is undefined

5. $\sqrt{96} = \sqrt{16 \cdot 6} = \sqrt{16}\sqrt{6} = 4\sqrt{6}$

7. $\sqrt{9x^9} = \sqrt{9x^8}\sqrt{x} = \sqrt{9}\sqrt{x^8}\sqrt{x} = 3x^4\sqrt{x}$

9. $\sqrt{\dfrac{4}{9}} = \dfrac{\sqrt{4}}{\sqrt{9}} = \dfrac{2}{3}$

11. $\sqrt{\dfrac{3}{5}} = \dfrac{\sqrt{3}}{\sqrt{5}} = \dfrac{\sqrt{3} \cdot \sqrt{5}}{\sqrt{5} \cdot \sqrt{5}} = \dfrac{\sqrt{15}}{5}$

13. $8\sqrt{7} - 5\sqrt{7} - \sqrt{7} = (8 - 5 - 1)\sqrt{7}$
$$= 2\sqrt{7}$$

15. $\sqrt{45} - \sqrt{20} = \sqrt{9}\sqrt{5} - \sqrt{4}\sqrt{5}$
$$= 3\sqrt{5} - 2\sqrt{5} = \sqrt{5}$$

17. $\sqrt{75x} + \sqrt{12x} = \sqrt{25}\sqrt{3x} + \sqrt{4}\sqrt{3x}$
$$= 5\sqrt{3x} + 2\sqrt{3x} = 7\sqrt{3x}$$

19. $\dfrac{\sqrt{12x^3y^2}}{xy} + \sqrt{27x} = \dfrac{\sqrt{4x^2y^2}\sqrt{3x}}{xy} + \sqrt{9}\sqrt{3x}$
$$= \dfrac{2\cancel{x}\cancel{y}\sqrt{3x}}{\cancel{x}\cancel{y}} + 3\sqrt{3x}$$
$$= 2\sqrt{3x} + 3\sqrt{3x} = 5\sqrt{3x}$$

21. $\sqrt{5}\left(3\sqrt{5} + \sqrt{2}\right) = 3\sqrt{5}\sqrt{5} + \sqrt{5}\sqrt{2}$
$$= 3 \cdot 5 + \sqrt{10}$$
$$= 15 + \sqrt{10}$$

23. $\left(3\sqrt{7} - 2\sqrt{3}\right)\left(2\sqrt{7} + 5\sqrt{3}\right)$

$$= \left(3\sqrt{7}\right)\left(2\sqrt{7}\right) + \left(3\sqrt{7}\right)\left(5\sqrt{3}\right) - \left(2\sqrt{3}\right)\left(2\sqrt{7}\right) - \left(2\sqrt{3}\right)\left(5\sqrt{3}\right)$$
$$= 3 \cdot 2\sqrt{7}\sqrt{7} + 3 \cdot 5\sqrt{7}\sqrt{3} - 2 \cdot 2\sqrt{3}\sqrt{7} - 2 \cdot 5\sqrt{3}\sqrt{3}$$
$$= 6 \cdot 7 + 15\sqrt{21} - 4\sqrt{21} - 10 \cdot 3$$
$$= 42 + 15\sqrt{21} - 4\sqrt{21} - 30$$
$$= 12 + 11\sqrt{21}$$

25. $\left(\sqrt{x} - 3\right)^2 = \left(\sqrt{x} - 3\right)\left(\sqrt{x} - 3\right)$

$$= \sqrt{x}\sqrt{x} - 3\sqrt{x} - 3\sqrt{x} + 9$$
$$= x - 6\sqrt{x} + 9$$

27. $\dfrac{7}{\sqrt{3}} = \dfrac{7\sqrt{3}}{\sqrt{3}\sqrt{3}} = \dfrac{7\sqrt{3}}{3}$

29. $\dfrac{x^2}{\sqrt{x}} = \dfrac{x^2\sqrt{x}}{\sqrt{x}\sqrt{x}} = \dfrac{\overset{x}{\cancel{x^2}}\sqrt{x}}{\cancel{x}} = x\sqrt{x}$

31. $\dfrac{18}{\sqrt{12}} = \dfrac{18}{\sqrt{4}\sqrt{3}} = \dfrac{\overset{9}{\cancel{18}}}{\cancel{2}\sqrt{3}} = \dfrac{9}{\sqrt{3}}$

$$= \dfrac{9\sqrt{3}}{\sqrt{3}\sqrt{3}} = \dfrac{\overset{3}{\cancel{9}}\sqrt{3}}{\underset{1}{\cancel{3}}} = 3\sqrt{3}$$

33. $\dfrac{14}{3 - \sqrt{2}} = \dfrac{14\left(3 + \sqrt{2}\right)}{\left(3 - \sqrt{2}\right)\left(3 + \sqrt{2}\right)}$

$$= \dfrac{14\left(3 + \sqrt{2}\right)}{9 - 2} = \dfrac{\overset{2}{\cancel{14}}\left(3 + \sqrt{2}\right)}{\cancel{7}}$$
$$= 2\left(3 + \sqrt{2}\right) \text{ or } 6 + 2\sqrt{2}$$

35. $\dfrac{2 + \sqrt{5}}{6 + \sqrt{5}} = \dfrac{\left(2 + \sqrt{5}\right)\left(6 - \sqrt{5}\right)}{\left(6 + \sqrt{5}\right)\left(6 - \sqrt{5}\right)}$

$$= \dfrac{12 - 2\sqrt{5} + 6\sqrt{5} - 5}{36 - 5}$$
$$= \dfrac{7 + 4\sqrt{5}}{31}$$

37. $\left(\sqrt{x + 7}\right)^2 - \left(\sqrt{x} + \sqrt{7}\right)^2 = x + 7 - \left(\sqrt{x} + \sqrt{7}\right)\left(\sqrt{x} + \sqrt{7}\right)$

$$= x + 7 - \left(x + \sqrt{x}\sqrt{7} + \sqrt{7}\sqrt{x} + 7\right)$$
$$= x + 7 - \left(x + \sqrt{7x} + \sqrt{7x} + 7\right)$$
$$= x + 7 - x - \sqrt{7x} - \sqrt{7x} - 7$$
$$= -2\sqrt{7x}$$

39. $\sqrt{x+5} = 7$   CHECK $x = 44$:

$\left(\sqrt{x+5}\right)^2 = (7)^2$   $\sqrt{x+5} = 7$

$x + 5 = 49$   $\sqrt{44+5} \overset{?}{=} 7$

$x = 44$   $\sqrt{49} \overset{?}{=} 7$

$7 \overset{\checkmark}{=} 7$

41. $10 - \sqrt{4u} = \quad 2$   CHECK $u = 16$:

$\underline{-10 \qquad\qquad -10}$   $10 - \sqrt{4u} = 2$

$-\sqrt{4u} = \quad -8$   $10 - \sqrt{4(16)} \overset{?}{=} 2$

$\sqrt{4u} = \quad 8$   $10 - \sqrt{64} \overset{?}{=} 2$

$\left(\sqrt{4u}\right)^2 = (8)^2$   $10 - 8 \overset{?}{=} 2$

$4u = \quad 64$   $2 \overset{\checkmark}{=} 2$

$u = \quad 16$

43. $8 - \sqrt{y-5} = 11$   CHECK $y = 14$:

$\underline{-8 \qquad\qquad -8}$   $8 - \sqrt{y-5} = 11$

$-\sqrt{y-5} = \quad 3$   $8 - \sqrt{14-5} \overset{?}{=} 11$

$\sqrt{y-5} = -3$   $8 - \sqrt{9} \overset{?}{=} 11$

$\left(\sqrt{y-5}\right)^2 = (-3)^2$   $8 - 3 \overset{?}{=} 11$

$y - 5 = \quad 9$   $5 \neq 11$

$y \quad = 14$   So the given equation has no solution.

45. $T = 2\pi\sqrt{\dfrac{L}{980}}; \; T = 1.5$

$1.5 = 2(3.14)\sqrt{\dfrac{L}{980}}$

$1.5 = 6.28\sqrt{\dfrac{L}{980}}$

$0.2388 = \sqrt{\dfrac{L}{980}}$

$(0.2388)^2 = \left(\sqrt{\dfrac{L}{980}}\right)^2$

$0.0570 = \dfrac{L}{980}$

$55.88 = L$

The length of the pendulum is 55.9 cm. (rounded to the nearest tenth).

## Chapter 10 Practice Test

1. $\sqrt{25x^{16}y^6} = \sqrt{25}\sqrt{x^{16}}\sqrt{y^6} = 5x^8 y^3$

3. $\sqrt{50x^3} - x\sqrt{32x} = \sqrt{25x^2}\sqrt{2x} - x\left(\sqrt{16}\sqrt{2x}\right)$

$$= 5x\sqrt{2x} - x\left(4\sqrt{2x}\right)$$

$$= 5x\sqrt{2x} - 4x\sqrt{2x} = x\sqrt{2x}$$

5. $\sqrt{20x^8y^9} + 3x^4y^4\sqrt{5y} = \sqrt{4x^8y^8}\sqrt{5y} + 3x^4y^4\sqrt{5y}$

$$= 2x^4y^4\sqrt{5y} + 3x^4y^4\sqrt{5y}$$

$$= 5x^4y^4\sqrt{5y}$$

7. $\left(2\sqrt{x} - \sqrt{5}\right)\left(\sqrt{x} + 3\sqrt{5}\right) = 2\sqrt{x}\sqrt{x} + \left(2\sqrt{x}\right)\left(3\sqrt{5}\right) - \sqrt{5}\sqrt{x} - 3\sqrt{5}\sqrt{5}$

$$= 2x + 6\sqrt{5x} - \sqrt{5x} - 3\cdot 5$$

$$= 2x + 5\sqrt{5x} - 15$$

9. $\left(\sqrt{x} - 4\right)^2 - \left(\sqrt{x-4}\right)^2 = \left(\sqrt{x} - 4\right)\left(\sqrt{x} - 4\right) - (x - 4)$

$$= \sqrt{x}\sqrt{x} - 4\sqrt{x} - 4\sqrt{x} + 16 - x + 4$$

$$= x - 8\sqrt{x} + 16 - x + 4$$

$$= -8\sqrt{x} + 20$$

11. $x^2 - 2x = 1$

$$\left(1 - \sqrt{2}\right)^2 - 2\left(1 - \sqrt{2}\right) \overset{?}{=} 1$$

$$\left(1 - \sqrt{2}\right)\left(1 - \sqrt{2}\right) - 2\left(1 - \sqrt{2}\right) \overset{?}{=} 1$$

$$1 - \sqrt{2} - \sqrt{2} + 2 - 2 + 2\sqrt{2} \overset{?}{=} 1$$

$$1 \overset{\checkmark}{=} 1$$

So $1 - \sqrt{2}$ is a solution to the equation.

13. $C = 1.8\sqrt{1000 - n};\ \ C = 18$

$$18 = 1.8\sqrt{1000 - n}$$

$$10 = \sqrt{1000 - n}$$

$$(10)^2 = \left(\sqrt{1000 - n}\right)^2$$

$$100 = 1000 - n$$

$$-900 = \qquad -n$$

$$900 = \qquad n$$

So 900 items are produced.

# Chapter 11
# Quadratic Equations

## Exercises 11.1

1.
$$x^2 + 4x = 32$$
$$x^2 + 4x - 32 = 0$$
$$(x + 8)(x - 4) = 0$$
$$x + 8 = 0 \quad \text{or} \quad x - 4 = 0$$
$$x = -8 \quad \text{or} \quad x = 4$$

3.
$$(x + 3)(x + 5) = 80$$
$$x^2 + 5x + 3x + 15 = 80$$
$$x^2 + 8x + 15 = 80$$
$$x^2 + 8x - 65 = 0$$
$$(x + 13)(x - 5) = 0$$
$$x + 13 = 0 \quad \text{or} \quad x - 5 = 0$$
$$x = -13 \quad \text{or} \quad x = 5$$

7.
$$b^2 - 16 = 0$$
$$b^2 = 16$$
$$b = \pm \sqrt{16} = \pm 4$$

9.
$$9b^2 - 16 = 0$$
$$9b^2 = 16$$
$$b^2 = \frac{16}{9}$$
$$b = \pm \sqrt{\frac{16}{9}} = \pm \frac{\sqrt{16}}{\sqrt{9}} = \pm \frac{4}{3}$$

11.
$$b^2 + 16 = 0$$
$$b^2 = -16$$
$$b = \pm \sqrt{-16}$$
Since $\sqrt{-16}$ is not a real number, the equation has no real solutions.

15.
$$36x^2 - 15 = 0$$
$$36x^2 = 15$$
$$x^2 = \frac{15}{36}$$
$$x = \pm \sqrt{\frac{15}{36}} = \pm \frac{\sqrt{15}}{\sqrt{36}} = \pm \frac{\sqrt{15}}{6}$$

17.
$$3b^2 = 11$$
$$b^2 = \frac{11}{3}$$
$$b = \pm \sqrt{\frac{11}{3}} = \pm \frac{\sqrt{11}}{\sqrt{3}} = \pm \frac{\sqrt{11}\sqrt{3}}{\sqrt{3}\sqrt{3}}$$
$$b = \pm \frac{\sqrt{33}}{3}$$

21.
$$9a^2 = 20$$
$$a^2 = \frac{20}{9}$$
$$a = \pm \sqrt{\frac{20}{9}} = \pm \frac{\sqrt{20}}{\sqrt{9}} = \pm \frac{\sqrt{20}}{3}$$
$$a = \pm \frac{\sqrt{4}\sqrt{5}}{3} = \pm \frac{2\sqrt{5}}{3}$$

25.
$$7y^2 - 4 = 5y^2 + 6$$
$$2y^2 - 4 = 6$$
$$2y^2 = 10$$
$$y^2 = 5$$
$$y = \pm \sqrt{5}$$

27.
$$5a^2 - 3a + 4 = 2a^2 - 3a + 13$$
$$3a^2 - 3a + 4 = -3a + 13$$
$$3a^2 + 4 = 13$$
$$3a^2 = 9$$
$$a^2 = 3$$
$$a = \pm \sqrt{3}$$

31. $(y-3)(y+4) = y$

$\qquad y^2 + y - 12 = y$

$\qquad\quad y^2 - 12 = 0$

$\qquad\qquad y^2 = 12$

$\qquad\qquad\ y = \pm \sqrt{12}$

$\qquad\qquad\ y = \pm \sqrt{4}\sqrt{3} = \pm 2\sqrt{3}$

35. $\qquad (x+2)^2 = 4(x+7)$

$\qquad x^2 + 4x + 4 = 4x + 28$

$\qquad\qquad x^2 + 4 = 28$

$\qquad\qquad\quad x^2 = 24$

$\qquad\qquad\quad\ x = \pm \sqrt{24}$

$\qquad\qquad\quad\ x = \pm \sqrt{4}\sqrt{6} = \pm 2\sqrt{6}$

37. $(t-2)^2 = 9$

$\qquad t - 2 = \pm \sqrt{9}$

$\qquad t - 2 = \pm 3$

$\quad t - 2 = 3 \quad \text{or} \quad t - 2 = -3$

$\qquad\ t = 5 \quad \text{or} \qquad\quad t = -1$

39. $(a+5)^2 = 7$

$\qquad a + 5 = \pm \sqrt{7}$

$\qquad\quad a = -5 \pm \sqrt{7}$

43. $(x+5)^2 = 10$

$\qquad x + 5 = \pm \sqrt{10}$

$\qquad\quad x = -5 \pm \sqrt{10}$

45. This exercise should read: $(x-6)^2 = -12x$. Then

$$(x-6)^2 = -12x$$

$$x^2 - 12x + 36 = -12x$$

$$x^2 + 36 = 0$$

$$x^2 = -36$$

$$x = \pm \sqrt{-36}$$

Since $\sqrt{-36}$ is not a real number, the equation has no real solutions.

47. $(2x-3)(x+4) = x(x+9)$

$\quad 2x^2 + 5x - 12 = x^2 + 9x$

$\qquad x^2 - 4x - 12 = 0$

$\qquad (x-6)(x+2) = 0$

$\quad x - 6 = 0 \quad \text{or} \quad x + 2 = 0$

$\qquad\ x = 6 \quad \text{or} \qquad\ x = -2$

49. $\left(m - \dfrac{2}{3}\right)^2 = \dfrac{4}{9}$

$\qquad m - \dfrac{2}{3} = \pm \sqrt{\dfrac{4}{9}}$

$\qquad m - \dfrac{2}{3} = \pm \dfrac{\sqrt{4}}{\sqrt{9}}$

$\qquad m - \dfrac{2}{3} = \pm \dfrac{2}{3}$

$\quad m - \dfrac{2}{3} = \dfrac{2}{3} \quad \text{or} \quad m - \dfrac{2}{3} = -\dfrac{2}{3}$

$\qquad\quad m = \dfrac{4}{3} \quad \text{or} \qquad\quad m = 0$

51. $\left(x + \dfrac{2}{5}\right)^2 = \dfrac{3}{25}$

$$x + \dfrac{2}{5} = \pm\sqrt{\dfrac{3}{25}}$$

$$x + \dfrac{2}{5} = \pm\dfrac{\sqrt{3}}{\sqrt{25}}$$

$$x + \dfrac{2}{5} = \pm\dfrac{\sqrt{3}}{5}$$

$$x = -\dfrac{2}{5} \pm \dfrac{\sqrt{3}}{5} = \dfrac{-2 \pm \sqrt{3}}{5}$$

55. $x + \dfrac{1}{x} = 2$

$$x\left(x + \dfrac{1}{x}\right) = x(2)$$

$$x^2 + 1 = 2x$$

$$x^2 - 2x + 1 = 0$$

$$(x - 1)^2 = 0$$

$$x - 1 = 0$$

$$x = 1$$

57. $\dfrac{x - 1}{x + 1} = \dfrac{x}{x + 3}$

$$(\cancel{x+1})(x + 3)\left(\dfrac{x - 1}{\cancel{x+1}}\right) = (x + 1)(\cancel{x+3})\left(\dfrac{x}{\cancel{x+3}}\right)$$

$$(x + 3)(x - 1) = (x + 1)(x)$$

$$x^2 + 2x - 3 = x^2 + x$$

$$2x - 3 = x$$

$$-3 = -x$$

$$3 = x$$

59. $\dfrac{2x}{x + 2} + 1 = x$

$$(x + 2)\left(\dfrac{2x}{x + 2} + 1\right) = (x + 2)x$$

$$(\cancel{x+2})\left(\dfrac{2x}{\cancel{x+2}}\right) + (x + 2)(1) = (x + 2)x$$

$$2x + x + 2 = x^2 + 2x$$

$$3x + 2 = x^2 + 2x$$

$$0 = x^2 - x - 2$$

$$0 = (x - 2)(x + 1)$$

$$0 = x - 2 \quad \text{or} \quad 0 = x + 1$$

$$2 = x \quad \text{or} \quad -1 = x$$

**61.**

$$a - \frac{5a}{a+1} = \frac{5}{a+1}$$

$$(a+1)\left(a - \frac{5a}{a+1}\right) = (a+1)\left(\frac{5}{a+1}\right)$$

$$(a+1)(a) - (a+1)\left(\frac{5a}{a+1}\right) = (a+1)\left(\frac{5}{a+1}\right)$$

$$(a+1)(a) - 5a = 5$$

$$a^2 + a - 5a = 5$$

$$a^2 - 4a = 5$$

$$a^2 - 4a - 5 = 0$$

$$(a-5)(a+1) = 0$$

$$a - 5 = 0 \quad \text{or} \quad a + 1 = 0$$

$$a = 5 \quad \text{or} \quad a = -1$$

Reject $a = -1$, since this value of $a$ produces denominators of 0 when it is checked in the original equation, and division by zero is undefined.

**63.**

$$\frac{3}{x-2} + \frac{7}{x+2} = \frac{x+1}{x-2}$$

$$(x-2)(x+2)\left(\frac{3}{x-2} + \frac{7}{x+2}\right) = (x-2)(x+2)\left(\frac{x+1}{x-2}\right)$$

$$(x-2)(x+2)\left(\frac{3}{x-2}\right) + (x-2)(x+2)\left(\frac{7}{x+2}\right) = (x-2)(x+2)\left(\frac{x+1}{x-2}\right)$$

$$3(x+2) + 7(x-2) = (x+2)(x+1)$$

$$3x + 6 + 7x - 14 = x^2 + 3x + 2$$

$$10x - 8 = x^2 + 3x + 2$$

$$0 = x^2 - 7x + 10$$

$$0 = (x-5)(x-2)$$

$$0 = x - 5 \quad \text{or} \quad 0 = x - 2$$

$$5 = x \quad \text{or} \quad 2 = x$$

Reject $x = 2$, since this value of $x$ produces denominators of 0 when it is checked in the original equation, and division by zero is undefined.

**65.**  
$$2x^2 + 7x - 5 = 3x^2 + 9x - 4$$
$$7x - 5 = x^2 + 9x - 4$$
$$-5 = x^2 + 2x - 4$$
$$0 = x^2 + 2x + 1$$
$$0 = (x+1)(x+1)$$
$$0 = x + 1 \quad \text{or} \quad 0 = x + 1$$
$$-1 = x \quad \text{or} \quad -1 = x$$
So $x = -1$.

**67.**  
$$(y-2)(y+3) = y + 10$$
$$y^2 + y - 6 = y + 10$$
$$y^2 - 6 = 10$$
$$y^2 = 16$$
$$y = \pm\sqrt{16} = \pm 4$$

69. $(y-2)(y+3) = (2y-7)(y+4)$

$y^2 + y - 6 = 2y^2 + y - 28$

$y - 6 = y^2 + y - 28$

$22 = y^2$

$\pm\sqrt{22} = y$

73.

$$4(x+1) = \frac{9}{x+1}$$

$$(x+1)[4(x+1)] = \frac{\cancel{x+1}}{1} \cdot \frac{9}{\cancel{x+1}}$$

$$4(x+1)^2 = 9$$

$$(x+1)^2 = \frac{9}{4}$$

$$x+1 = \pm\sqrt{\frac{9}{4}} = \pm\frac{\sqrt{9}}{\sqrt{4}}$$

$$x+1 = \pm\frac{3}{2}$$

$$x+1 = \frac{3}{2} \quad \text{or} \quad x+1 = -\frac{3}{2}$$

$$x = \frac{1}{2} \quad \text{or} \qquad x = -\frac{5}{2}$$

75. $x = \pm 2.65$

77. $k = \pm 2.58$

79. $x = \pm 2.22$

81. $a = 1.52, a = -1.22$

83. $y = 0.23, y = -1.19$

85. Step 1: We can add the same quantity, 7, to both sides of an equation to obtain an equivalent equation.

Step 2: We can add the same quantity, 4, to both sides of an equation to obtain an equivalent equation.

Step 3: $7 + 4 = 11$

Step 4: $x^2 + 4x + 4$ factors as the square of the sum $x + 2$.

Step 5: $u^2 = d$ implies that $u = \pm\sqrt{d}$.

Step 6: We can subtract the same quantity, 2, from both sides of an equation to obtain an equivalent equation.

86. $\left(2 + \sqrt{3}\right)^2 - 4\left(2 + \sqrt{3}\right) + 1 = \left(2 + \sqrt{3}\right)\left(2 + \sqrt{3}\right) - 4\left(2 + \sqrt{3}\right) + 1$

$= 4 + 2\sqrt{3} + 2\sqrt{3} + 3 - 8 - 4\sqrt{3} + 1$

$= 0$

So $2 + \sqrt{3}$ is a solution to the equation $x^2 - 4x + 1 = 0$.

$\left(2 - \sqrt{3}\right)^2 - 4\left(2 - \sqrt{3}\right) + 1 = \left(2 - \sqrt{3}\right)\left(2 - \sqrt{3}\right) - 4\left(2 - \sqrt{3}\right) + 1$

$= 4 - 2\sqrt{3} - 2\sqrt{3} + 3 - 8 + 4\sqrt{3} + 1$

$= 0$

So $2 - \sqrt{3}$ is also a solution to the equation $x^2 - 4x + 1 = 0$. (This is no coincidence. In fact, whenever $p + \sqrt{d}$ satisfies the quadratic equation $ax^2 + bx + c = 0$, where $a$, $b$, and $c$ are integers, $a \neq 0$, it must be true that $p - \sqrt{d}$ also satisfies the equation.)

87. $\begin{cases} 2x + y = 6 \\ 3x - 2y = 23 \end{cases}$ $\xrightarrow{\text{multiply by 2}}$ $4x + 2y = 12$ $\qquad$ $2x + y = 6$

$\qquad\qquad\qquad\qquad\qquad\;\xrightarrow{\text{as is}}$ $\dfrac{3x - 2y = 23}{7x \quad\;\; = 35}$ Add. $\quad$ $2(5) + y = 6$

$\qquad\qquad\qquad\qquad\qquad\qquad\qquad\qquad\qquad\; x \quad\;\; = 5$ $\qquad\qquad 10 + y = 6$

$\qquad\qquad\qquad\qquad\qquad\qquad\qquad\qquad\qquad\qquad\qquad\qquad\qquad\qquad y = -4$

Solution: $(5, -4)$
CHECK: $3x - 2y = 23$

$3(5) - 2(-4) \overset{?}{=} 23$

$\qquad\quad 15 + 8 \overset{?}{=} 23$

$\qquad\qquad\quad 23 \overset{\checkmark}{=} 23$

89. $\begin{cases} 6x - 4y = 9 \\ 3x - 2y = 2 \end{cases}$ $\xrightarrow{\text{as is}}$ $\qquad$ $6x - 4y = \quad 9$

$\qquad\qquad\qquad\qquad\;\xrightarrow{\text{multiply by } -2}$ $\dfrac{-6x + 4y = -4}{0 = \quad 5}$ Add

$\qquad\qquad\qquad\qquad\qquad\qquad\qquad\qquad$ Contradiction. This system has no solution.

91. Let $x =$ number of hours needed if they work together.

$$\frac{x}{4} + \frac{x}{3} = 1$$

$$12\left(\frac{x}{4} + \frac{x}{3}\right) = 12(1)$$

$$3x + 4x = 12$$

$$7x = 12$$

$$x = \frac{12}{7} \text{ or } 1\frac{5}{7}$$

To prepare the meal together, Lorraine and Renaldo need $1\frac{5}{7}$ hours.

## Exercises 11.2

3. $\quad x^2 + 8x + \;\; 6 = \qquad 0$

$\qquad x^2 + 8x \qquad\; = \;\; -6$ $\qquad\qquad 2p = 8; \; p = \dfrac{8}{2} = 4; \; p^2 = 4^2 = 16$

$\qquad\qquad\quad \dfrac{+16 \quad +16}{}$

$\qquad x^2 + 8x + 16 = \qquad 10$

$\qquad\qquad (x + 4)^2 = \qquad 10$

$\qquad\qquad\quad\; x + 4 = \; \pm\sqrt{10}$

$\qquad\qquad\qquad\quad\; x = -4 \pm \sqrt{10}$

5. $x^2 - 10x \quad = \quad 15$

$\phantom{x^2 - 10x} +25 \quad +25$

$\overline{x^2 - 10x + 25 = \quad 40}$

$(x - 5)^2 = \quad 40$

$x - 5 = \pm \sqrt{40}$

$x - 5 = \pm 2\sqrt{10}$

$x = 5 \pm 2\sqrt{10}$

$2p = -10; \; p = \dfrac{-10}{2} = -5; \; p^2 = (-5)^2 = 25$

7. $a^2 - 8a - 20 = \quad 0$

$a^2 - 8a \quad\quad = \quad 20$

$\phantom{a^2 - 8a} +16 \quad +16$

$\overline{a^2 - 8a + 16 = \quad 36}$

$(a - 4)^2 = \quad 36$

$a - 4 = \pm \sqrt{36}$

$a - 4 = \pm 6$

$a - 4 = 6 \quad \text{or} \quad a - 4 = -6$

$a = 10 \quad \text{or} \quad a = -2$

$2p = -8; \; p = \dfrac{-8}{2} = -4; \; p^2 = (-4)^2 = 16$

11. $2z^2 - 12z + 4 = \quad 0$

$2z^2 - 12z \quad\quad = -4$

$z^2 - 6z \quad\quad = -2$

$\phantom{z^2 - 6z} +9 \quad +9$

$\overline{z^2 - 6z + 9 = \quad 7}$

$(z - 3)^2 = \quad 7$

$z - 3 = \pm \sqrt{7}$

$z = 3 \pm \sqrt{7}$

$2p = -6; \; p = \dfrac{-6}{2} = -3; \; p^2 = (-3)^2 = 9$

15. $u^2 + 5u - 2 = \quad 0$

$u^2 + 5u \quad\quad = \quad 2$

$\phantom{u^2 + 5u} +\dfrac{25}{4} \quad +\dfrac{25}{4}$

$\overline{u^2 + 5u + \dfrac{25}{4} = 2 + \dfrac{25}{4}}$

$\left(u + \dfrac{5}{2}\right)^2 = \dfrac{8}{4} + \dfrac{25}{4}$

$\left(u + \dfrac{5}{2}\right)^2 = \dfrac{33}{4}$

$u + \dfrac{5}{2} = \pm \sqrt{\dfrac{33}{4}} = \pm \dfrac{\sqrt{33}}{2}$

$u = -\dfrac{5}{2} \pm \dfrac{\sqrt{33}}{2}$

$u = \dfrac{-5 \pm \sqrt{33}}{2}$

$2p = 5; \; p = \dfrac{5}{2}; \; p^2 = \left(\dfrac{5}{2}\right)^2 = \dfrac{25}{4}$

19. 
$$w^2 - 3w = 2w^2 - 7w + 2$$
$$-w^2 - 3w = -7w + 2$$
$$-w^2 + 4w = 2$$
$$w^2 - 4w = -2$$
$$\underline{\phantom{w^2 - 4w} +4 \qquad +4}$$
$$w^2 - 4w + 4 = 2$$
$$(w - 2)^2 = 2$$
$$w - 2 = \pm \sqrt{2}$$
$$w = 2 \pm \sqrt{2}$$

$$2p = -4; \; p = \frac{-4}{2} = -2; \; p^2 = (-2)^2 = 4$$

23. 
$$(x - 3)(x + 2) = 9x - 1$$
$$x^2 - x - 6 = 9x - 1$$
$$x^2 - 10x - 6 = -1$$
$$x^2 - 10x = 5$$
$$\underline{\phantom{x^2 - 10x} +25 \qquad +25}$$
$$x^2 - 10x + 25 = 30$$
$$(x - 5)^2 = 30$$
$$x - 5 = \pm \sqrt{30}$$
$$x = 5 \pm \sqrt{30}$$

$$2p = -10; \; p = -5; \; p^2 = (-5)^2 = 25$$

25. 
$$(a - 2)(a + 1) = 6$$
$$a^2 - a - 2 = 6$$
$$a^2 - a = 8$$
$$\underline{\phantom{a^2 - a} +\frac{1}{4} \qquad +\frac{1}{4}}$$
$$a^2 - a + \frac{1}{4} = 8 + \frac{1}{4}$$
$$\left(a - \frac{1}{2}\right)^2 = \frac{32}{4} + \frac{1}{4}$$
$$\left(a - \frac{1}{2}\right)^2 = \frac{33}{4}$$
$$a - \frac{1}{2} = \pm \sqrt{\frac{33}{4}} = \pm \frac{\sqrt{33}}{2}$$
$$a = \frac{1}{2} \pm \frac{\sqrt{33}}{2}$$
$$a = \frac{1 \pm \sqrt{33}}{2}$$

$$2p = -1; \; p = -\frac{1}{2}; \; p^2 = \left(-\frac{1}{2}\right)^2 = \frac{1}{4}$$

29. 
$$2x^2 \qquad + 3 = 6x$$
$$2x^2 - 6x + 3 = 0$$
$$2x^2 - 6x \qquad = -3$$
$$x^2 - 3x \qquad = -\frac{3}{2}$$

$$2p = -3; \quad p = -\frac{3}{2}; \quad p^2 = \left(-\frac{3}{2}\right)^2 = \frac{9}{4}$$

$$+\frac{9}{4} \quad +\frac{9}{4}$$
$$\overline{\rule{3cm}{0.4pt}}$$
$$x^2 - 3x + \frac{9}{4} = -\frac{3}{2} + \frac{9}{4}$$
$$\left(x - \frac{3}{2}\right)^2 = -\frac{6}{4} + \frac{9}{4}$$
$$\left(x - \frac{3}{2}\right)^2 = \frac{3}{4}$$
$$x - \frac{3}{2} = \pm\sqrt{\frac{3}{4}} = \pm\frac{\sqrt{3}}{2}$$
$$x = \frac{3}{2} \pm \frac{\sqrt{3}}{2} = \frac{3 \pm \sqrt{3}}{2}$$

33. 
$$4z^2 + 20z + 19 = 0$$
$$4z^2 + 20z \qquad = -19$$
$$z^2 + 5z \qquad = -\frac{19}{4}$$

$$2p = 5; \quad p = \frac{5}{2}; \quad p^2 = \left(\frac{5}{2}\right)^2 = \frac{25}{4}$$

$$+\frac{25}{4} \quad +\frac{25}{4}$$
$$\overline{\rule{3cm}{0.4pt}}$$
$$z^2 + 5z + \frac{25}{4} = \frac{-19}{4} + \frac{25}{4}$$
$$\left(z + \frac{5}{2}\right)^2 = \frac{6}{4}$$
$$z + \frac{5}{2} = \pm\sqrt{\frac{6}{4}} = \pm\frac{\sqrt{6}}{2}$$
$$z = -\frac{5}{2} \pm \frac{\sqrt{6}}{2}$$
$$z = \frac{-5 \pm \sqrt{6}}{2}$$

35. 
$$(x + 3)^2 = 6$$
$$x + 3 = \pm\sqrt{6}$$
$$x = -3 \pm \sqrt{6}$$

37. 
$$(x + 3)^2 = 6x$$
$$x^2 + 6x + 9 = 6x$$
$$x^2 + 9 = 0$$
$$x^2 = -9$$
This has no real solution.

**39.**
$$x^2 + 8x - 9 = 0$$
$$(x + 9)(x - 1) = 0$$
$$x + 9 = 0 \quad \text{or} \quad x - 1 = 0$$
$$x = -9 \quad \text{or} \quad x = 1$$

**41.**
$$3x^2 + 4 = 8x$$
$$3x^2 - 8x + 4 = 0$$
$$(3x - 2)(x - 2) = 0$$
$$3x - 2 = 0 \quad \text{or} \quad x - 2 = 0$$
$$x = \frac{2}{3} \quad \text{or} \quad x = 2$$

**43.** The constant of a perfect square is the square of one-half of the coefficient of the middle term, provided that the leading coefficient is equal to 1.

**44.**
$$\frac{x}{x-1} = \frac{2}{x-2}$$
$$\frac{(x-1)(x-2)}{1} \cdot \frac{x}{x-1} = \frac{(x-1)(x-2)}{1} \cdot \frac{2}{x-2}$$
$$x(x-2) = 2(x-1)$$
$$x^2 - 2x = 2x - 2$$
$$x^2 - 4x = -2$$
$$\frac{\phantom{x^2 - 4x} +4 \qquad +4}{x^2 - 4x + 4 = \phantom{xx} 2}$$
$$(x-2)^2 = 2$$
$$x - 2 = \pm\sqrt{2}$$
$$x = 2 \pm \sqrt{2}$$

CHECK: $x = 2 + \sqrt{2}$:
$$\frac{x}{x-1} = \frac{2}{x-2}$$
$$\frac{2+\sqrt{2}}{2+\sqrt{2}-1} \overset{?}{=} \frac{2}{2+\sqrt{2}-2}$$
$$\frac{2+\sqrt{2}}{1+\sqrt{2}} \overset{?}{=} \frac{2}{\sqrt{2}}$$
$$\frac{2+\sqrt{2}}{1+\sqrt{2}} \cdot \frac{1-\sqrt{2}}{1-\sqrt{2}} \overset{?}{=} \frac{2}{\sqrt{2}} \cdot \frac{\sqrt{2}}{\sqrt{2}}$$
$$\frac{2-2\sqrt{2}+\sqrt{2}-2}{1-2} \overset{?}{=} \frac{2\sqrt{2}}{2}$$
$$\frac{-\sqrt{2}}{-1} \overset{?}{=} \sqrt{2}$$
$$\sqrt{2} \overset{\checkmark}{=} \sqrt{2}$$

CHECK: $x = 2 - \sqrt{2}$:
$$\frac{x}{x-1} = \frac{2}{x-2}$$
$$\frac{2-\sqrt{2}}{2-\sqrt{2}-1} \overset{?}{=} \frac{2}{2-\sqrt{2}-2}$$
$$\frac{2-\sqrt{2}}{1-\sqrt{2}} \overset{?}{=} \frac{2}{-\sqrt{2}}$$
$$\frac{2-\sqrt{2}}{1-\sqrt{2}} \cdot \frac{1+\sqrt{2}}{1+\sqrt{2}} \overset{?}{=} \frac{2}{-\sqrt{2}} \cdot \frac{\sqrt{2}}{\sqrt{2}}$$
$$\frac{2+2\sqrt{2}-\sqrt{2}-2}{1-2} \overset{?}{=} \frac{2\sqrt{2}}{-2}_{-1}$$
$$\frac{\sqrt{2}}{-1} \overset{?}{=} -\sqrt{2}$$
$$-\sqrt{2} \overset{\checkmark}{=} -\sqrt{2}$$

**45.**
$$\sqrt{48} - \sqrt{75} = \sqrt{16 \cdot 3} - \sqrt{25 \cdot 3}$$
$$= \sqrt{16}\sqrt{3} - \sqrt{25}\sqrt{3}$$
$$= 4\sqrt{3} - 5\sqrt{3} = -\sqrt{3}$$

**47.**
$$\sqrt{\frac{2}{7}} = \frac{\sqrt{2}}{\sqrt{7}} = \frac{\sqrt{2}\sqrt{7}}{\sqrt{7}\sqrt{7}} = \frac{\sqrt{14}}{7}$$

49. $\dfrac{3}{2-\sqrt{2}} = \dfrac{3\left(3+\sqrt{2}\right)}{\left(3-\sqrt{2}\right)\left(3+\sqrt{2}\right)}$

$\phantom{\dfrac{3}{2-\sqrt{2}}} = \dfrac{3\left(3+\sqrt{2}\right)}{9-2} = \dfrac{3\left(3+\sqrt{2}\right)}{7}$

51. $\dfrac{6+4\sqrt{3}}{2} = \dfrac{\cancel{6}^{\,3}}{\cancel{2}} + \dfrac{\cancel{4}^{\,2}\sqrt{3}}{\cancel{2}} = 3+2\sqrt{3}$

53. Let $x$ = number of hours Yvonne needs to finish the puzzle.

$$\dfrac{3}{10} + \dfrac{x}{8} = 1$$
$$40\left(\dfrac{3}{10} + \dfrac{x}{8}\right) = 40(1)$$
$$12 + 5x = 40$$
$$5x = 28$$
$$x = \dfrac{28}{5} \text{ or } 5\dfrac{3}{5}$$

It takes Yvonne $5\frac{3}{5}$ hours or 5 hours and 36 minutes to finish the puzzle.

## Exercises 11.3

1. $x^2 + 3x - 5 = 0$
   $a = 1, b = 3, c = -5$

5. $\quad 2u^2 = 8u$
   $\quad 2u^2 - 8u = 0$
   $\quad a = 2, b = -8, c = 0$

9. $x^2 + 3x - 5 = 0$
   $a = 1, b = 3, c = -5$
   $x = \dfrac{-b \pm \sqrt{b^2 - 4ac}}{2a}$
   $x = \dfrac{-(3) \pm \sqrt{(3)^2 - 4(1)(-5)}}{2(1)}$
   $x = \dfrac{-3 \pm \sqrt{9 + 20}}{2}$
   $x = \dfrac{-3 \pm \sqrt{29}}{2}$

13. $u^2 - 2u + 3 = 0$
    $a = 1, b = -2, c = 3$
    $u = \dfrac{-b \pm \sqrt{b^2 - 4ac}}{2a}$
    $u = \dfrac{-(-2) \pm \sqrt{(-2)^2 - 4(1)(3)}}{2(1)}$
    $u = \dfrac{2 \pm \sqrt{4 - 12}}{2}$
    $u = \dfrac{2 \pm \sqrt{-8}}{2}$

    No real solutions, since the answer involves the square root of a negative number.

**17.** $2x^2 - 3x - 1 = 0$

$a = 2, b = -3, c = -1$

$x = \dfrac{-b \pm \sqrt{b^2 - 4ac}}{2a}$

$x = \dfrac{-(-3) \pm \sqrt{(-3)^2 - 4(2)(-1)}}{2(2)}$

$x = \dfrac{3 \pm \sqrt{9 + 8}}{4}$

$x = \dfrac{3 \pm \sqrt{17}}{4}$

**19.** $5x^2 - x = 2$

$5x^2 - x - 2 = 0$

$a = 5, b = -1, c = -2$

$x = \dfrac{-b \pm \sqrt{b^2 - 4ac}}{2a}$

$x = \dfrac{-(-1) \pm \sqrt{(-1)^2 - 4(5)(-2)}}{2(5)}$

$x = \dfrac{1 \pm \sqrt{1 + 40}}{10}$

$x = \dfrac{1 \pm \sqrt{41}}{10}$

**21.** $t^2 - 3t + 4 = 2t^2 + 4t - 3$

$-3t + 4 = t^2 + 4t - 3$

$4 = t^2 + 7t - 3$

$0 = t^2 + 7t - 7$

$a = 1, b = 7, c = -7$

$t = \dfrac{-b \pm \sqrt{b^2 - 4ac}}{2a}$

$t = \dfrac{-(7) \pm \sqrt{(7)^2 - 4(1)(-7)}}{2(1)}$

$t = \dfrac{-7 \pm \sqrt{49 + 28}}{2}$

$t = \dfrac{-7 \pm \sqrt{77}}{2}$

**23.** $(5w + 2)(w - 1) = 3w + 1$

$5w^2 - 3w - 2 = 3w + 1$

$5w^2 - 6w - 2 = 1$

$5w^2 - 6w - 3 = 0$

$a = 5, b = -6, c = -3$

$w = \dfrac{-b \pm \sqrt{b^2 - 4ac}}{2a}$

$w = \dfrac{-(-6) \pm \sqrt{(-6)^2 - 4(5)(-3)}}{2(5)}$

$w = \dfrac{6 \pm \sqrt{36 + 60}}{10} = \dfrac{6 \pm \sqrt{96}}{10}$

$w = \dfrac{6 \pm 4\sqrt{6}}{10} = \dfrac{\cancel{2}\left(3 \pm 2\sqrt{6}\right)}{\underset{5}{\cancel{10}}}$

$w = \dfrac{3 \pm 2\sqrt{6}}{5}$

**27.** $3x^2 - 5x + 7 = 2x(x - 5) + 9x + 5$

$3x^2 - 5x + 7 = 2x^2 - 10x + 9x + 5$

$3x^2 - 5x + 7 = 2x^2 - x + 5$

$x^2 - 5x + 7 = -x + 5$

$x^2 - 4x + 7 = 5$

$x^2 - 4x + 2 = 0$

$a = 1, b = -4, c = 2$

$x = \dfrac{-b \pm \sqrt{b^2 - 4ac}}{2a}$

$x = \dfrac{-(-4) \pm \sqrt{(-4)^2 - 4(1)(2)}}{2(1)}$

$x = \dfrac{4 \pm \sqrt{16 - 8}}{2} = \dfrac{4 \pm \sqrt{8}}{2}$

$x = \dfrac{4 \pm 2\sqrt{2}}{2} = \dfrac{\cancel{2}\left(2 \pm \sqrt{2}\right)}{\cancel{2}}$

$x = 2 \pm \sqrt{2}$

**31.** $x^2(x - 1) = (x - 1)^3$

$x^3 - x^2 = x^3 - 3x^2 + 3x - 1$

$-x^2 = -3x^2 + 3x - 1$

$0 = -2x^2 + 3x - 1$

$0 = 2x^2 - 3x + 1$

$0 = (2x - 1)(x - 1)$

$0 = 2x - 1 \quad \text{or} \quad 0 = x - 1$

$1 = 2x \qquad \text{or} \quad 1 = x$

$\dfrac{1}{2} = x \qquad \text{or} \quad 1 = x$

33. $4x = 9x^2$

$\phantom{33.}0 = 9x^2 - 4x$

$\phantom{33.}0 = x(9x - 4)$

$\phantom{33.}0 = x \quad \text{or} \quad 0 = 9x - 4$

$\phantom{33.}0 = x \quad \text{or} \quad 4 = 9x$

$\phantom{33.}0 = x \quad \text{or} \quad \dfrac{4}{9} = x$

37. $\dfrac{w}{2} = \dfrac{3}{w+2}$

$\dfrac{2(w+2)}{1} \cdot \dfrac{w}{2} = \dfrac{2(w+2)}{1} \cdot \dfrac{3}{w+2}$

$w(w+2) = 6$

$w^2 + 2w = 6$

$w^2 + 2w - 6 = 0$

$a = 1, b = 2, c = -6$

$w = \dfrac{-b \pm \sqrt{b^2 - 4ac}}{2a}$

$w = \dfrac{-(2) \pm \sqrt{(2)^2 - 4(1)(-6)}}{2(1)}$

$w = \dfrac{-2 \pm \sqrt{4 + 24}}{2} = \dfrac{-2 \pm \sqrt{28}}{2}$

$w = \dfrac{-2 \pm 2\sqrt{7}}{2} = \dfrac{2\left(-1 \pm \sqrt{7}\right)}{2} = -1 \pm \sqrt{7}$

41. $x = 1.62, x = -0.62$

43. $t = 8.22, t = -1.22$

45. $w = 1.72, w = -4.06$

47. $x = 3.06, x = -1.17$

49. $R = s^2 - 250s + 600$

$50,000 = s^2 - 250s + 600$

$\phantom{50,00}0 = s^2 - 250s - 49,400$

$\phantom{50,00}0 = (s - 380)(s + 130)$

$0 = s - 380 \quad \text{or} \quad 0 = s + 130$

$380 = s \quad \text{or} \quad -130 = s$

Reject, since it makes no sense to manufacture a negative number of square feet of plastic.

So the firm should maufacture 380 sq. ft. of plastic.

51.
$$h = 120 + 40t - 16t^2$$
$$140 = 120 + 40t - 16t^2$$
$$16t^2 - 40t + 20 = 0$$
$$4t^2 - 10t + 5 = 0$$
$$a = 4, b = -10, c = 5$$
$$t = \frac{-b \pm \sqrt{b^2 - 4ac}}{2a}$$
$$t = \frac{-(-10) \pm \sqrt{(-10)^2 - 4(4)(5)}}{2(4)}$$
$$t = \frac{10 \pm \sqrt{100 - 80}}{8} = \frac{10 \pm \sqrt{20}}{8}$$
$$t = \frac{10 \pm 2\sqrt{5}}{8} = \frac{\cancel{2}\left(5 \pm \sqrt{5}\right)}{\underset{4}{\cancel{8}}} = \frac{5 \pm \sqrt{5}}{4}$$
$$t = \frac{5 + \sqrt{5}}{4} \quad \text{or} \quad t = \frac{5 - \sqrt{5}}{4}$$
$$t = 1.8 \quad \text{or} \quad t = 0.7$$
(rounded to the nearest tenth)

The ball is 140 ft. above the ground after 0.7 seconds (when it is on the way up) and again after 1.8 seconds (when it is on the way down).

53. $2x^2 + 7x + 4 = 0$
$a = 2, b = 7, c = 4$
$$x = \frac{-b \pm \sqrt{b^2 - 4ac}}{2a}$$
$$x = \frac{-(7) \pm \sqrt{(7)^2 - 4(2)(4)}}{2(2)}$$
$$x = \frac{-7 \pm \sqrt{49 - 32}}{4}$$
$$x = \frac{-7 \pm \sqrt{17}}{4}$$

Using the quadratic formula is easier than the method of completing the square.

54. The factoring method, when it works, is usually the easiest method to use. However, it does not always work. Completing the square and the quadratic formula work for any quadratic equation. Generally, completing the square is the most complicated of these two methods.

55. (a) Since $b = -3$ and $c = -1$,
$$x = \frac{-(-3) \pm \sqrt{9 - 4(-1)}}{2} = \frac{3 \pm \sqrt{9 + 4}}{2} = \frac{3 \pm \sqrt{13}}{2}.$$

(b) The minus under the square root should be a plus sign.

(c) The "5" should be divided by 2 as well. That is,
$$x = \frac{5 \pm \sqrt{25 - 12}}{2} = \frac{5 \pm \sqrt{13}}{2}.$$

(d) This is correct up until the last step. Then

$$x = \frac{6 \pm 4\sqrt{3}}{2} = \frac{\cancel{2}\left(3 \pm 2\sqrt{3}\right)}{\cancel{2}} = 3 \pm 2\sqrt{3}, \text{ not } 6 \pm 2\sqrt{3}.$$

57. $3\sqrt{2}\left(\sqrt{2} + \sqrt{7}\right) + \sqrt{7}\left(5\sqrt{2} - \sqrt{7}\right) = 3\sqrt{2}\sqrt{2} + 3\sqrt{2}\sqrt{7} + 5\sqrt{2}\sqrt{7} - \sqrt{7}\sqrt{7}$

$$= 3(2) + 3\sqrt{14} + 5\sqrt{14} - 7$$
$$= 6 + 8\sqrt{14} - 7$$
$$= 8\sqrt{14} - 1$$

59. $\left(2\sqrt{x} - \sqrt{5}\right)\left(3\sqrt{x} - 4\sqrt{5}\right) = (2\sqrt{x})(3\sqrt{x}) - (2\sqrt{x})(4\sqrt{5}) - (\sqrt{5})(3\sqrt{x}) + (\sqrt{5})(4\sqrt{5})$

$$= 6x - 8\sqrt{5x} - 3\sqrt{5x} + 20$$
$$= 6x - 11\sqrt{5x} + 20$$

61. Let $p =$ price of a pastrami sandwich.
Let $t =$ price of a tuna sandwich.

$$\begin{cases} 5p + 6t = 42.00 \\ 4p + 9t = 47.25 \end{cases}$$

$\xrightarrow{\text{multiply by 3}}$   $15p + 18t = 126.00$      $5p + 6t = 42.00$

$\xrightarrow{\text{multiply by } -2}$   $\begin{array}{r} -8p - 18t = \phantom{0}94.50 \text{ Add.} \\ \hline 7p \phantom{-18t} = \phantom{0}31.50 \\ p = 4.50 \end{array}$      $\begin{array}{r} 5(4.50) + 6t = 42.00 \\ 22.50 + 6t = 42.00 \\ 6t = 19.50 \\ t = 3.25 \end{array}$

So a pastrami sandwich costs \$4.50 and a tuna sandwich costs \$3.25.

## Exercises 11.4

1.    $x^2 + 6x + 5 = 0$          $x^2 + 6x \phantom{+5} = -5$

    $(x + 1)(x + 5) = 0$            $\underline{\phantom{x^2 + 6x} +9 \phantom{=} +9}$

    $x + 1 = 0$   or   $x + 5 = 0$     $x^2 + 6x + 9 = 4$

       $x = -1$   or      $x = -5$      $(x + 3)^2 = 4$

                               $x + 3 = \pm\sqrt{4} = \pm 2$

               $x + 3 = 2$   or   $x + 3 = -2$

                   $x = -1$   or      $x = -5$

5.        $2r^2 + 1 = 3r$       $a = 2, b = -3, c = 1$

     $2r^2 - 3r + 1 = 0$        $r = \dfrac{-b \pm \sqrt{b^2 - 4ac}}{2a}$

   $(2r - 1)(r - 1) = 0$

    $2r - 1 = 0$   or   $r - 1 = 0$    $r = \dfrac{-(-3) \pm \sqrt{(-3)^2 - 4(2)(1)}}{2(2)}$

      $2r = 1$   or       $r = 1$

                            $r = \dfrac{3 \pm \sqrt{9 - 8}}{4} = \dfrac{3 \pm \sqrt{1}}{4} = \dfrac{3 \pm 1}{4}$

        $r = \dfrac{1}{2}$   or       $r = 1$

                      $r = \dfrac{3 + 1}{4} = \dfrac{4}{4} = 1$   or   $t = \dfrac{3 - 1}{4} = \dfrac{2}{4} = \dfrac{1}{2}$

7.
$$w^2 = 4w + 5$$
$$w^2 - 4w = 5$$
$$w^2 - 4w - 5 = 0$$
$$(w - 5)(w + 1) = 0$$
$$w - 5 = 0 \quad \text{or} \quad w + 1 = 0$$
$$w = 5 \quad \text{or} \quad w = -1$$

$$w^2 \qquad = 4w + 5$$
$$w^2 - 4w \quad = \quad 5$$
$$\underline{\qquad +4 \qquad\qquad +4}$$
$$w^2 - 4w + 4 = \qquad 9$$
$$(w - 2)^2 = 9$$
$$w - 2 = \pm\sqrt{9} = \pm 3$$
$$w - 2 = 3 \quad \text{or} \quad w - 2 = -3$$
$$w = 5 \quad \text{or} \quad w = -1$$

13.
$$4x^2 \qquad = 16x - 28$$
$$4x^2 - 16x \quad = \quad -28$$
$$x^2 - 4x \quad = \quad -7$$
$$\underline{\qquad +4 \qquad\qquad +4}$$
$$x^2 - 4x + 4 = \qquad -3$$
$$(x - 2)^2 = -3$$
$$x - 2 = \pm\sqrt{-3}$$

Thus, there are no real solutions, since square roots of negative numbers are not real.

$$4x^2 = 16x - 28$$
$$4x^2 - 16x = -28$$
$$4x^2 - 16x + 28 = 0$$
$$x^2 - 4x + 7 = 0$$
$$a = 1, b = -4, c = 7$$
$$x = \frac{-b \pm \sqrt{b^2 - 4ac}}{2a}$$
$$x = \frac{-(-4) \pm \sqrt{(-4)^2 - 4(1)(7)}}{2(1)}$$
$$x = \frac{4 \pm \sqrt{16 - 28}}{2} = \frac{4 \pm \sqrt{-12}}{2}, \text{ which leads}$$

to the same conclusion for the same reason.

15.
$$(x - 1)^2 = 5$$
$$x - 1 = \pm\sqrt{5}$$
$$x = 1 \pm \sqrt{5}$$

$$(x - 1)^2 = 5$$
$$x^2 - 2x + 1 = 5$$
$$x^2 - 2x - 4 = 0$$
$$a = 1, b = -2, c = -4$$
$$x = \frac{-b \pm \sqrt{b^2 - 4ac}}{2a}$$
$$x = \frac{-(-2) \pm \sqrt{(-2)^2 - 4(1)(-4)}}{2(1)}$$
$$x = \frac{2 \pm \sqrt{4 + 16}}{2} = \frac{2 \pm \sqrt{20}}{2}$$
$$x = \frac{2 \pm 2\sqrt{5}}{2} = \frac{\cancel{2}\left(1 \pm \sqrt{5}\right)}{\cancel{2}}$$
$$x = 1 \pm \sqrt{5}$$

17.

$$(x-1)^2 = 5x$$
$$x^2 - 2x + 1 = 5x$$
$$x^2 - 7x + 1 = 0$$
$$a = 1, b = -7, c = 1$$
$$x = \frac{-b \pm \sqrt{b^2 - 4ac}}{2a}$$
$$x = \frac{-(-7) \pm \sqrt{(-7)^2 - 4(1)(1)}}{2(1)}$$
$$x = \frac{7 \pm \sqrt{49 - 4}}{2} = \frac{7 \pm \sqrt{45}}{2}$$
$$x = \frac{7 \pm 3\sqrt{5}}{2}$$

$$(x-1)^2 = 5x$$
$$x^2 - 2x + 1 = 5x$$
$$x^2 - 7x + 1 = 0$$
$$x^2 - 7x = -1$$
$$\phantom{x^2 - 7x} +\frac{49}{4} \quad +\frac{49}{4}$$
$$\overline{x^2 - 7x + \frac{49}{4} = -1 + \frac{49}{4}}$$
$$\left(x - \frac{7}{2}\right)^2 = \frac{-4}{4} + \frac{49}{4}$$
$$\left(x - \frac{7}{2}\right)^2 = \frac{45}{4}$$
$$x - \frac{7}{2} = \pm\sqrt{\frac{45}{4}} = \pm\frac{\sqrt{45}}{2}$$
$$x - \frac{7}{2} = \pm\frac{3\sqrt{5}}{2}$$
$$x = \frac{7}{2} \pm \frac{3\sqrt{5}}{2} = \frac{7 \pm 3\sqrt{5}}{2}$$

21.

$$y^2 - 4y + 10 = 5(y + 2)$$
$$y^2 - 4y + 10 = 5y + 10$$
$$y^2 - 9x + 10 = 10$$
$$y^2 - 9y = 0$$
$$y(y - 9) = 0$$
$$y = 0 \quad \text{or} \quad y - 9 = 0$$
$$y = 0 \quad \text{or} \qquad y = 9$$

$$y^2 - 9y = 0$$
$$a = 1, b = -9, c = 0$$
$$y = \frac{-b \pm \sqrt{b^2 - 4ac}}{2a}$$
$$y = \frac{-(-9) \pm \sqrt{(-9)^2 - 4(1)(0)}}{2(1)}$$
$$y = \frac{9 \pm \sqrt{81 - 0}}{2} = \frac{9 \pm \sqrt{81}}{2}$$
$$y = \frac{9 \pm 9}{2}$$
$$y = \frac{9 + 9}{2} = \frac{18}{2} = 9 \text{ or } y = \frac{9 - 9}{2} = \frac{0}{2} = 0$$

23.

$$(t + 4)(t - 8) = 13$$
$$t^2 - 4t - 32 = 13$$
$$t^2 - 4t - 45 = 0$$
$$(t - 9)(t + 5) = 0$$
$$t - 9 = 0 \quad \text{or} \quad t + 5 = 0$$
$$t = 9 \quad \text{or} \qquad t = -5$$

$$(t + 4)(t - 8) = 13$$
$$t^2 - 4t - 32 = 13$$
$$t^2 - 4t = 45$$
$$\phantom{t^2 - 4t} +4 \quad +4$$
$$\overline{t^2 - 4t + 4 = 49}$$
$$(t - 2)^2 = 49$$
$$t - 2 = \pm\sqrt{49} = \pm 7$$
$$t - 2 = 7 \quad \text{or} \quad t - 2 = -7$$
$$t = 9 \quad \text{or} \qquad t = -5$$

27.

$$z^2 - 3z \quad = 3z - 9$$
$$z^2 - 6z \quad = -9$$
$$z^2 - 6z + 9 = 0$$
$$(z-3)^2 = 0$$
$$z - 3 = 0$$
$$z = 3$$

$$z^2 - 3z \quad = 3z - 9$$
$$z^2 - 6z \quad = -9$$
$$z^2 - 6z + 9 = 0$$
$$a = 1, b = -6, c = 9$$
$$z = \frac{-b \pm \sqrt{b^2 - 4ac}}{2a}$$
$$z = \frac{-(-6) \pm \sqrt{(-6)^2 - 4(1)(9)}}{2(1)}$$
$$z = \frac{6 \pm \sqrt{36 - 36}}{2} = \frac{6 \pm \sqrt{0}}{2}$$
$$z = \frac{6 \pm 0}{2} = \frac{6}{2} = 3$$

33.

$$x^2 + 1 = \frac{5}{2}x$$
$$2(x^2 + 1) = 2\left(\frac{5}{2}x\right)$$
$$2x^2 + 2 = \frac{\cancel{2}}{1} \cdot \frac{5}{\cancel{2}}x$$
$$2x^2 + 2 = 5x$$
$$2x^2 - 5x + 2 = 0$$
$$(2x - 1)(x - 2) = 0$$
$$2x - 1 = 0 \quad \text{or} \quad x - 2 = 0$$
$$2x = 1 \quad \text{or} \qquad x = 2$$
$$x = \frac{1}{2} \quad \text{or} \qquad x = 2$$

$$2x^2 - 5x + 2 = 0$$
$$a = 2, b = -5, c = 2$$
$$x = \frac{-b \pm \sqrt{b^2 - 4ac}}{2a}$$
$$x = \frac{-(-5) \pm \sqrt{(-5)^2 - 4(2)(2)}}{2(2)}$$
$$x = \frac{5 \pm \sqrt{25 - 16}}{4}$$
$$x = \frac{5 \pm \sqrt{9}}{4} = \frac{5 \pm 3}{4}$$
$$x = \frac{5 + 3}{4} = \frac{8}{4} = 2 \text{ or } x = \frac{5 - 3}{4} = \frac{2}{4} = \frac{1}{2}$$

35.

$$\frac{x}{x + 1} = \frac{4}{x + 4}$$
$$\frac{(x+1)(x+4)}{1} \cdot \frac{x}{x+1} = \frac{(x+1)(x+4)}{1} \cdot \frac{4}{x+4}$$
$$x(x + 4) = 4(x + 1)$$
$$x^2 + 4x = 4x + 4$$
$$x^2 = 4$$
$$x = \pm \sqrt{4}$$
$$x = \pm 2$$

$$x^2 = 4$$
$$x^2 - 4 = 0$$
$$(x - 2)(x + 2) = 0$$
$$x - 2 = 0 \quad \text{or} \quad x + 2 = 0$$
$$x = 2 \quad \text{or} \qquad x = -2$$

39.
$$\frac{3x}{x+1} + \frac{2}{x-1} = 4$$

$$\frac{(x+1)(x-1)}{1} \cdot \frac{3x}{x+1} + \frac{(x+1)(x-1)}{1} \cdot \frac{2}{x-1} = (x+1)(x-1)4$$

$$3x(x-1) + 2(x+1) = 4(x+1)(x-1)$$

$$3x^2 - 3x + 2x + 2 = 4(x^2 - 1)$$

$$3x^2 - x + 2 = 4x^2 - 4$$

$$-x + 2 = x^2 - 4$$

$$2 = x^2 + x - 4$$

$$0 = x^2 + x - 6$$

$$0 = (x+3)(x-2)$$

$$0 = x+3 \quad \text{or} \quad 0 = x-2$$

$$-3 = x \quad \quad \text{or} \quad 2 = x$$

$x^2 + x - 6 = 0$

$a = 1, b = 1, c = -6$

$$x = \frac{-b \pm \sqrt{b^2 - 4ac}}{2a}$$

$$x = \frac{-(1) \pm \sqrt{(1)^2 - 4(1)(-6)}}{2(1)}$$

$$x = \frac{-1 \pm \sqrt{1 + 24}}{2}$$

$$x = \frac{-1 \pm \sqrt{25}}{2} = \frac{-1 \pm 5}{2}$$

$$x = \frac{-1 + 5}{2} = \frac{4}{2} = 2 \text{ or } x = \frac{-1 - 5}{2} = \frac{-6}{2} = -3$$

41. $2x^2 + 3x = 20$

Method 1—factoring:

$2x^2 + 3x - 20 = 0$

$(2x - 5)(x + 4) = 0$

$2x - 5 = 0 \quad \text{or} \quad x + 4 = 0$

$\quad 2x = 5 \quad \text{or} \quad \quad x = -4$

$$x = \frac{5}{2} \quad \text{or} \quad \quad x = -4$$

Method 2—completing the square:

$$2x^2 + 3x \qquad = 20$$

$$x^2 + \frac{3}{2}x \qquad = 10$$

$$\underline{\qquad +\frac{9}{16} \qquad\qquad +\frac{9}{16}\qquad}$$

$$x^2 + \frac{3}{2}x + \frac{9}{16} = 10 + \frac{9}{16}$$

$$\left(x + \frac{3}{4}\right)^2 = \frac{160}{16} + \frac{9}{16}$$

$$\left(x + \frac{3}{4}\right)^2 = \frac{169}{16}$$

$$x + \frac{3}{4} = \pm\sqrt{\frac{169}{16}} = \pm\frac{\sqrt{169}}{\sqrt{16}}$$

$$x + \frac{3}{4} = \pm\frac{13}{4}$$

$$x = -\frac{3}{4} \pm \frac{13}{4}$$

$$x = \frac{-3 \pm 13}{4}$$

$$x = \frac{-3 + 13}{4} = \frac{10}{4} = \frac{5}{2} \quad \text{or} \quad x = \frac{-3 - 13}{4} = \frac{-16}{4} = -4$$

Method 3—quadratic formula

$$2x^2 + 3x - 20 = 0$$

$$a = 2, b = 3, c = -20$$

$$x = \frac{-b \pm \sqrt{b^2 - 4ac}}{2a}$$

$$x = \frac{-(3) \pm \sqrt{(3)^2 - 4(2)(-20)}}{2(2)}$$

$$x = \frac{-3 \pm \sqrt{9 + 160}}{4}$$

$$x = \frac{-3 \pm \sqrt{169}}{4} = \frac{-3 \pm 13}{4}$$

$$x = \frac{-3 + 13}{4} = \frac{10}{4} = \frac{5}{2} \quad \text{or} \quad x = \frac{-3 - 13}{4} = \frac{-16}{4} = -4$$

The easiest of these methods is the first, while the second one appears to be the most difficult.

42.  $3x^2 - 5x - 1 = 0$

$a = 3, b = -5, c = -1$

$x = \dfrac{-b \pm \sqrt{b^2 - 4ac}}{2a}$

$x = \dfrac{-(-5) \pm \sqrt{(-5)^2 - 4(3)(-1)}}{2(3)}$

$x = \dfrac{5 \pm \sqrt{25 + 12}}{6} = \dfrac{5 \pm \sqrt{37}}{6}$

CHECK  $x = \dfrac{5 + \sqrt{37}}{6}$:

$$3x^2 - 5x - 1 = 0$$

$$3\left(\frac{5 + \sqrt{37}}{6}\right)^2 - 5\left(\frac{5 + \sqrt{37}}{6}\right) - 1 \stackrel{?}{=} 0$$

$$3\left(\frac{5 + \sqrt{37}}{6}\right)\left(\frac{5 + \sqrt{37}}{6}\right) - 5\left(\frac{5 + \sqrt{37}}{6}\right) - 1 \stackrel{?}{=} 0$$

$$\frac{\cancel{3}}{1}\left(\frac{25 + 10\sqrt{37} + 37}{\underset{12}{\cancel{36}}}\right) - \frac{5}{1}\left(\frac{5 + \sqrt{37}}{6}\right) - 1 \stackrel{?}{=} 0$$

$$\frac{62 + 10\sqrt{37}}{12} - \frac{25 + 5\sqrt{37}}{6} - 1 \stackrel{?}{=} 0$$

$$\frac{\cancel{2}(31 + 5\sqrt{37})}{\underset{6}{\cancel{12}}} - \frac{25 + 5\sqrt{37}}{6} - 1 \stackrel{?}{=} 0$$

$$\frac{31 + 5\sqrt{37} - (25 + 5\sqrt{37})}{6} - 1 \stackrel{?}{=} 0$$

$$\frac{31 + 5\sqrt{37} - 25 - 5\sqrt{37}}{6} - 1 \stackrel{?}{=} 0$$

$$\frac{6}{6} - 1 \stackrel{?}{=} 0$$

$$1 - 1 \stackrel{?}{=} 0$$

$$0 \stackrel{\checkmark}{=} 0$$

$$3x^2 - 5x - 1 = 0$$

$$\underline{\phantom{3x^2 - 5x} +1 \quad +1}$$

$$3x^2 - 5x = 1$$

$$x^2 - \frac{5}{3}x = \frac{1}{3}$$

$$\underline{\phantom{x^2 - \frac{5}{3}x} +\frac{25}{36} \qquad +\frac{25}{36}}$$

$$x^2 - \frac{5}{3}x + \frac{25}{36} = \frac{1}{3} + \frac{25}{36}$$

$$\left(x - \frac{5}{6}\right)^2 = \frac{12}{36} + \frac{25}{36}$$

$$\left(x - \frac{5}{6}\right)^2 = \frac{37}{36}$$

$$x - \frac{5}{6} = \pm\sqrt{\frac{37}{36}} = \pm\frac{\sqrt{37}}{\sqrt{36}}$$

$$x - \frac{5}{6} = \pm\frac{\sqrt{37}}{6}$$

$$x = \frac{5}{6} \pm \frac{\sqrt{37}}{6} = \frac{5 \pm \sqrt{37}}{6}$$

The second method of checking was by far the easier in this example. The first check was more complicated than the actual solution. If the first check happened to fail, we really would not know whether an error was made in the solution to the problem or in the check of that solution. Further, the first check only verified that $x = \dfrac{5 + \sqrt{37}}{6}$ is a valid solution. We would still have to examine $x = \dfrac{5 - \sqrt{37}}{6}$ and verify that it satisfied the equation as well.

43. (a) $m = \dfrac{y_2 - y_1}{x_2 - x_1} = \dfrac{-4 - 3}{3 - (-1)} = \dfrac{-7}{4}$

   (b) $m = 0$, since the points have the same $y$-coordinate.

   (c) $m$ is undefined, since the points have the same $x$-coordinate.

45. $m = \dfrac{y_2 - y_1}{x_2 - x_1} = \dfrac{-1 - (-5)}{1 - (-2)} = \dfrac{-1 + 5}{1 + 2} = \dfrac{4}{3}$

   $(x_1, y_1) = (-2, -5)$

   $y - y_1 = m(x - x_1)$

   $y - (-5) = \dfrac{4}{5}[x - (-2)]$

   $y + 5 = \dfrac{4}{3}(x + 2)$ or $y = \dfrac{4}{3}x - \dfrac{7}{3}$

## Exercises 11.5

3. Let $x =$ one of the numbers.

Then $20 - x =$ the other number.
$$x(20 - x) = 96$$
$$20x - x^2 = 96$$
$$x^2 - 20x + 96 = 0$$
$$(x - 8)(x - 12) = 0$$
$$x - 8 = 0 \quad \text{or} \quad x - 12 = 0$$
$$x = 8 \quad \text{or} \qquad x = 12$$

If $x = 8$, then $20 - x = 20 - 8 = 12$. If $x = 12$, then $20 - x = 20 - 12 = 8$. In both cases, we conclude that the numbers are 8 and 12.

CHECK:

$8 + 12 \overset{\checkmark}{=} 20$

$8(12) \overset{\checkmark}{=} 96$

5. Let $W =$ width of the rectangle.

Then $2W + 3 =$ length of the rectangle.
$$W(2W + 3) = 90$$
$$2W^2 + 3W = 90$$
$$2W^2 + 3W - 90 = 0$$
$$(2W + 15)(W - 6) = 0$$
$$2W + 15 = 0 \qquad \text{or} \quad W - 6 = 0$$
$$W = -\frac{15}{2} \quad \text{or} \qquad W = 6$$

Reject $W = -\dfrac{15}{2}$, since width cannot be negative.

So $W = 6$. Then $2W + 3 = 2(6) + 3 = 15$. Thus, the rectangle has a width of 6 meters and a length of 15 meters.

CHECK: 15 is 3 more than twice 6, and $15(6) = 90$.

9. Let $d =$ length of the diagonal. From the Pythagorean Theorem, we get
$$8^2 + 8^2 = d^2$$
$$64 + 64 = d^2$$
$$128 = d^2$$
$$\pm \sqrt{128} = d$$
$$\pm 8\sqrt{2} = d$$

Reject $d = -8\sqrt{2}$, since we cannot have a negative length. So $d = 8\sqrt{2}$. Thus, the length of the diagonal is $8\sqrt{2}$ inches.

CHECK:

$8^2 + 8^2 \overset{?}{=} \left(8\sqrt{2}\right)^2$

$64 + 64 \overset{?}{=} 64 \cdot 2$

$128 \overset{\checkmark}{=} 128$

13. Let $x$ = length of the shortest side.
Then $x + 1$ = length of the middle side and
$x + 2$ = length of the longest side.
Using the Pythagorean Theorem, it follows that
$$x^2 + (x+1)^2 = (x+2)^2$$
$$x^2 + x^2 + 2x + 1 = x^2 + 4x + 4$$
$$2x^2 + 2x + 1 = x^2 + 4x + 4$$
$$x^2 - 2x - 3 = 0$$
$$(x-3)(x+1) = 0$$
$$x + 3 = 0 \quad \text{or} \quad x + 1 = 0$$
$$x = 3 \quad \text{or} \quad x = -1$$
Reject $x = -1$, since the side of a triangle cannot have negative length. So $x = 3$. Then
$x + 1 = 3 + 1 = 4$ and $x + 2 = 3 + 2 = 5$. Thus, the sides of the triangle have lengths of 3, 4, and 5.
CHECK: 3, 4, and 5 are three consecutive integers and $3^2 + 4^2 = 9 + 16 = 25 = 5^2$.

15. Let $n$ = numerator of the original fraction.
Then $n + 1$ = denominator of the original fraction.
$$\frac{n+3}{n+1} = \frac{n}{n+1} + 1$$
$$(n+1)\left(\frac{n+3}{n+1}\right) = (n+1)\left(\frac{n}{n+1} + 1\right)$$
$$\frac{n+1}{1} \cdot \frac{n+3}{n+1} = \frac{n+1}{1} \cdot \frac{n}{n+1} + (n+1) \cdot 1$$
$$n + 3 = n + n + 1$$
$$n + 3 = 2n + 1$$
$$3 = n + 1$$
$$2 = n$$
Then $n + 1 = 2 + 1 = 3$, Thus, the original fraction is $\frac{2}{3}$.
CHECK: $\frac{2+3}{3} = \frac{5}{3}$, which is one more than $\frac{2}{3}$.

17. Let $s$ = number of seats in each row.
Then $s - 8$ = number of rows of seats.
$$s(s-8) = 768$$
$$s^2 - 8s = 768$$
$$s^2 - 8s - 768 = 0$$
$$(s-32)(s+24) = 0$$
$$s - 32 = 0 \quad \text{or} \quad s + 24 = 0$$
$$s = 32 \quad \text{or} \quad s = -24$$
Reject $s = -24$, since the number of seats in a row cannot be negative. So $x = 32$. Then
$s - 8 = 32 - 8 = 24$. Thus, there are 24 rows of seats in the concert hall, and each row has 32 seats in it.
CHECK: 24 is eight less than 32, and $24 \cdot 32 = 768$.

19. Let $x = $ speed of the motorist for the first part of the trip.
    Then $x - 20 = $ speed of the motorist for the second part of the trip.

$$\frac{120}{x} + \frac{30}{x - 20} = 2$$

$$x(x - 20)\left(\frac{120}{x} + \frac{30}{x - 20}\right) = x(x - 20) \cdot 2$$

$$\frac{\cancel{x}(x - 20)}{1} \cdot \frac{120}{\cancel{x}} + \frac{x\cancel{(x - 20)}}{1} \cdot \frac{30}{\cancel{x - 20}} = 2x(x - 20)$$

$$120(x - 20) + 30x = 2x(x - 20)$$

$$120x - 2400 + 30x = 2x^2 - 40x$$

$$150x - 2400 = 2x^2 - 40x$$

$$0 = 2x^2 - 190x + 2400$$

$$0 = x^2 - 95x + 1200$$

$$0 = (x - 15)(x - 80)$$

$$x - 15 = 0 \quad \text{or} \quad x - 80 = 0$$

$$x = 15 \quad \text{or} \quad x = 80$$

We reject $x = 15$, since this would mean that $x - 20 = 15 - 20 = -5$, and a negative speed is impossible. So $x = 80$. Thus, the motorist's speed for the first part of the trip is 80 kph.

CHECK: At 80 kph, the motorist covers 120 kilometers in $\frac{120}{80} = \frac{3}{2}$ hours. At $(80 - 20) = 60$ kph, the motorist covers the remaining 30 kilometers in $\frac{30}{60} = \frac{1}{2}$ hour. Then $\frac{3}{2} + \frac{1}{2} = \frac{4}{2} = 2$, as required.

23. Let $n = $ distance from $P$ to $R$ (in meters).
    By the Pythagorean Theorem,
    $x^2 = (8.6)^2 + (4.9)^2$
    $\quad = 73.96 + 24.01$
    $\quad = 97.97$
    So $x = 9.9$
    The distance between $P$ and $R$ is 9.9 meters.

27. (a) $\quad h = 1000 - 16t^2$
    $\quad\quad 700 = 1000 - 16t^2$
    $\quad -300 = -16t^2$
    $\quad\quad \dfrac{75}{4} = t^2$
    $\quad 18.75 = t^2$
    $\quad\quad 4.3 = t$

    The object takes 4.3 seconds to reach a height of 700 feet.

(b) $\quad h = 1000 - 16t^2$
    $\quad\quad 0 = 1000 - 16t^2$
    $\quad 16t^2 = 1000$
    $\quad\quad t^2 = \dfrac{125}{2}$
    $\quad\quad t^2 = 62.5$
    $\quad\quad t = 7.9$

    The object hits the ground 7.9 seconds after it is dropped.

31. Let $d$ = distance from home plate to second base.
   By the Pythagorean Theorem.
   $$90^2 + 90^2 = d^2$$
   $$8100 + 8100 = d^2$$
   $$16200 = d^2$$
   $$\sqrt{16200} = d$$
   $$127 = d$$
   The distance from home plate to second base is 127 ft. (rounded to the nearest foot).

33. Let $w$ = width of the path.
   $$(15 + 2w)(10 + w) - 150 = 100$$
   $$150 + 35w + 2w^2 - 150 = 100$$
   $$2w^2 + 35w = 100$$
   $$2w^2 + 35w - 100 = 0$$
   $$(2w - 5)(w + 20) = 0$$
   $$2w - 5 = 0 \quad \text{or} \quad w + 20 = 0$$
   $$2w = 5 \quad \text{or} \quad w = -20$$
   $$w = \frac{5}{2} \quad \text{or} \quad w = -20$$
   Reject $w = -20$, since the width of the path cannot be negative. So the path should be $\frac{5}{2}$ or $2\frac{1}{2}$ ft.

35. Let $x$ = number of days that the father needs to paint the house alone.
   Then $x + 9$ = number of days that the son needs to paint the house alone.
   $$\frac{6}{x} + \frac{6}{x + 9} = 1$$
   $$x(x + 9)\left(\frac{6}{x} + \frac{6}{x + 9}\right) = x(x + 9)(1)$$
   $$\not{x}(x + 9)\left(\frac{6}{\not{x}}\right) + x(x + 9)\left(\frac{6}{x + 9}\right) = x(x + 9)$$
   $$6(x + 9) + 6x = x(x + 9)$$
   $$6x + 54 + 6x = x^2 + 9x$$
   $$0 = x^2 - 3x - 54$$
   $$0 = (x - 9)(x + 6)$$
   $$0 = x - 9 \quad \text{or} \quad 0 = x + 6$$
   $$9 = x \quad \text{or} \quad -6 = x$$
   Reject $x = -6$, since the number of days needed cannot be negative. Thus, the father needs 9 days to paint the alone, while the son needs $x + 9 = 9 + 9 = 18$ days to paint the house alone.

37. (a)
$$\frac{1}{l} = \frac{l}{1+l}$$

$$\frac{l(1+l)}{1} \cdot \frac{1}{l} = \frac{l(1+l)}{1} \cdot \frac{l}{1+l}$$

$$1 + l = l^2$$

$$0 = l^2 - l - 1$$

$$a = 1, b = -1, c = -1$$

$$l = \frac{-b \pm \sqrt{b^2 - 4ac}}{2a}$$

$$l = \frac{-(-1) \pm \sqrt{(-1)^2 - 4(1)(-1)}}{2(1)}$$

$$l = \frac{1 \pm \sqrt{1+4}}{2} = \frac{1 \pm \sqrt{5}}{2}$$

So $l = \dfrac{1+\sqrt{5}}{2}$ or $l = \dfrac{1-\sqrt{5}}{2}$.

Since $\sqrt{5}$ is greater than 1, $\dfrac{1-\sqrt{5}}{2}$ is negative, and thus must be rejected. So $l = \dfrac{1+\sqrt{5}}{2}$ inches.

(b)
$$\frac{w}{l} = \frac{l}{w+l}$$

$$\frac{l(w+l)}{1} \cdot \frac{w}{l} = \frac{l(w+l)}{1} \cdot \frac{l}{w+l}$$

$$w(w+l) = l^2$$

$$w^2 + wl = l^2$$

$$0 = l^2 - wl - w^2$$

$$a = 1, b = -w, c = -w^2$$

$$l = \frac{-b \pm \sqrt{b^2 - 4ac}}{2a}$$

$$l = \frac{-(-w) \pm \sqrt{(-w)^2 - 4(1)(-w^2)}}{2(1)}$$

$$l = \frac{w \pm \sqrt{w^2 + 4w^2}}{2} = \frac{w \pm \sqrt{5w^2}}{2}$$

So $l = \dfrac{w \pm \sqrt{5}w}{2} = \dfrac{\left(1 \pm \sqrt{5}\right)w}{2}$.

As in part (a), we reject $\dfrac{\left(1 - \sqrt{5}\right)w}{2}$ because it is negative. Thus, $l = \dfrac{\left(1+\sqrt{5}\right)w}{2}$. (If $w = 1$, we get the golden ratio of part (a).)

38.
$$\frac{1}{x-1} = \frac{x}{1}$$

$$\frac{x-1}{1} \cdot \frac{1}{x-1} = \frac{x-1}{1} \cdot \frac{x}{1}$$

$$1 = x(x-1)$$
$$1 = x^2 - x$$
$$0 = x^2 - x - 1$$

Proceed with the quadratic formula to find that $x = \dfrac{1 + \sqrt{5}}{2}$. (This quadratic equation is the same as the one encountered in part (a) of exercise (37), except for the letter used to represent the variable.)

39. Let $x =$ length of the shortest side.
Then $x + 2 =$ length of the middle side and
$x + 4 =$ length of the longest side.
Use the Pythagorean Theorem to find that $x^2 + (x+2)^2 = (x+4)^2$. Then
$$x^2 + x^2 + 4x + 4 = x^2 + 8x + 16$$
$$2x^2 + 4x + 4 = x^2 + 8x + 16$$
$$x^2 - 4x - 12 = 0$$
$$(x-6)(x+2) = 0$$
$$x - 6 = 0 \quad \text{or} \quad x + 2 = 0$$
$$x = 6 \quad \text{or} \quad x = -2$$

Reject $x = 6$, since it is not odd. Reject $x = -2$, since we cannot have a negative length. Since both possible solutions to the problem have been rejected, we conclude that it is impossible to find three consecutive odd integers that are the sides of a right triangle.

## Exercises 11.6

3. $y = x^2 + 2$

There are no $x$-intercepts.

The $y$-intercept is 2.

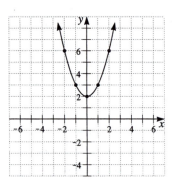

7. $y = -x^2 + 1$

$0 = -x^2 + 1$
$x^2 = 1$
$x = \pm 1$
The $x$-intercepts are 1 and $-1$.

The $y$-intercept is 1.

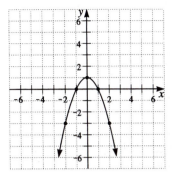

11. $y = 6x - 2x^2$

$0 = 6x - 2x^2$
$0 = 2x(3 - x)$
$0 = 2x$ or $0 = 3 - x$

$0 = x$ or $3 = x$
The $x$-intercepts are 0 and 3.

The $y$-intercept is 0.

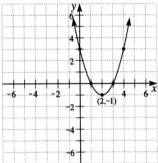

17. $y = x^2 - 4x + 3$

$0 = x^2 - 4x + 3$
$0 = (x - 1)(x - 3)$
$0 = x - 1$ or $0 = x - 3$
$1 = x$ or $3 = x$
The $x$-intercepts are 1 and 3.

The $y$-intercept is 3.

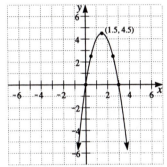

23. $y = x^2 + 4x + 4$

$0 = x^2 + 4x + 4$
$0 = (x + 2)(x + 2)$
$0 = x + 2$ or $0 = x + 2$
$-2 = x$ or $-2 = x$
The $x$-intercept is $-2$.

The $y$-intercept is 4.

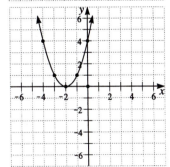

29. $y = -x^2 + x + 6$

$0 = -x^2 + x + 6$
$0 = x^2 - x - 6$
$0 = (x - 3)(x + 2)$
$0 = x - 3$    or    $0 = x + 2$
$3 = x$     or    $-2 = x$
The $x$-intercepts are $3$ and $-2$.

The $y$-intercept is $6$.

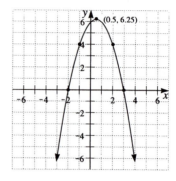

33. $y = -x^2 - 8x - 5$

$0 = -x^2 - 8x - 5$
$0 = x^2 + 8x + 5$
$a = 1, b = 8, c = 5$
$$x = \frac{-b \pm \sqrt{b^2 - 4ac}}{2a}$$
$$= \frac{-8 \pm \sqrt{64 - 20}}{2} = \frac{-8 \pm \sqrt{44}}{2}$$
$$= \frac{-8 \pm 2\sqrt{11}}{2} = -4 \pm \sqrt{11}$$
The $x$-intercepts are $-4 + \sqrt{11} \approx -0.7$ and $-4 - \sqrt{11} \approx -7.3$.

The $y$-intercept is $-5$.

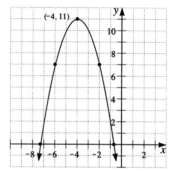

39. $y = 4x^2 - 8x - 5$

$0 = 4x^2 - 8x - 5$
$0 = (2x + 1)(2x - 5)$
   $0 = 2x + 1$    or    $0 = 2x - 5$
$-1 = 2x$     or    $5 = 2x$
$-\dfrac{1}{2} = x$     or    $\dfrac{5}{2} = x$
The $x$-intercepts are $-\dfrac{1}{2}$ and $\dfrac{5}{2}$.

The $y$-intercept is $-5$.

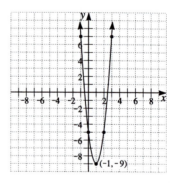

41. (a)

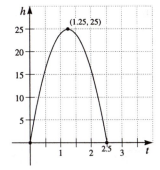

(b) At its highest point, the object is 25 ft above the ground.

(c) The object reaches its highest point 1.25 seconds after it is thrown.

43. (a) $(3, 6)$

(b) $x = 3$

(c) $x \leq 3$

(d) $x \geq 3$

45. (a) $(-4, 0)$

(b) $x = -4$

(c) $x \geq -4$

(d) $x \leq -4$

47.

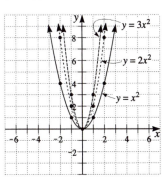

As the coefficient of $x^2$ increases from 1 to 2 to 3 the corresponding parabolas become narrower.

48.

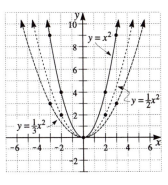

As the coefficient of $x^2$ decreases from 1 to $\dfrac{1}{2}$ to $\dfrac{1}{3}$ the corresponding parabolas become wider.

49.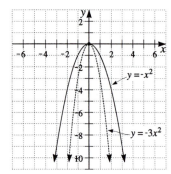

The coefficient of $-3$ narrows the graph of $y = x$ and then "flips" the resulting graph over the $x$-axis.

50.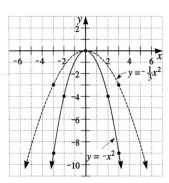

The coefficient of $-\dfrac{1}{3}$ widens the graph of $y = x^2$ and then "flips" the resulting graph over the $x$-axis.

51. $(x-5)^2 = (x-5)(x-5)$
$$= x^2 - 10x + 25$$

53. $(x+5)(x-5) = x^2 - 25$

55. $(x+p)^2 = (x+p)(x+p)$
$$= x^2 + 2px + p^2$$

57. Let $t$ = number of hours the assistant works.
Then $t - 1$ = number of hours the handyman works.
$$11t + 19(t - 1) = 146$$
$$11t + 19t - 19 = 146$$
$$30t - 19 = 146$$
$$30t = 165$$
$$t = 5.5$$

Since the assistant started at 10:00 A.M. and worked for 5.5 hours to finish the job, the job was completed at 3:30 P.M.

CHECK: The assistant worked for 5.5 hours and was paid \$11 per hour, so he earned $5.5(\$11) = \$60.50$. The handyman worked for 4.5 hours and was paid \$19 per hour, so he earned $4.5(\$19) = \$85.50$. $\$60.50 + \$85.50 \overset{\checkmark}{=} \$146$.

## Chapter 11 Review Exercises

1. $x^2 - 7x - 6 = 0$
   $a = 1, b = -7, c = -6$
   $$x = \frac{-b \pm \sqrt{b^2 - 4ac}}{2a}$$
   $$x = \frac{-(-7) \pm \sqrt{(-7)^2 - 4(1)(-6)}}{2(1)}$$
   $$x = \frac{7 \pm \sqrt{49 + 24}}{2} = \frac{7 \pm \sqrt{73}}{2}$$

3. $x^2 \quad\quad + 5 = 4x$
   $x^2 - 4x + 5 = \quad 0$
   $x^2 - 4x \quad\quad = -5$
   $$\underline{\quad\quad\quad +4 \quad +4}$$
   $x^2 - 4x + 4 = -1$
   $(x - 2)^2 = -1$
   $x - 2 = \pm\sqrt{-1}$
   Therefore, there are no real solutions.

5. $(u - 6)^2 = 13$
   $u - 6 = \pm\sqrt{13}$
   $u = 6 \pm \sqrt{13}$

7. $2y^2 + 7y = 15$
   $2y^2 + 7y - 15 = 0$
   $(2y - 3)(y + 5) = 0$
   $2y - 3 = 0 \quad$ or $\quad y + 5 = 0$
   $y = \dfrac{3}{2} \quad$ or $\quad y = -5$

9. $18x^2 - 24x + 6 = 0$
   $3x^2 - 4x + 1 = 0$
   $(3x - 1)(x - 1) = 0$
   $3x - 1 = 0 \quad$ or $\quad x - 1 = 0$
   $x = \dfrac{1}{3} \quad$ or $\quad x = 1$

11. $(x - 6)^2 = (x + 3)(x - 5)$
    $x^2 - 12x + 36 = x^2 - 2x - 15$
    $-12x + 36 = -2x - 15$
    $36 = 10x - 15$
    $51 = 10x$
    $\dfrac{51}{10} = x$

13. $u^2 + 1 = \dfrac{13u}{6}$
    $6(u^2 + 1) = \dfrac{\cancel{6}}{1} \cdot \dfrac{13u}{\cancel{6}}$
    $6u^2 + 6 = 13u$
    $6u^2 - 13u + 6 = 0$
    $(3u - 2)(2u - 3) = 0$
    $3u - 2 = 0 \quad$ or $\quad 2u - 3 = 0$
    $u = \dfrac{2}{3} \quad$ or $\quad u = \dfrac{3}{2}$

15. $3x(x - 2) = (x - 3)^2$
    $3x^2 - 6x = x^2 - 6x + 9$
    $2x^2 - 6x = -6x + 9$
    $2x^2 = 9$
    $x^2 = \dfrac{9}{2}$
    $x = \pm\sqrt{\dfrac{9}{2}} = \pm\dfrac{\sqrt{9}}{\sqrt{2}} = \pm\dfrac{3}{\sqrt{2}}$
    $= \pm\dfrac{3\sqrt{2}}{\sqrt{2}\sqrt{2}} = \pm\dfrac{3\sqrt{2}}{2}$

17.
$$\frac{x+3}{x+6} = \frac{x+2}{x+4}$$

$$\frac{(x+6)(x+4)}{1} \cdot \frac{x+3}{x+6} = \frac{(x+6)(x+4)}{1} \cdot \frac{x+2}{x+4}$$

$$(x+4)(x+3) = (x+6)(x+2)$$

$$x^2 + 7x + 12 = x^2 + 8x + 12$$

$$7x + 12 = 8x + 12$$

$$7x = 8x$$

$$0 = x$$

19. $x^2 - 7x + 3 = 0$

$x^2 - 7x \qquad = -3$

$$\begin{array}{cc} +\dfrac{49}{4} & +\dfrac{49}{4} \end{array}$$

$$x^2 - 7x + \frac{49}{4} = -3 + \frac{49}{4}$$

$$\left(x - \frac{7}{2}\right)^2 = -\frac{12}{4} + \frac{49}{4}$$

$$\left(x - \frac{7}{2}\right)^2 = \frac{37}{4}$$

$$x - \frac{7}{2} = \pm\sqrt{\frac{37}{4}} = \pm\frac{\sqrt{37}}{\sqrt{4}} = \pm\frac{\sqrt{37}}{2}$$

$$x = \frac{7}{2} \pm \frac{\sqrt{37}}{2} = \frac{7 \pm \sqrt{37}}{2}$$

21. $y = x^2 - 6x$

$0 = x^2 - 6x$

$0 = x(x - 6)$

$0 = x$   or   $0 = x - 6$

$0 = x$   or   $6 = x$

The $x$-intercepts are 0 and 6.

The $y$-intercept is 0.

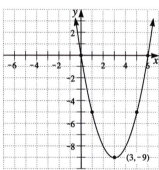

23. $y = -x^2 - 2x + 8$

$0 = -x^2 - 2x + 8$

$0 = x^2 + 2x - 8$

$0 = (x + 4)(x - 2)$

$0 = x + 4$   or   $0 = x - 2$

$-4 = x$      or   $2 = x$

The $x$-intercepts are $-4$ and 2.

The $y$-intercept is 8.

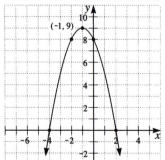

25. $y = 2x^2 - 4x + 3$

There are no $x$-intercepts.

The $y$-intercept is 3.

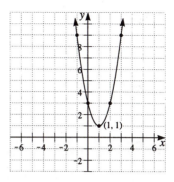

## Chapter 11 Practice Test

1. $(x + 5)(x - 2) = 18$
$$x^2 + 3x - 10 = 0$$
$$x^2 + 3x - 28 = 0$$
$$(x + 7)(x - 4) = 0$$
$$x + 7 = 0 \quad \text{or} \quad x - 4 = 0$$
$$x = -7 \quad \text{or} \qquad x = 4$$

3. $x^2 - x + 14 = 2x(x - 3)$
$$x^2 - x + 14 = 2x^2 - 6x$$
$$0 = x^2 - 5x - 14$$
$$0 = (x - 7)(x + 2)$$
$$0 = x - 7 \quad \text{or} \quad 0 = x + 2$$
$$7 = x \qquad \text{or} \quad -2 = x$$

5. $(x + 5)^2 = 10x$
$$x^2 + 10x + 25 = 10x$$
$$x^2 + 25 = 0$$
$$x^2 = -25$$
$$x = \pm\sqrt{-25}$$

Since square roots of negative number are not real, the equation has no real solution.

7. $5x^2 + 15 = 30x$
$$5x^2 - 30x + 15 = 0$$
$$5(x^2 - 6x + 3) = 0$$
$$x^2 - 6x + 3 = 0$$
$$a = 1, b = -6, c = 3$$
$$x = \frac{-b \pm \sqrt{b^2 - 4ac}}{2a}$$
$$x = \frac{-(-6) \pm \sqrt{(-6)^2 - 4(1)(3)}}{2(1)}$$
$$x = \frac{6 \pm \sqrt{36 - 12}}{2} = \frac{6 \pm \sqrt{24}}{2}$$
$$x = \frac{6 \pm 2\sqrt{6}}{2} = \frac{\cancel{2}\left(3 \pm \sqrt{6}\right)}{\cancel{2}}$$
$$x = 3 \pm \sqrt{6}$$

9.
$$3x^2 - 12x \quad = 7$$
$$x^2 - 4x \quad = \frac{7}{3}$$
$$\underline{\qquad +4 \qquad +4 \qquad}$$
$$x^2 - 4x + 4 = \frac{7}{3} + 4$$
$$(x-2)^2 = \frac{19}{3}$$
$$x - 2 = \pm\sqrt{\frac{19}{3}}$$
$$x - 2 = \pm\frac{\sqrt{19}}{\sqrt{3}} = \pm\frac{\sqrt{19}\sqrt{3}}{\sqrt{3}\sqrt{3}}$$
$$x - 2 = \pm\frac{\sqrt{57}}{3}$$
$$x = 2 \pm\frac{\sqrt{57}}{3} = \frac{6 \pm \sqrt{57}}{3}$$

$$3x^2 - 12x = 7$$
$$3x^2 - 12x - 7 = 0$$
$$a = 3, b = -12, c = -7$$
$$x = \frac{-b \pm \sqrt{b^2 - 4ac}}{2a}$$
$$x = \frac{-(-12) \pm \sqrt{(-12)^2 - 4(3)(-7)}}{2(3)}$$
$$x = \frac{12 \pm \sqrt{144 + 84}}{6} = \frac{12 \pm \sqrt{228}}{6}$$
$$x = \frac{12 \pm 2\sqrt{57}}{6} = \frac{\cancel{2}\left(6 \pm \sqrt{57}\right)}{\cancel{6}_{3}}$$
$$x = \frac{6 \pm \sqrt{57}}{3}$$

11.   $y = 6x - x^2$

$$0 = 6x - x^2$$
$$0 = x(6 - x)$$
$$0 = x \quad \text{or} \quad 0 = 6 - x$$
$$0 = x \quad \text{or} \quad 6 = x$$
The $x$-intercepts are 0 and 6.

The $y$-intercept is 0.

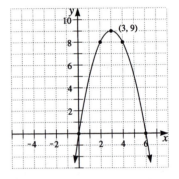

# Chapters 10 - 11
# Cumulative Review

1. $\sqrt{36x^{16}y^{12}} = \sqrt{36}\sqrt{x^{16}}\sqrt{y^{12}} = 6x^8y^6$

3. $\dfrac{7}{\sqrt{6}} = \dfrac{7\sqrt{6}}{\sqrt{6}\sqrt{6}} = \dfrac{7\sqrt{6}}{6}$

5. $\dfrac{20}{3-\sqrt{5}} = \dfrac{20\left(3+\sqrt{5}\right)}{\left(3-\sqrt{5}\right)\left(3+\sqrt{5}\right)}$

$= \dfrac{20\left(3+\sqrt{5}\right)}{9-5} = \dfrac{\overset{5}{\cancel{20}}\left(3+\sqrt{5}\right)}{\cancel{4}}$

$= 5\left(3+\sqrt{5}\right)$

7. $\sqrt{120} = \sqrt{4}\sqrt{30} = 2\sqrt{30}$

9. $\sqrt{45t} - \sqrt{20t} = \sqrt{9 \cdot 5t} - \sqrt{4 \cdot 5t}$

$= \sqrt{9}\sqrt{5t} - \sqrt{4}\sqrt{5t}$

$= 3\sqrt{5t} - 2\sqrt{5t} = \sqrt{5t}$

11. $\sqrt{\dfrac{3}{7}} + \sqrt{21} = \dfrac{\sqrt{3}}{\sqrt{7}} + \sqrt{21}$

$= \dfrac{\sqrt{3}\sqrt{7}}{\sqrt{7}\sqrt{7}} + \sqrt{21} = \dfrac{\sqrt{21}}{7} + \sqrt{21}$

$= \left(\dfrac{1}{7} + 1\right)\sqrt{21} = \dfrac{8}{7}\sqrt{21}$

13. $\sqrt{27x^6y^5} - xy\sqrt{12x^4y^3} = \sqrt{9x^6y^4 \cdot 3y} - xy\sqrt{4x^4y^2 \cdot 3y}$

$= \sqrt{9x^6y^4}\sqrt{3y} - xy\sqrt{4x^4y^2}\sqrt{3y}$

$= 3x^3y^2\sqrt{3y} - xy(2x^2y)\sqrt{3y}$

$= 3x^3y^2\sqrt{3y} - 2x^3y^2\sqrt{3y}$

$= x^3y^2\sqrt{3y}$

15.  $\left(3\sqrt{2}-4\sqrt{5}\right)\left(4\sqrt{2}-2\sqrt{5}\right)=(3\sqrt{2})(4\sqrt{2})-(3\sqrt{2})(2\sqrt{5})-(4\sqrt{5})(4\sqrt{2})+(4\sqrt{5})(2\sqrt{5})$

$$=3\cdot4\sqrt{2}\sqrt{2}-3\cdot2\sqrt{2}\sqrt{5}-4\cdot4\sqrt{5}\sqrt{2}+4\cdot2\sqrt{5}\sqrt{5}$$

$$=12\cdot2-6\sqrt{10}-16\sqrt{10}+8\cdot5$$

$$=24-6\sqrt{10}-16\sqrt{10}+40$$

$$=64-22\sqrt{10}$$

17.  $\dfrac{15}{\sqrt{7}-\sqrt{2}}-\dfrac{10}{\sqrt{2}}=\dfrac{15\left(\sqrt{7}+\sqrt{2}\right)}{\left(\sqrt{7}-\sqrt{2}\right)\left(\sqrt{7}+\sqrt{2}\right)}-\dfrac{10\sqrt{2}}{\sqrt{2}\sqrt{2}}$

$$=\dfrac{15\left(\sqrt{7}+\sqrt{2}\right)}{7-2}-\dfrac{\overset{5}{\cancel{10}}\sqrt{2}}{\cancel{2}}$$

$$=\dfrac{\overset{3}{\cancel{15}}\left(\sqrt{7}+\sqrt{2}\right)}{\cancel{5}}-5\sqrt{2}$$

$$=3\left(\sqrt{7}+\sqrt{2}\right)-5\sqrt{2}$$

$$=3\sqrt{7}+3\sqrt{3}-5\sqrt{2}$$

$$=3\sqrt{7}-2\sqrt{2}$$

19.  $\qquad x^2-3x=10$

$$x^2-3x-10=0$$

$$(x-5)(x+2)=0$$

$$x-5=0\quad\text{or}\quad x+2=0$$

$$x=5\quad\text{or}\qquad x=-2$$

21.  $\qquad\dfrac{x}{x-2}=\dfrac{x+1}{x-3}$

$$\dfrac{(x-2)(x-3)}{1}\left(\dfrac{x}{x-2}\right)=\dfrac{(x-2)(x-3)}{1}\left(\dfrac{x+1}{x-3}\right)$$

$$(x-3)x=(x-2)(x+1)$$

$$x^2-3x=x^2+x-2x-2$$

$$x^2-3x=x^2-x-2$$

$$-3x=-x-2$$

$$-2x=-2$$

$$x=1$$

23. $3x^2 - 6x - 2 = x^2 - x - 4$

$2x^2 - 6x - 2 = -x - 4$

$2x^2 - 5x - 2 = -4$

$2x^2 - 5x + 2 = 0$

$(2x - 1)(x - 2) = 0$

$2x - 1 = 0 \quad \text{or} \quad x - 2 = 0$

$2x = 1 \quad \text{or} \qquad x = 2$

$x = \dfrac{1}{2} \quad \text{or} \qquad x = 2$

25. $(x + 3)(x - 4) = 8$

$x^2 - 4x + 3x - 12 = 8$

$x^2 - x - 12 = 8$

$x^2 - x - 20 = 0$

$(x - 5)(x + 4) = 0$

$x - 5 = 0 \quad \text{or} \quad x + 4 = 0$

$x = 5 \quad \text{or} \qquad x = -4$

27. $(2x - 3)(x - 4) = (x - 2)(2x - 1)$

$2x^2 - 8x - 3x + 12 = 2x^2 - x - 4x + 2$

$2x^2 - 11x + 12 = 2x^2 - 5x + 2$

$-11x + 12 = -5x + 2$

$12 = 6x + 2$

$10 = 6x$

$\dfrac{5}{3} = x$

29. $(s - 2)^2 = 10$

$s - 2 = \pm\sqrt{10}$

$s = 2 \pm \sqrt{10}$

31. $\sqrt{x} - 3 = 4$

$\sqrt{x} = 7$

$\left(\sqrt{x}\right)^2 = 7^2$

$x = 49$

33. $3\sqrt{t + 4} - 1 = 7$

$3\sqrt{t + 4} = 8$

$\sqrt{t + 4} = \dfrac{8}{3}$

$\left(\sqrt{t + 4}\right)^2 = \left(\dfrac{8}{3}\right)^2$

$t + 4 = \dfrac{64}{9}$

$t = \dfrac{28}{9}$

35. $2x^2 - 10x + 6 = 0$

$2x^2 - 10x = -6$

$x^2 - 5x = -3$

$$\begin{array}{r} +\dfrac{25}{4} \quad +\dfrac{25}{4} \\ \hline \end{array}$$

$x^2 - 5x + \dfrac{25}{4} = -3 + \dfrac{25}{4}$

$\left(x - \dfrac{5}{2}\right)^2 = \dfrac{13}{4}$

$x - \dfrac{5}{2} = \pm\sqrt{\dfrac{13}{4}} = \pm\dfrac{\sqrt{13}}{\sqrt{4}}$

$x - \dfrac{5}{2} = \pm\dfrac{\sqrt{13}}{2}$

$x = \dfrac{5}{2} \pm \dfrac{\sqrt{13}}{2}$

$x = \dfrac{5 \pm \sqrt{13}}{2}$

37. $y = (2x - 3)^2$

$0 = (2x - 3)^2$

$0 = 2x - 3$

$3 = 2x$

$\dfrac{3}{2} = x$

The $x$-intercept is $\dfrac{3}{2}$.

The $y$-intercept is 9.

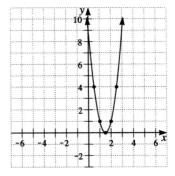

39. $y = x^2 - x - 1$

$0 = x^2 - x - 1$

$a = 1, b = -1, c = -1$

$x = \dfrac{-b \pm \sqrt{b^2 - 4ac}}{2a}$

$x = \dfrac{-(-1) \pm \sqrt{(-1)^2 - 4(1)(-1)}}{2(1)}$

$x = \dfrac{1 \pm \sqrt{5}}{2}$

The $x$-intercepts are $\dfrac{1 + \sqrt{5}}{2} \approx 1.6$ and

$\dfrac{1 - \sqrt{5}}{2} \approx -0.6$.

The $y$-intercept is $-1$.

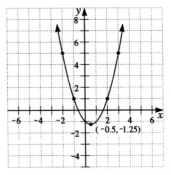

41. Let $x = $ length of the third side of the triangle.

By the Pythagorean Theorem,

$8^2 + x^2 = 17^2$

$64 + x^2 = 289$

$x^2 = 225$

$x = \pm 15$

We reject $x = -15$, since length cannot be negative. So $x = 15$. Since the length of the third side is 15, the perimeter of the triangle is $8 + 15 + 17 = 40$.

43. Let $x = $ Kim's average speed on the way there (in mph).

Then $x - 30 = $ Kim's average speed on the way back (in mph).

$$\frac{280}{x} + \frac{280}{x - 30} = 11$$

$$x(x - 30)\left(\frac{280}{x} + \frac{280}{x - 30}\right) = x(x - 30)(11)$$

$$\frac{\not{x}(x - 30)}{1} \cdot \frac{280}{\not{x}} + \frac{x(\not{x - 30})}{1} \cdot \frac{280}{\not{x - 30}} = 11x(x - 30)$$

$$280(x - 30) + 280x = 11x^2 - 330x$$

$$280x - 8400 + 280x = 11x^2 - 330x$$

$$560x - 8400 = 11x^2 - 330x$$

$$0 = 11x^2 - 890x + 8400$$

$$0 = (x - 70)(11x - 120)$$

$$0 = x - 70 \quad \text{or} \quad 0 = 11x - 120$$

$$70 = x \quad \text{or} \quad 120 = 11x$$

$$70 = x \quad \text{or} \quad \frac{120}{11} = x$$

$$70 = x \quad \text{or} \quad 10.9 = x$$

We reject $x = 10.9$, since this value would make Kim's average speed negative on the way back. So $x = 70$. Therefore, Kim's average speed on the way there is 70 mph.

# Chapters 10 - 11
# Cumulative Practice Test

1. $\sqrt{72} = \sqrt{36 \cdot 2} = \sqrt{36}\sqrt{2}$

$\qquad = 6\sqrt{2}$

3. $3\sqrt{20} - 4\sqrt{45} = 3\sqrt{4 \cdot 5} - 4\sqrt{9 \cdot 5}$

$\qquad\qquad\qquad = 3\sqrt{4}\sqrt{5} - 4\sqrt{9}\sqrt{5}$

$\qquad\qquad\qquad = 3\left(2\sqrt{5}\right) - 4\left(3\sqrt{5}\right)$

$\qquad\qquad\qquad = 6\sqrt{5} - 12\sqrt{5}$

$\qquad\qquad\qquad = -6\sqrt{5}$

5. $\dfrac{8}{\sqrt{5}} = \dfrac{8\sqrt{5}}{\sqrt{5}\sqrt{5}} = \dfrac{8\sqrt{5}}{5}$

7. $\dfrac{24}{4 - \sqrt{7}} = \dfrac{24\left(4 + \sqrt{7}\right)}{\left(4 - \sqrt{7}\right)\left(4 + \sqrt{7}\right)}$

$\qquad = \dfrac{24\left(4 + \sqrt{7}\right)}{16 - 7} = \dfrac{\overset{8}{\cancel{24}}\left(4 + \sqrt{7}\right)}{\underset{3}{\cancel{9}}} = \dfrac{8\left(4 + \sqrt{7}\right)}{3}$

9. $\left(5\sqrt{t} + \sqrt{3}\right)\left(2\sqrt{t} - 4\sqrt{3}\right) = (5\sqrt{t})(2\sqrt{t}) - (5\sqrt{t})\left(4\sqrt{3}\right) + \sqrt{3}(2\sqrt{t}) - \sqrt{3}\left(4\sqrt{3}\right)$

$\qquad\qquad\qquad\qquad\qquad\qquad = 10t - 20\sqrt{3t} + 2\sqrt{3t} - 12$

$\qquad\qquad\qquad\qquad\qquad\qquad = 10t - 18\sqrt{3t} - 12$

11. $\qquad (x + 3)(x - 5) = 9$

$\qquad x^2 - 5x + 3x - 15 = 9$

$\qquad\qquad x^2 - 2x - 15 = 9$

$\qquad\qquad x^2 - 2x - 24 = 0$

$\qquad\qquad (x - 6)(x + 4) = 0$

$\qquad x - 6 = 0 \quad \text{or} \quad x + 4 = 0$

$\qquad\qquad x = 6 \quad \text{or} \qquad x = -4$

13.
$$\frac{x}{2x-5} = \frac{x+4}{2x}$$

$$\frac{(2x-5)(2x)}{1}\left(\frac{x}{2x-5}\right) = \frac{(2x-5)(2x)}{1}\left(\frac{x+4}{2x}\right)$$

$$(2x)(x) = (2x-5)(x+4)$$

$$2x^2 = 2x^2 + 8x - 5x - 20$$

$$2x^2 = 2x^2 + 3x - 20$$

$$0 = 3x - 20$$

$$20 = 3x$$

$$\frac{20}{3} = x$$

15.
$$2x^2 + 1 = 5x$$
$$2x^2 - 5x + 1 = 0$$
$$a = 2, b = -5, c = 1$$

$$x = \frac{-b \pm \sqrt{b^2 - 4ac}}{2a}$$

$$x = \frac{-(-5) \pm \sqrt{(-5)^2 - 4(2)(1)}}{2(2)}$$

$$x = \frac{5 \pm \sqrt{17}}{4}$$

17.
$$(2x+5)^2 = 4x(x-3)$$
$$4x^2 + 20x + 25 = 4x^2 - 12x$$
$$20x + 25 = -12x$$
$$25 = -32x$$
$$-\frac{25}{32} = x$$

19.  Let $x$ = length of the rectangle (in inches).
By the Pythagorean Theorem,
$$x^2 + 5^2 = 10^2$$
$$x^2 + 25 = 100$$
$$x^2 = 75$$
$$x = \pm\sqrt{75}$$
$$x = \pm 5\sqrt{3}$$
We reject $x = -5\sqrt{3}$, since length cannot be negative. So $x = 5\sqrt{3}$. Then the area of the rectangle is $(length) \times (width) = (5\sqrt{3})(5) = 25\sqrt{3}$ sq. in.

21.  $y = x^2 + 4x - 5$

$$0 = x^2 + 4x - 5$$
$$0 = (x+5)(x-1)$$
$$0 = x+5 \quad \text{or} \quad 0 = x-1$$
$$-5 = x \qquad \text{or} \quad 1 = x$$

The $x$-intercepts are $-5$ and $1$.

The $y$-intercept is $-5$.

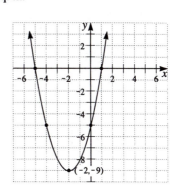

# Supplementary Chapter on Geometry

## Exercises G.1

1. $\angle AEB$ and $\angle BED$
   $\angle EAD$ and $\angle DAB$

3. (a) $60°$      (b) $30°$      (c) $45°$      (d) $72°$      (e) $1°$

5. Let $x =$ the angle.
   Then $180 - x =$ its supplement.
   $$x = 3(180 - x) - 28$$
   $$x = 540 - 3x - 28$$
   $$x = 512 - 3x$$
   $$4x = 512$$
   $$x = 128$$
   The angle is $128°$.

7. Let $x =$ the angle.
   Then $90 - x =$ its complement.
   $$x = 2(90 - x) + 12$$
   $$x = 180 - 2x + 12$$
   $$x = 192 - 2x$$
   $$3x = 192$$
   $$x = 64$$
   The angle is $64°$.

9. $$x + (3x + 10) = 90$$
   $$4x + 10 = 90$$
   $$4x = 80$$
   $$x = 20$$

11. $$5x + 90 + 10x + 30 = 180$$
    $$15x + 120 = 180$$
    $$15x = 60$$
    $$x = 4$$

13. $\angle 1 = 145°$ (supplement of $\angle 4$)
    $\angle 2 = 35°$ ($\angle 2$ and $\angle 4$ are vertical angles)
    $\angle 3 = 145°$ (supplement of $\angle 4$)

15. $\angle 1 = 65°$ ($\angle 1 + 90 + 25 = 180$)
    $\angle 2 = 115°$ (supplement of $\angle 1$)
    $\angle 3 = 65°$ ($\angle 1$ and $\angle 3$ are vertical angles)

17. $\angle 1 = \angle 3 = \angle 6 = \angle 8 = 55°$
    $\angle 2 = \angle 5 = \angle 7 = 125°$

19. $$2x + (x + 12) = 180$$
    $$3x + 12 = 180$$
    $$3x = 168$$
    $$x = 56$$

21. $\angle 1 = \angle 4 = \angle 5 = \angle 7 = \angle 9 = \angle 11 = \angle 14 = \angle 16 = 140°$
    $\angle 3 = \angle 6 = \angle 8 = \angle 10 = \angle 12 = \angle 13 = \angle 15 = 40°$

23. $\angle 1 = 50°, \angle 2 = 130°, \angle 3 = 50°, \angle 4 = 130°$

## Exercises G.2

1. $\angle C = 180 - (34 + 23) = 180 - 57 = 123°$

3. Since $AC = BC$, $\angle A = \angle B$. Let $x =$ the number of degrees in $\angle A$ and in $\angle B$.

$$x + x + 40 = 180$$
$$2x + 40 = 180$$
$$2x = 140$$
$$x = 70$$

So $\angle A = 70°$ and $\angle B = 70°$.

5. $\quad x + 4x + (2x - 16) = 180$
$$7x - 16 = 180$$
$$7x = 196$$
$$x = 28$$

7. $\quad x + (2x + 20) + (2x + 20) = 180$
$$5x + 40 = 180$$
$$5x = 140$$
$$x = 28$$

9. $\quad x + 23 + 90 = 180$
$$x + 113 = 180$$
$$x = 67$$

11. $\quad x + (2x + 12) + 90 = 180$
$$3x + 102 = 180$$
$$3x = 78$$
$$x = 26$$

13. $\quad \dfrac{|\overline{DE}|}{15} = \dfrac{8}{5}$, so $|\overline{DE}| = 24$

$\dfrac{|\overline{EF}|}{15} = \dfrac{4}{5}$, so $|\overline{EF}| = 12$

15. $\quad \dfrac{|\overline{AC}|}{24} = \dfrac{10}{12}$, so $|\overline{AC}| = 20$

17. $\quad \dfrac{|\overline{EA}|}{|\overline{EC}|} = \dfrac{|\overline{DB}|}{|\overline{DC}|}$

$\dfrac{|\overline{EA}|}{18} = \dfrac{6}{10}$, so $|\overline{EA}| = 10.8$

19. $\quad |\overline{AB}|^2 = 3^2 + 4^2$
$$|\overline{AB}|^2 = 9 + 16$$
$$|\overline{AB}|^2 = 25$$
$$|\overline{AB}| = 5$$

21. $\quad |\overline{AC}|^2 + 4^2 = 9^2$
$$|\overline{AC}|^2 + 16 = 81$$
$$|\overline{AC}|^2 = 65$$
$$|\overline{AC}| = \sqrt{65}$$

23. Let $x = |\overline{AC}| = |\overline{BC}|$.
$$x^2 + x^2 = 8^2$$
$$2x^2 = 64$$
$$x^2 = 32$$
$$x = \sqrt{32} = 4\sqrt{2}$$
So $|\overline{AC}| = 4\sqrt{2}$ and $|\overline{BC}| = 4\sqrt{2}$.

25. $\quad \dfrac{|\overline{ED}|}{|\overline{DC}|} = \dfrac{|\overline{BA}|}{|\overline{AC}|}$

$\dfrac{|\overline{ED}|}{8} = \dfrac{6}{4}$, so $|\overline{ED}| = 12$

27. $\quad |\overline{AC}|^2 = 4^2 + 8^2$
$$|\overline{AC}|^2 = 16 + 64$$
$$|\overline{AC}|^2 = 80$$
$$|\overline{AC}| = \sqrt{80} = 4\sqrt{5}$$

29. Let $x =$ distance between top of the ladder and the ground (in feet).
$$x^2 + 10^2 = 30^2$$
$$x^2 + 100 = 900$$
$$x^2 = 800$$
$$x = \sqrt{800} = 20\sqrt{2}$$

The top of the ladder is $20\sqrt{2}$ (approximately 28.3) feet above the ground.

31. $x + 63 + 90 = 180$
$$x + 153 = 180$$
$$x = 27$$

33. Referring to the sketch, $a = 50$, since it is the supplement of 130. Then
$$50 + b + b = 180$$
$$50 + 2b = 180$$
$$2b = 130$$
$$b = 65$$
Since the lines are parallel, $x + b = 180$
$$x + 65 = 180$$
$$x = 115$$

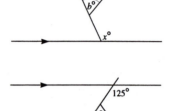

35. Referring to the sketch, $a + 125 = 180$, (since the lines are parallel). So $a = 55$. Then
$$55 + 55 + x = 180$$
$$110 + x = 180$$
$$x = 70$$

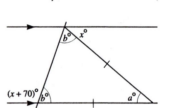

37. Referring to the sketch, $a = x$, (since the lines are parallel) and $b + x = x + 70$ (for the same reason). From the second equation, we get $b = 70$. Then $a + 70 + 70 = 180$, so $a = 40$. Therefore, $x = 40$.

39. Let $x =$ height of the flagpole (in feet).
$$\frac{x}{18} = \frac{6}{4}$$
$$\frac{x}{18} = \frac{3}{2}$$
$$\cancel{18} \cdot \frac{x}{\cancel{18}} = \overset{9}{\cancel{18}} \left( \frac{3}{\cancel{2}_{1}} \right)$$
$$x = 27$$
The flagpole is 27 ft high.

41. We recognize that since $\triangle ABC$ is a right triangle, $\angle 1 + \angle 2 = 90°$.

    (a) The entire figure is a square since the angles are all right angles and the sides are all of equal length (each side is of length $a + b$).
    Looking at the inside figure, we can see that $\angle 1 + \angle 2 + \angle 3 = 180°$. Since we know that $\angle 1 + \angle 2 = 90°$, it follows that $\angle 3 = 90°$. Similarly, all of the angles of the inside figure are right angles and all its sides are of equal length. Therefore, the inside figure is a square.

    (b) Since the side of the outer square is of length $a + b$, its area is $(a + b)^2$.

    (c) Since the side of the inner square is of length $c$, its area is $c^2$.

(d)   Since the area of a right triangle is $\dfrac{1}{2}$ the product of its legs, the area of each triangle is $\dfrac{1}{2}ab$.

(e)   The figure clearly shows that the area of the outer square is equal to the area of the inner square plus the area of the four inner triangles.

(f)   Using the area relationship described in part (e), we have

$$(a+b)^2 = c^2 + 4 \cdot \frac{1}{2}(ab) \quad \leftarrow \quad \textit{This says that the area of the outer square equals}$$
$$a^2 + 2ab + b^2 = c^2 + 2ab \qquad \qquad \textit{the area of the inner square plus the area of the}$$
$$a^2 + b^2 = c^2 \qquad \qquad \qquad \textit{four inner triangles.}$$

Thus we have proven the Pythagorean Theorem which says that in a right $\triangle ABC$, we have $a^2 + b^2 = c^2$.

## Exercises G.3

1.   $\angle C = 360 - (110 + 130 + 40) = 360 - 280 = 80°$

3.   $\angle C = 60°$ (opposite angles in a parallelogram are equal)
     $\angle B = 120°$ (adjacent angles in a parallelogram are supplementary)
     $\angle D = 120°$ (opposite angles in a parallelogram are equal)

5.   $2x + 3 = x + 8$
     $x + 3 = 8$
     $x = 5$

7.   $5x - 18 + 6x = 180$
     $11x - 18 = 180$
     $11x = 198$
     $x = 18$

9.   Let $x$ = length of the diagonal (in cm).
     $x^2 = 5^2 + 8^2$
     $x^2 = 25 + 64$
     $x^2 = 89$
     $x = \sqrt{89}$

     The length of the diagonal is $\sqrt{89}$ (approximately 9.4) cm.

11.   Let $x$ = length of the diagonal (in inches).
     $x^2 = 4^2 + 4^2$
     $x^2 = 16 + 16$
     $x^2 = 32$
     $x = \sqrt{32} = 4\sqrt{2}$

     The length of the diagonal is $4\sqrt{2}$ (approximately 5.7) in.

13.   Let $x$ = length of the side (in cm).
     $x^2 + x^2 = 5^2$
     $2x^2 = 25$
     $x^2 = \dfrac{25}{2}$
     $x = \sqrt{\dfrac{25}{2}} = \dfrac{\sqrt{25}}{\sqrt{2}} = \dfrac{5}{\sqrt{2}} = \dfrac{5\sqrt{2}}{2}$

     The length of the side is $\dfrac{5\sqrt{2}}{2}$ (approximately 3.5) cm.

15.   $p = 2(5) + 2(8) = 10 + 16 = 26$ ft.
     $A = (5)(8) = 40$ sq. ft.

17.   $p = 2(6) + 2(12) = 12 + 24 = 36$ in.
     $A = (6)(12) = 72$ sq. in.

19.   $p = 4(3) = 12$ in.
     $A = 3^2 = 9$ sq. in.

21.   $A = (8)(6) = 48$

23. $A = (10)(6) = 60$

25. $A = \dfrac{1}{2}(8)(6) = 24$

27. $A = \dfrac{1}{2}(12)(4) = 24$

29. $A = \dfrac{1}{2}(6)(4) = 12$

31. $A = \dfrac{1}{2}(5 + 12)(6) = 51$

33. $\left|\overline{AD}\right|^2 = 5^2 + 12^2$
    $\left|\overline{AD}\right|^2 = 25 + 144$
    $\left|\overline{AD}\right|^2 = 169$
    $\left|\overline{AD}\right| = 13$
    Then $p = 2(13) + 2(15) = 56$
    $A = (15)(12) = 180$

35. Let $x =$ length of the hypotenuse.
    $x^2 = 5^2 + 8^2$
    $x^2 = 25 + 64$
    $x^2 = 89$
    $x = \sqrt{89}$
    Then $p = 5 + 8 + \sqrt{89} = 13 + \sqrt{89}$
    (approximately 22.4)
    $A = \dfrac{1}{2}(5)(8) = 20$

37. Let $x =$ length of the missing leg.
    $x^2 + 7^2 = 9^2$
    $x^2 + 49 = 81$
    $x^2 = 32$
    $x = \sqrt{32} = 4\sqrt{2}$
    Then $p = 7 + 9 + 4\sqrt{2} = 16 + 4\sqrt{2}$
    (approximately 21.7)
    $A = \dfrac{1}{2}\left(4\sqrt{2}\right)(7) = 14\sqrt{2}$
    (approximately 19.8)

39. Let $x =$ height of the rectangle (in cm).
    $x^2 + 8^2 = 12^2$
    $x^2 + 64 = 144$
    $x^2 = 80$
    $x = \sqrt{80} = 4\sqrt{5}$
    Then $p = 2\left(4\sqrt{5}\right) + 2(8) = 8\sqrt{5} + 16$
    (approximately 33.9) cm.
    $A = (8)\left(4\sqrt{5}\right) = 32\sqrt{5}$
    (approximately 71.6) sq. cm.

41. Let $x =$ side of the square (in mm).
    $x^2 + x^2 = 10^2$
    $2x^2 = 100$
    $x^2 = 50$
    $x = \sqrt{50} = 5\sqrt{2}$
    Then $p = 4\left(5\sqrt{2}\right) = 20\sqrt{2}$ (approximately 28.3) mm.
    $A = \left(5\sqrt{2}\right)^2 = 50$ sq. mm.

43. Drop a perpendicular from $D$ to $AB$. Call its foot $E$. Then
    $\left|\overline{EB}\right| = 5$, so $\left|\overline{AE}\right| = 3$. Using the Pythagorean Theorem,
    $\left|\overline{AD}\right|^2 = 3^2 + 4^2 = 9 + 16 = 25$, so $\left|\overline{AD}\right| = 5$. Then
    $p = 8 + 4 + 5 + 5 = 22$ and $A = \dfrac{1}{2}(5 + 8)(4) = 26$.

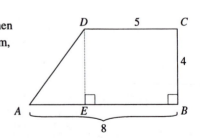

45.  $p = 10 + 10 + 10 + 26 = 56$

Drop a perpendicular from both $D$ and $C$ to $AB$, calling their feet $E$ and $F$, respectively. Since $|\overline{EF}| = 10$ and $|\overline{AE}| = |\overline{FB}|$, it follows that $|\overline{AE}| = 8$. Then

$$|\overline{DE}|^2 + 8^2 = 10^2$$
$$|\overline{DE}|^2 + 64 = 100$$
$$|\overline{DE}|^2 = 36$$
$$|\overline{DE}| = 6$$
$$A = \frac{1}{2}(10 + 26)(6) = 108$$

47.  Area of $ABCDE$ = (Area of rectangle $ABDE$) − (Area of $\triangle BCD$)

$|\overline{BD}|^2 = 6^2 + 6^2$

$|\overline{BD}|^2 = 36 + 36$

$|\overline{BD}|^2 = 72$

$|\overline{BD}| = \sqrt{72} = 6\sqrt{2}$

Then area of rectangle $ABDE = 9\left(6\sqrt{2}\right) = 54\sqrt{2}$ and

Area of $\triangle BCD = \frac{1}{2}(6)(6) = 18$.

So Area of $ABCDE = 54\sqrt{2} - 18$ (approximately 58.4)

49.  From the sketch, $|\overline{DG}| = 4$ and $|\overline{FC}| = 2$.

Area of $AEFG$ = (Area of rectangle $ABCD$) − (Area of $\triangle ADG$)
$$- (\text{Area of } \triangle GCF) - (\text{Area of } \triangle FBE)$$

Area of rectangle $ABCD = (10)(8) = 80$

Area of $\triangle ADG = \frac{1}{2}(8)(4) = 16$

Area of $\triangle GCF = \frac{1}{2}(6)(2) = 6$

Area of $\triangle FBE = \frac{1}{2}(6)(7) = 21$

So Area of $AEFG = 80 - 16 - 6 - 21 = 37$.

51.  From the sketch, $|\overline{DF}| = 6$ and $|\overline{DG}| = 9$.

Area of the figure = (Area of rectangle $ABCD$) + (Area of $\triangle FDG$) + (Area of $\triangle ECG$)

Area of rectangle $ABCD = (12)(8) = 96$

Area of $\triangle FDG = \frac{1}{2}(6)(9) = 27$

Area of $\triangle ECG = \frac{1}{2}(3)(6) = 9$

So Area of the figure = $96 + 27 + 9 = 132$.

53.  From the sketch, $|\overline{DE}| = 9$.

Area of shaded region = (Area of rectangle $ACDF$) − (Area of trapezoid $BCDE$)
$$- (\text{Area of } \triangle EFG)$$

Area of rectangle $ACDF = (12)(6) = 72$

Area of trapezoid $BCDE = \frac{1}{2}(4 + 9)(6) = 39$

Area of $\triangle EFG = \frac{1}{2}(2)(3) = 3$

So Area of shaded region = $72 - 39 - 3 = 30$.

## Exercises G.4

1. $\overset{\frown}{AB} = 35°$

3. $x + (2x - 30) = 360$
$\phantom{x + (}3x - 30 = 360$
$\phantom{x + (3x - }3x = 390$
$\phantom{x + (3x - 3}x = 130$
So $\overset{\frown}{AB} = 130°$

5. $6\pi$ inches

7. $8\pi$ feet

9. Length of arc $= \dfrac{30}{360}(18\pi) = \dfrac{3\pi}{2}$ inches

11. Length of $\overset{\frown}{AB} = \dfrac{65}{360}(12\pi) = \dfrac{13\pi}{6}$
Then perimeter of sector $= 6 + 6 + \dfrac{13\pi}{6} = 12 + \dfrac{13\pi}{6}$

13. Length of $\overset{\frown}{AB} = \dfrac{10}{360}(10\pi) = \dfrac{5\pi}{18}$
Then perimeter of sector $= 5 + 5 + \dfrac{5\pi}{18} = 10 + \dfrac{5\pi}{18}$

15. $9\pi$ sq. in.

17. $36\pi$ sq. ft.

19. Length of $\overset{\frown}{AB} = \dfrac{1}{2}(20\pi) = 10\pi$.
Length of $\overline{AB} = 20$.
So perimeter of semicircle $= (10\pi + 20)$ inches.
Area of semicircle $= \dfrac{1}{2}(100\pi) = 50\pi$ sq. in.

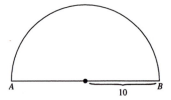

21. $\dfrac{80}{360}(25\pi) = \dfrac{50\pi}{9}$ sq. in.

23. Radius of the semicircle $= 3$, so the area of the semicircle is $\dfrac{1}{2}(9\pi) = \dfrac{9}{2}\pi$. Since the area of the rectangle $= (6)(8) = 48$, the area of the figure $= 48 + \dfrac{9}{2}\pi$. Length of the semicircular arc $= \frac{1}{2}(6\pi) = 3\pi$. So the perimeter of the figure $= 8 + 6 + 8 + 3\pi = 22 + 3\pi$.

25. Area of shaded region $=$ (Area of square $ABCD$) $-$ (Area of circle).
Area of square $ABCD = 8^2 = 64$.
Area of circle $= \pi(4)^2 = 16\pi$.
So Area of shaded region $= (64 - 16\pi)$ sq. cm.

27. Radius of large semicircle $= 6 + 8 = 14$, so area of large semicircle $= \dfrac{1}{2} \cdot \pi(14)^2 = 98\pi$.
Radius of small semicircle $= 6$, so area of small semicircle $= \dfrac{1}{2} \cdot \pi(6)^2 = 18\pi$.
So area of figure $= 98\pi - 18\pi = 80\pi$.

29. $\left|\overline{AB}\right|^2 = 6^2 + 8^2$

   $\left|\overline{AB}\right|^2 = 36 + 64$

   $\left|\overline{AB}\right|^2 = 100$

   $\left|\overline{AB}\right| = 10$            Then length of $\overparen{AB} = \dfrac{1}{2}(10\pi) = 5\pi$.

31. Perimeter $= \left|\overline{AE}\right| + \left|\overline{ED}\right| + \left|\overline{DC}\right| + \left|\overline{CB}\right| + $ length of $\overparen{AB} = 20 + 10 + 30 + 10 + $ length of $\overparen{AB}$.

   Since $\left|\overline{AB}\right| = \left|\overline{EB}\right| - \left|\overline{EA}\right| = 30 - 20 = 10$,   $\overparen{AB}$ is a semicircle with diameter $= 10$.

   So length of $\overparen{AB} = \dfrac{1}{2}(10\pi) = 5\pi$. Then perimeter $= (70 + 5\pi)$ cm.

   Area $=$ (Area of rectangle $BCDE$) $+$ (area of semicircle) $= (30)(10) + \dfrac{1}{2} \cdot \pi(5)^2$

        $= \left(300 + \dfrac{25\pi}{2}\right)$ sq. cm.

33. A round 12-inch pie has a radius of 6 inches and therefore has an area of $\pi(6)^2 = 36\pi$ sq. in. Each round 8-inch pie has a radius of 4 inches and therefore has an area of $\pi(4)^2 = 16\pi$ sq. in. This means that the total area of the two 8-inch pies is $2(16\pi) = 32\pi$ sq. in. As a result, one 12-inch pie contains more pizza than two 8-inch pies. Since the 12-inch pie costs \$9, each dollar buys $\dfrac{36\pi}{9} = 4\pi$ sq. in. of pizza. Since the two 8-inch pies costs $2(\$4)$ or \$8, each dollar buys $\dfrac{32\pi}{8} = 4\pi$ sq. in. of pizza. Therefore, both options have the same value.

## Exercises G.5

1. $S.A. = 396$ sq. cm.     $V = 504$ cu. cm.     3. $S.A. = 40\pi$ sq. ft.     $V = 32\pi$ cu. ft.

5. $S.A. = 24\pi$ sq. m.     $V = 12\pi$ cu. m.     7. $S.A. = 275.6$ sq. ft.     $V = 762.4$ cu. ft.

9. $S.A. = 113.0$ sq. ft.     $V = 113.0$ cu. ft.     11. $S.A. = 226.2$ sq. cm.     $V = 251.3$ cu. cm.

13. $S.A. = 134.6$ sq. m.     $V = 64.8$ cu. m.     15. $S.A. = 502.7$ sq. ft.     $V = 599.5$ cu. ft.

17. $S.A. = 4,712.4$ sq. mm.     $V = 29,452.4$ cu. mm.

19. $V = 144\pi$ cu. m.               21. $V = \dfrac{500\pi}{3}$ cu. cm.

23. Lateral $S.A. = 60\pi$ sq. in.        25. 15 cu. yds.

27. \$377.75        29. 859        31. \$35.60

33. 790.3 sq. cm.        35. 11,196.6 cu. ft.        37. $360\pi$ cu. cm.

39. (a) Each of the six faces of a cube is a square of side $x$. Since the area of each square face is $x^2$, the surface area of the cube is $6x^2$.

     (b) If the edge of a cube is doubled in length, the surface area is multiplied by 4. The volume is multiplied by 8.

     (c) If the edge of a cube is tripled in length, the surface area is multiplied by 9. The volume is multiplied by 27.

     (d) If the edge of a cube has its length multiplied by $k$, the surface area is multiplied by $k^2$. The volume is multiplied by $k^3$.

# Appendix A
# A Review of Arithmetic

## Exercises A.1

1. $\dfrac{6}{9} = \dfrac{2}{3} = \dfrac{60}{90}$

5. $\dfrac{33}{44} = \dfrac{3}{4} = \dfrac{66}{88}$

9. $1 = \dfrac{7}{7} = \dfrac{12}{12}$

13. $\dfrac{6}{42} = \dfrac{6 \cdot 1}{6 \cdot 7} = \dfrac{1}{7}$

17. $\dfrac{20}{49}$ cannot be reduced.

21. $\dfrac{90}{15} = \dfrac{15 \cdot 6}{15 \cdot 1} = \dfrac{6}{1} = 6$

25. $\dfrac{81}{100}$ cannot be reduced.

29. $\dfrac{15}{7} = \dfrac{15 \cdot 3}{7 \cdot 3} = \dfrac{45}{21}$ So ? = 21.

33. $\dfrac{1}{18} = \dfrac{1 \cdot 6}{18 \cdot 6} = \dfrac{6}{108}$ So ? = 6.

37. $\dfrac{54}{60} = \dfrac{\cancel{6} \cdot 9}{\cancel{6} \cdot 10} = \dfrac{9}{10}$ hit the bull's-eye

$\dfrac{6}{60} = \dfrac{\cancel{6} \cdot 1}{\cancel{6} \cdot 10} = \dfrac{1}{10}$ missed the bull's-eye

39. $\dfrac{24}{28} = \dfrac{\cancel{4} \cdot 6}{\cancel{4} \cdot 7} = \dfrac{6}{7}$ was correct

$\dfrac{4}{28} = \dfrac{\cancel{4} \cdot 1}{\cancel{4} \cdot 7} = \dfrac{1}{7}$ was incorrect

## Exercises A.2

1. $\dfrac{2}{3} \cdot \dfrac{5}{7} = \dfrac{2 \cdot 5}{3 \cdot 7} = \dfrac{10}{21}$

5. $\dfrac{\overset{2}{\cancel{8}}}{\underset{3}{\cancel{9}}} \cdot \dfrac{\cancel{3}}{\cancel{4}} = \dfrac{2}{3}$

9. $\dfrac{5}{6} \div \dfrac{4}{15} = \dfrac{5}{\underset{2}{\cancel{6}}} \cdot \dfrac{\overset{5}{\cancel{15}}}{4} = \dfrac{25}{8}$

13. $\dfrac{15}{18} \div \dfrac{24}{25} = \dfrac{\overset{5}{\cancel{15}}}{18} \cdot \dfrac{25}{\underset{8}{\cancel{24}}} = \dfrac{125}{144}$

15. $\dfrac{\cancel{3}}{\cancel{5}} \cdot \dfrac{\cancel{5}}{\cancel{3}} = \dfrac{1}{1} = 1$

17. $\dfrac{3}{5} \div \dfrac{5}{3} = \dfrac{3}{5} \cdot \dfrac{3}{5} = \dfrac{9}{25}$

21. $18 \div \dfrac{3}{2} = \dfrac{\overset{6}{\cancel{18}}}{1} \cdot \dfrac{2}{\cancel{3}} = \dfrac{12}{1} = 12$

25. $\dfrac{\overset{7}{\cancel{28}}}{\underset{5}{\cancel{45}}} \cdot \dfrac{\overset{7}{\cancel{63}}}{\underset{10}{\cancel{40}}} = \dfrac{49}{50}$

29. $\dfrac{12}{25} \cdot \dfrac{5}{2} \div \dfrac{6}{5} = \left( \dfrac{\overset{6}{\cancel{12}}}{\underset{5}{25}} \cdot \dfrac{\cancel{5}}{\cancel{2}} \right) \div \dfrac{6}{5}$

$\qquad\qquad = \dfrac{6}{5} \div \dfrac{6}{5}$

$\qquad\qquad = \dfrac{\cancel{6}}{\cancel{5}} \cdot \dfrac{\cancel{5}}{\cancel{6}} = \dfrac{1}{1} = 1$

31. The reciprocal of $\dfrac{5}{9}$ is $\dfrac{9}{5}$.

33. $\dfrac{4}{7}$ is the reciprocal of $\dfrac{7}{4}$.

35. $\dfrac{4}{\cancel{5}} \cdot \dfrac{\overset{4,500}{\cancel{22,500}}}{1} = \dfrac{18,000}{1} = 18,000$

Allison earns $18,000.

39. $\dfrac{1}{\cancel{4}} \cdot \dfrac{\overset{75}{\cancel{300}}}{1} = \dfrac{75}{1} = 75$

$\qquad \dfrac{3}{\cancel{5}} \cdot \dfrac{\overset{60}{\cancel{300}}}{1} = \dfrac{180}{1} = 180$

So 75 hours are for assembly and 180 hours are for testing. Therefore, $300 - (75 + 180) = 300 - 255 = 45$ hours are left for packing.

## Exercises A.3

1. $\dfrac{2}{7} + \dfrac{3}{7} = \dfrac{2+3}{7} = \dfrac{5}{7}$

5. $\dfrac{5}{12} + \dfrac{1}{12} = \dfrac{5+1}{12} = \dfrac{6}{12}$

$\qquad\qquad\qquad = \dfrac{\cancel{6} \cdot 1}{\cancel{6} \cdot 2} = \dfrac{1}{2}$

9. $\dfrac{1}{2} + \dfrac{1}{4} = \dfrac{1 \cdot 2}{2 \cdot 2} + \dfrac{1}{4}$

$\qquad\quad = \dfrac{2}{4} + \dfrac{1}{4} = \dfrac{2+1}{4} = \dfrac{3}{4}$

11. $\dfrac{5}{6} - \dfrac{3}{8} = \dfrac{5 \cdot 4}{6 \cdot 4} - \dfrac{3 \cdot 3}{8 \cdot 3}$

$\qquad\quad = \dfrac{20}{24} - \dfrac{9}{24} = \dfrac{20-9}{24} = \dfrac{11}{24}$

15. $\dfrac{7}{12} + \dfrac{1}{3} = \dfrac{7}{12} + \dfrac{1 \cdot 4}{3 \cdot 4}$

$\qquad\quad = \dfrac{7}{12} + \dfrac{4}{12} = \dfrac{7+4}{12} = \dfrac{11}{12}$

17. $\dfrac{1}{2} + \dfrac{1}{3} + \dfrac{1}{4} = \dfrac{1 \cdot 6}{2 \cdot 6} + \dfrac{1 \cdot 4}{3 \cdot 4} + \dfrac{1 \cdot 3}{4 \cdot 3}$

$\qquad\qquad\quad = \dfrac{6}{12} + \dfrac{4}{12} + \dfrac{3}{12}$

$\qquad\qquad\quad = \dfrac{6+4+3}{12} = \dfrac{13}{12}$

21. $\dfrac{3}{28} + \dfrac{2}{35} = \dfrac{3 \cdot 5}{28 \cdot 5} + \dfrac{2 \cdot 4}{35 \cdot 4}$

$\qquad = \dfrac{15}{140} + \dfrac{8}{140}$

$\qquad = \dfrac{15 + 8}{140} = \dfrac{23}{140}$

25. $5 \div \dfrac{3}{5} = \dfrac{5}{1} \cdot \dfrac{5}{3} = \dfrac{25}{3}$

29. $12\dfrac{4}{5} = 12 + \dfrac{4}{5}$

$\qquad = \dfrac{12}{1} + \dfrac{4}{5}$

$\qquad = \dfrac{12 \cdot 5}{1 \cdot 5} + \dfrac{4}{5}$

$\qquad = \dfrac{60}{5} + \dfrac{4}{5} = \dfrac{60 + 4}{5} = \dfrac{64}{5}$

33. $\dfrac{43}{6} = \dfrac{42 + 1}{6}$

$\qquad = \dfrac{42}{6} + \dfrac{1}{6}$

$\qquad = 7 + \dfrac{1}{6} = 7\dfrac{1}{6}$

37. $\dfrac{5}{12} = \dfrac{5 \cdot 5}{12 \cdot 5} = \dfrac{25}{60}$

$\dfrac{7}{15} = \dfrac{7 \cdot 4}{15 \cdot 4} = \dfrac{28}{60}$

$\dfrac{9}{20} = \dfrac{9 \cdot 3}{20 \cdot 3} = \dfrac{27}{60}$

Therefore, $\dfrac{5}{12}$ is the smallest of the three fractions.

41. $240 \div 22\dfrac{1}{2} = \dfrac{240}{1} \div \dfrac{45}{2}$

$\qquad = \dfrac{\overset{16}{240}}{1} \cdot \dfrac{2}{\underset{3}{45}}$

$\qquad = \dfrac{32}{3} \text{ or } 10\dfrac{2}{3}$

On a 240-mile trip, the car uses $10\dfrac{2}{3}$ gallons of gasoline.

45. $27\dfrac{1}{2} \div 4\dfrac{1}{2} = \dfrac{55}{2} \div \dfrac{9}{2}$

$\qquad = \dfrac{55}{\cancel{2}} \cdot \dfrac{\cancel{2}}{9}$

$\qquad = \dfrac{55}{9} \text{ or } 6\dfrac{1}{9}$

Raul's rate of speed is $6\dfrac{1}{9}$ miles per hour.

## Exercises A.4

1.
```
    4.7
    3.5
 +21.7
 ─────
  29.9
```

5.
```
   21.620
    4.100
 +57.236
 ───────
  82.956
```

9.
```
   9.27
 −7.85
 ─────
  1.42
```

15.
```
   13.05
 ×  2.63
 ──────
   3915
   7830
   2610
 ───────
 34.3215
```

19.
```
      7.92
   5)39.60
     35
    ───
     46
     45
    ───
     10
     10
    ───
      0
```

23. $0.16\overline{)3.28}$ becomes
```
         20.5
    16)328.0
       32
      ───
        80
        80
       ───
         0
```

27.
```
    0.032
 ×   0.05
 ────────
  0.00160
```

33. $0.015\overline{)3.900}$ becomes
```
          260
    15)3900
       30
      ───
        90
        90
       ───
        00
        00
       ───
         0
```

37. $0.12\overline{)102}$ becomes
```
            850
    12)10200
       96
      ───
        60
        60
       ───
        00
        00
       ───
         0
```

The customer used 850 kilowatt hours.

41.    $18.65
   ×    35.4
    ───────
       7460
       9325
      5 595
    ───────
    $660.210

The mixture used for this process costs $660.21.


## Exercises A.5

3.  $0.78 = 78\%$

7.  $9\% = 0.09$

11.  $28\% = 0.28$

15.  $2 = 200\%$

19.  $0.007 = 0.7\%$

21.  $62.4\% = 0.624$

25.  $30\%$ of $70 = (0.30)(70) = 21$

27.  $7.2\%$ of $35 = (0.072)(35) = 2.52$

29.  $0.8\%$ of $5 = (0.008)(5) = 0.04$

33.  $\dfrac{5}{8} = 0.625 = 62.5\%$

35.  $\dfrac{8}{20} = \dfrac{2}{5} = 0.4 = 40\%$

39.  $\dfrac{24.6}{0.30}$ becomes $0.30\overline{)24.6}$, which becomes

$$
\begin{array}{r}
82 \\
3\overline{)246} \\
\underline{24}\phantom{0} \\
6 \\
\underline{6} \\
0
\end{array}
$$

43.  $6\%$ of $12,500 = (0.06)(12,500)$
$$= 750$$

So in 5 years, the population of the town will be $12,500 + 750 = 13,250$.

47.  $52\%$ of $8,600 = (0.52)(8,600)$
$$= 4,472$$
$48\%$ of $9,250 = (0.48)(9,250)$
$$= 4,440$$

Therefore, more votes were cast in this town in 1988.

# Appendix B
# Using a Scientific Calculator

Appendix B Exercises

1. 861.55

5. 65.5

9. 4,847.04

13. 30.45

17. 27.62

21. 0.06

25. $(186,000)(60) = 11,160,000$ miles.

27. 15% of $12 = (0.15)($12) = $1.80$.

The wholesaler marks up the price to $12 + $1.80 = $13.80$.

12% of $13.80 = (0.12)($13.80) = $1.66$.

So the consumer pays $13.80 + $1.66 = $15.46$.

# Chapter Tests, Cumulative Reviews and Cumulative Tests, and Practice Final

## Chapter 1 Test A

1. Let $A = \{5, 10, 15, 20, \ldots, 40\}$ and $B = \{x \mid x$ is an even integer between 13 and 31$\}$. Answer parts (a) through (d) "true" or "false".

   (a) $24 \in A$

   (b) $24 \in B$

   (c) $12 \in B$

   (d) $A$ has more elements than $B$.

   (e) List the set $C = \{x \mid x \in A$ and $x \in B\}$

In each of the following problems, compute the given expression.

2. $-8 + 1 - 2 - 4 + 5$

3. $|2 - 7| + |7 - 2|$

4. $2 + (-5) - (-3) - 4$

5. $\dfrac{(-2)(-5)(-3)}{-4 - 6}$

6. $\dfrac{(+1) - (-2)(-3)}{(+1)(-2) - (-3)}$

7. $(3 + 2)(5 - 9)$

8. $3 + 2 \cdot 5 - 9$

9. $-9[3 + 2(5 - 9)]$

10. $7 - 2[7 + 2(7 - 2)]$

11. Factor the number 72 into its prime factors.

## Chapter 1 Test B

1. Let $A = \{3, 6, 9, 12, \ldots, 27\}$ and $B = \{x \mid x$ is a prime less than 27$\}$. Answer parts (a) through (d) "true" or "false".

   (a) $21 \in A$

   (b) $21 \in B$

   (c) $2 \in B$

   (d) $A$ has fewer elements than $B$.

   (e) List the set $C = \{x \mid x \in A$ and $x \in B\}$

In each of the following problems, compute the given expression.

2. $6 - 3 - 7 + 1 - 2$

3. $|3 - 5| - |3 + 5|$

4. $-5 - (-1) + 6 + (-2)$

5. $\dfrac{(-4)(-3)(+2)}{5 - (-1)}$

6. $\dfrac{-1 - (-3)(-5)}{(-1)(-3) - (-5)}$

7. $(2 - 4 \cdot 3) - 6$

8. $2 - (4 \cdot 3 - 6)$

9. $5[(2 - 4)(3 - 6)]$

10. $1 + 3[1 - 3(1 + 3)]$

11. Factor the number 60 into its prime factors.

## Chapter 2 Test A

In problems 1−4, evaluate each of the given expressions.

1. $(-2)^6$

2. $-2^6$

3. $(-8 + 5 - (-2))^7$

4. $3 - 5(3 - 5)^3$

In problems 5−8, evaluate the given expression for $x = -1$, $y = 4$, and $z = -2$.

5. $x + y - z$

6. $x(y - z)$

7. $x^2 y - xz^3$

8. $|xy - z| + z$

In problems 9−13, perform the indicated operations and simplify as completely as possible.

9. $x^4 - 2xy^3 + x^3 y - 3xy^3 + 4x^4 - 6x^3 y$

10. $-5x^3 y^2 (-2xy)(xy^4)$

11. $x(xy - 3y^2) + 3y(xy + x) - xy(x + 1)$

12. $4x^2 y^2 (x + 2y^3) - 8xy^3 (xy^2)$

13. $3 + [x - 2(x + 1)]$

14. If we let $n$ stand for the "number," translate each of the following phrases:

   (a) three less than four times a number

   (b) seven more than three times a number is equal to one less than the number

15. Tin plates cost \$4 each and copper plates cost \$7 each. A customer bought a certain number of tin plates and one less than three times that many copper plates.

   (a) How many tin plates did the customer buy?

   (b) How much does a copper plate cost?

   (c) How many copper plates did the customer buy?

(d) How much does a tin plate cost?

(e) How much did the customer spend on the copper plates?

(f) How much did the customer spend on the tin plates?

(g) How much did the customer spend altogether?

## Chapter 2 Test B

In problems 1−4, evaluate each of the given expressions.

1. $(-4)^3$

2. $-4^3$

3. $(5 - 8 - (-1))^4$

4. $1 - 2(1 - 2)^7$

In problems 5−8, evaluate the given expression for $x = 2$, $y = -3$, and $z = -1$.

5. $2x - y + z$

6. $2(x - y) + z$

7. $x^3 - xyz^2$

8. $|x + y| + |xy|z$

In problems 9−13, perform the indicated operations and simplify as completely as possible.

9. $2x^2y^2 - xy - 3x^3 + 5xy - x^3 + 4x^2y^2$

10. $3xy^4(-x^3y^2)(-2xy^3)$

11. $y(x^3y + 2x) - x(xy + x^2y^2) - 2xy(1 - x)$

12. $5x^3y^3(y - 4x) + 2xy^2(10x^3y)$

13. $4 - [x + 3(x - 2)]$

14. If we let $n$ stand for the "number," translate each of the following phrases:

(a) four times three less than a number

(b) six more than five times a number is equal to ten more than the number

15. Walnuts cost $5 a pound and cashews cost $6 a pound. Mrs. O'Brien bought a certain number of pounds of walnuts and one more than four times that many pounds of cashews.

(a) How much does a pound of cashews cost?

(b) How many pounds of walnuts did Mrs. O'Brien buy?

(c) How much does a pound of walnuts cost?

(d) How many pounds of cashews did Mrs. O'Brien buy?

(e) How much did Mrs. O'Brien spend on walnuts?

(f) How much did Mrs. O'Brien spend on cashews?

(g) How much did Mrs. O'Brien spend altogether?

## Chapter 3 Test A

1. Determine whether the given equation is conditional, an identity, or a contradiction.

   (a) $2x - 4(2x - 1) = -6x + 4$

   (b) $2x - 4(2x - 1) = 6x + 4$

   (c) $2x - 4(2x - 1) = -6x - 4$

2. Determine whether or not the given value is a solution to the equation or inequality.

   (a) $x(x - 2) - (x + 1) = -2 - x; \quad x = -1$

   (b) $s(s + 5) + 2 = (s + 1)(s + 4) - 2; \quad s = 2$

   (c) $5 + 4(u + 3) > 9; \quad u = -2$

3. Solve each of the following equations or inequalities.

   (a) $4 + 2a = -2a + 16$

   (b) $3(2z + 1) - (2 - z) = 4(z - 1) - 1$

   (c) $t^2 - t(1 - t) = 2t(t - 2) + 12$

   (d) Solve and sketch the solution set on a number line: $\quad 5 - 3(x + 1) \leq -4$

   (e) Solve and sketch the solution set on a number line: $\quad -7 < 1 - 4x < 13$

4. One number is 7 more than 4 times another number. Find the numbers if their sum is 32.

5. A purse contains only dimes and quarters. If there are 16 coins altogether, and their total value is $2.50, how many of each type of coin are there in the purse?

6. A man leaves his home and travels at the rate of 40 mph. One hour later, his wife leaves and travels along the same road at the rate of 60 mph. How far from their home will she catch up to her husband?

## Chapter 3 Test B

1. Determine whether the given equation is conditional, an identity, or a contradiction.

   (a) $4x - 3(1 - x) = x - 3$

   (b) $4x - 3(1 - x) = 7x + 3$

   (c) $4x - 3(1 - x) = 7x - 3$

2. Determine whether or not the given value is a solution to the equation or inequality.

   (a) $(x + 3)(x - 1) - (3 - x) = 2x; \quad x = 2$

   (b) $b(b - 4) + 3 = b^2 + 7b; \quad b = -1$

   (c) $-2 - 5(t - 1) \leq 13; \quad t = -2$

3. Solve each of the following equations or inequalities.

(a) $-2 + 5n = -2n + 5$

(b) $3(3 - y) - 4(y + 4) = 2 - 4y$

(c) $4t^2 + 10 = 2t(2t + 1)$

(d) Solve and sketch the solution set on a number line: $2 + 3(4 - x) < 2$

(e) Solve and sketch the solution set on a number line: $-4 \leq 2 - 3x \leq -1$

4. Admission to a play costs $3 for a child's ticket and $5 for an adult's ticket. If 100 people attend the play and $380 is collected, how many children's tickets were sold?

5. Two sides of a triangle are equal in length. The third side is three less than the sum of the other two. If the perimeter of the triangle is 25 inches, find the lengths of its sides.

6. If 11 more than 3 times a number is 1 less than 5 times the number, what is the number?

## Cumulative Review: Chapters 1–3

In exercises 1–24, perform the indicated operations and simplify as completely as possible.

1. $-2 - 3 - 4$

2. $(-2)(-3)(-4)$

3. $(8 - 2 \cdot 3) - 1$

4. $8 - (2 \cdot 3 - 1)$

5. $(-3)^4$

6. $-3^4$

7. $a^2 a^4 a^6$

8. $3bb^5 b^2 + 2b^3 b^4 b$

9. $3r^3 s^5 - 5r^5 s^3 + r^5 s^3 - r^3 s^5$

10. $6y^2 + 3y - 4 - 5y^2 - 6y + 2$

11. $3x^2(2y - 5z)$

12. $3x^2(2y)(-5z)$

13. $-t^3(u^2)(-6t)$

14. $-t^3(u^2 - 6t)$

15. $8(p + 2q) - 4(p - 4q)$

16. $6(3c^5 - 2d^4) + 4(3d^4 - 2c^5)$

17. $v^2 w^3(5vw + 6v^3 w^2) - v^3 w^2(4w^2 - 3v^2 w^3)$

18. $2xz^3(3yz - x^2 y) + 3yz^2(x^3 z - 2xz^2)$

19. $5a^2 b^2 - 3ab - (5ab - 3a^2 b^2)$

20. $r(s - 2t) - (rs - 2t)$

21. $h + 6\{h - 5(h + 4)\}$

22. $n - [n^2 - n(n^2 - n)]$

23. $u(u - 2v) + 2v(u - 2v) - (u^2 - 4v^2)$

24. $c^3 - c(c^2 - cd + d^2) - d(c^2 - cd + d^2)$

In exercises 25–32, evaluate the given expression for $x = 4$, $y = -2$, and $z = -3$.

25. $x - y^2$

26. $(x - y)^2$

27. $x^2 - y^2$

28. $xy + (x + y)z$

29. $|xy + z - x(y + z)|$

30. $|xy + z| - |x(y + z)|$

31. $(y + 1)^3 + (z + 1)^2$

32. $(x - 4)(4y^4)(3z^3)$

In exercises 33–44, solve the equation or inequality. In the case of an inequality, sketch the solution set on a number line.

33. $4x + 1 = 9x - 9$

34. $8 - 3s = 3s - 10$

35. $3(b + 5) + 2(2b - 1) = -1$

36. $4(3a + 2) - 3(2a - 1) = 11$

37. $x(6x + 5) - 3x(2x - 1) = -40$

38. $y - 3(y - 3) > 3$

39. $3(2z - 5) + 2(3z + 5) \leq 7z$

40. $4(k + 2) + 2(k + 4) = 5(k + 1) - 3(k + 3)$

41. $-2 \leq 3(c - 2) + 4 < 7$

42. $1 < 1 + 3(4 - 2y) < 7$

43. $3\{x - 3(x - 3)\} = -x - 3$

44. $x - 3\{x - 2(x - 3)\} = 2$

Solve each of the following algebraically. Be sure to clearly label what the variable represents.

45. One number is 3 more than 2 times another. If their difference is 11, find the numbers.

46. The length of a rectangle is 2 less than 5 times its width. If the perimeter is 11 times the width, find the dimensions of the rectangle.

47. Sara and David line 51 miles apart. Sara can pedal her bicycle at the rate of 6 mph, while David can pedal his bicycle at the rate of 7 mph. If Sara leaves her house at 9:00 a.m. and David leaves his house two hours later, when will they meet?

48. A florist sell red geranium plants for $2 apiece and white geranium plants for $3 apiece. If he sell 50 plants and earns $113, how many white geranium plants did he sell?

## Cumulative Test: Chapters 1–3

1. Evaluate each of the following:
   (a) $-2^3 + 3(-3)^2$
   (b) $(-3 + 5 - 6)^3$

2. Evaluate each of the following for $a = 2, b = -3$:
   (a) $(b - a)^2 - ab^2$
   (b) $|3a + 2b| + |2a + 3b|$

3. Perform the indicated operations and simplify as completely as possible.

   (a) $3x^3 - 8x^2 + 5x - x^3 + 4x^2 + 2x$

   (b) $3(2x + y) - 5(x - y)$

   (c) $2ab(a^3 + ab) - 3a^2(a^2b - 2b^2)$

   (d) $5(m - 2n) - 3(m - 4n) - 2(n + m)$

   (e) $3u^2v^4(u^3 + v^2) - 3uv(2u^3v)(2uv^2)$

   (f) $3 - 2b(3 - 2b(3 - 2b))$

4. Solve each of the following equations or inequalities.

   (a) $4 - 5x = 10 - 7x$

   (b) $3(z - 4) + 4(z - 3) = 11$

   (c) $2 + 3(4 - c) > -10$

   (d) $4(s + 1) - 3(s - 1) = 2(s + 2) - (s - 3)$

   (e) $2(t - 3) + 3(t - 2) = 12 - t$

   (f) $3t + 7(5 - t) = 4(8 - t)$

5. Solve the following inequalities and sketch the solution set on a number line.

   (a) $5 - 2(1 + x) < -7$

   (b) $-3 < 5x + 2 \leq 2$

6. Solve each of the following algebraically.

   (a) One number is 6 more than 4 times another number. If their difference is 36, find the numbers.

   (b) Cities $A$ and $B$ are 2200 miles apart. A train leaves city $A$ at 10:00 a.m. and travels toward city $B$ at the rate of 200 mph. A second train leaves city $B$ at 11:00 a.m. and travels toward city $A$ at the rate of 300 mph. At what time will the two trains pass one another?

   (c) A charity car wash collects $156 by washing 40 cars. If it costs $3 to wash a compact car and $5 to wash a full-sized car, how many full-sized cars were washed?

## Chapter 4 Test A

1. Reduce each of the following to lowest terms:

   (a) $\dfrac{12}{-21}$

   (b) $\dfrac{y^9}{y^4}$

   (c) $\dfrac{-12x^8}{-8x^7}$

   (d) $\dfrac{-28a^3b^4}{8a^6b}$

2. Perform the indicated operations and express your answer in lowest terms.

(a) $\dfrac{3x^2y}{4xy^3} \cdot \dfrac{6x^2y^2}{xy}$

(b) $\dfrac{3x^2y}{4xy^3} \div \dfrac{6x^2y^2}{xy}$

(c) $\dfrac{5s^2t}{8s^3} \cdot \dfrac{-4st^3}{15t^2} \cdot \dfrac{-6}{t^2}$

(d) $\dfrac{3a^4b^3}{7a^5} \div 6b^2$

(e) $\dfrac{c}{4} + \dfrac{c}{4}$

(f) $\dfrac{c}{4} \cdot \dfrac{c}{4}$

(g) $\dfrac{5}{3x^2} + \dfrac{1}{3x^2}$

(h) $\dfrac{u}{2} - \dfrac{2u}{5}$

(i) $\dfrac{3}{4y} + \dfrac{2}{3y}$

(j) $\dfrac{2}{5xy^3} - \dfrac{5}{2x^2y^2}$

(k) $\dfrac{x^2 - 3x + 4}{6x^2} - \dfrac{x^2 + 3x + 4}{6x^2}$

3. Solve each of the following equations or inequalities.

(a) $\dfrac{5x}{6} - \dfrac{x}{4} = 7$

(b) $\dfrac{2x - 1}{3} + \dfrac{x + 1}{8} \le -1$

(c) $\dfrac{5 - y}{5} - \dfrac{4 - y}{4} = \dfrac{1}{10}$

(d) $0.01z + 0.1z = 1.32$

Solve each of the following algebraically.

4. If there are 39.37 inches in a meter, how many meters are there in 100 inches?

5. When $\dfrac{3}{5}$ of a number is subtracted from 5 more than the number, the result is one less than the number. Find the number.

6. A bank contains pennies, nickels, and dimes. There are 15 more pennies than nickels, and 3 times as many dimes as pennies. If the value of the coins is $6.45, how many of each type of coin are in the bank?

7. An amount of money is invested at 7% and $1000 more than that amount is invested at 10%. If the total interest on the two investments is $950, how much was invested at each rate?

8. The length of a rectangle is 15 cm more than $\dfrac{5}{6}$ of its width. If the perimeter is 10 cm less than 7 times the width, find the dimensions of the rectangle.

9. How many liters of a 24% salt solution must be mixed with 40 liters of a 30% salt solution in order to produce a 26% salt solution?

## Chapter 4 Test B

1. Reduce each of the following to lowest terms:

   (a) $\dfrac{-15}{-40}$

   (b) $\dfrac{z^{11}}{z^3}$

   (c) $\dfrac{15x^9}{-9x^6}$

   (d) $\dfrac{-30p^5q^2}{12p^4q^3}$

2. Perform the indicated operations and express your answer in lowest terms.

   (a) $\dfrac{2xy^3}{3x^2y^2} \cdot \dfrac{x^3y^2}{12xy}$

   (b) $\dfrac{2xy^3}{3x^2y^2} \div \dfrac{x^3y^2}{12xy}$

   (c) $\dfrac{3u^3v}{-7uv^3} \cdot \dfrac{14v}{15u} \cdot \dfrac{-5v^2}{2uv}$

   (d) $\dfrac{5c^2d^4}{6d^3} \div 10c^3$

   (e) $\dfrac{m}{6} + \dfrac{m}{6}$

   (f) $\dfrac{m}{6} \cdot \dfrac{m}{6}$

   (g) $\dfrac{13}{5x^3} - \dfrac{3}{5x^3}$

   (h) $\dfrac{v}{6} + \dfrac{4v}{9}$

   (i) $\dfrac{4}{5r} - \dfrac{1}{4r}$

   (j) $\dfrac{3}{4a^2b} + \dfrac{5}{3ab^3}$

   (k) $\dfrac{x^2 + 8x - 7}{2x^3} + \dfrac{7 - 4x - x^2}{2x^3}$

3. Solve each of the following equations or inequalities.

   (a) $\dfrac{3x}{8} + \dfrac{x}{6} = 13$

   (b) $\dfrac{x-3}{4} - \dfrac{3x+1}{5} > -2$

   (c) $\dfrac{6-y}{6} + \dfrac{5-y}{5} = \dfrac{8}{15}$

   (d) $0.2t + 0.04t = 14.4$

Solve each of the following algebraically.

4. If there are 35.31 cubic feet in a cubic meter, how many cubic meters are there in 200 cubic feet?

5. When two times a number is added to $\dfrac{2}{3}$ of one more than the number, the result is one less than three times the number. Find the number.

6. A man has 60 coins in his pocket, consisting of nickels, dimes, and quarters. He has three times as many dimes as quarters and the total value of his coins is $6.50. How many of each type of coin are in the man's pocket?

7. Two people leave from the same point and travel in opposite directions. One leaves at 1:00 p.m. and drives at the rate of 50 kph. The other leaves 15 minutes later and drives at the rate of 60 kph. At what time will they be 205 kilometers apart?

8. A woman invest a sum of money at 8% and twice as much at 9%. If her annual interest from the two investments is $390, how much was invested at each rate?

9. A chemist forms 20 gallons of a 35% hydrochloric acid solution by mixing a 25% solution and a 50% solution together. How many gallons of each are used to form the mixture?

## Chapter 5 Test A

1. Determine whether or not the point $(2, -3)$ satisfies the equation $2x + 3y = 5(5y + 7x)$.

2. Find the missing coordinates for the ordered pair of the given equation.
   (a) $4x + y = 9$ $\quad$ $(3, \ )$
   (b) $x + 2y = 8$ $\quad$ $( \ , -2)$
   (c) $2x - 3y = 12$ $\quad$ $(0, \ )$

3. Sketch the graphs of each of the following in a rectangular coordinate system using the intercept method.
   (a) $x + 3y = 6$
   (b) $2x - 5y = 10$
   (c) $3x + 4y = 0$
   (d) $x = -1$
   (e) $y = 5$

4. Find the slope of the line passing through the points $(4, -3)$ and $(-3, 4)$.

5. Write an equation of the line with slope $= -\dfrac{2}{5}$ which passes through the point $(6, -1)$.

6. Write an equation of the line passing through the points $(4, -1)$ and $(0, 7)$.

7. Write equations of the horizontal and vertical lines which pass through the point $(-2, 3)$.

8. What is the slope of a line whose equation is $2x + 3y + 4 = 0$?

9. Write an equation of the line passing through the point $(-1, 2)$ which is parallel to the line whose equation is $y = 3x + 7$.

10. Write an equation of the line whose graph appears in the following figure.

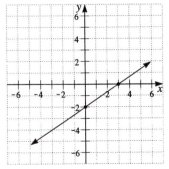

## Chapter 5 Test B

1. Determine whether or not the point $(-3, 1)$ satisfies the equation $3(x + 4y) = 3y + 2x$.

2. Find the missing coordinates for the ordered pair of the given equation.

   (a)  $5x - y = 7$     $(\ , 3)$

   (b)  $x - 3y = -2$    $(-11, \ )$

   (c)  $4x + 5y = 40$   $(\ , 0)$

3. Sketch the graphs of each of the following in a rectangular coordinate system using the intercept method.

   (a)  $3x + 2y = -6$

   (b)  $4x - 3y = 12$

   (c)  $x + 3y = 0$

   (d)  $x = 5$

   (e)  $y = -1$

4. Find the slope of the line passing through the points $(-5, 2)$ and $(-2, 5)$.

5. Write an equation of the line with slope $= -\dfrac{3}{4}$ which passes through the point $(1, -4)$.

6. Write an equation of the line passing through the points $(3, -2)$ and $(0, 10)$.

7. Write equations of the horizontal and vertical lines which pass through the point $(4, -2)$.

8. What is the slope of a line whose equation is $3x - 2y + 8 = 0$?

9. Write an equation of the line passing through the point $(2, -1)$ which is parallel to the line whose equation is $y = 5x - 4$.

10. Write an equation of the line whose graph appears in the following figure.

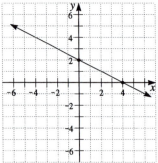

## Chapter 6 Test A

1. Solve the following system of equations by the graphical method:

$$\begin{cases} x + 3y = 6 \\ 2x - y = 5 \end{cases}$$

Solve each of the following systems of equations algebraically.

2. $\begin{cases} 2x + 3y = 5 \\ x - 2y = 6 \end{cases}$

3. $\begin{cases} 4x - 5y = -10 \\ 5x + 6y = 12 \end{cases}$

4. $\begin{cases} x + \dfrac{3}{4}y = 1 \\ \dfrac{4}{3}x + y = 2 \end{cases}$

Solve the following algebraically.

5. Five pounds of plums and four pounds of peaches cost $5.00. Two pounds of plums and three pounds of peaches cost $2.70. Find the cost of a pound of plums and a pound of peaches.

## Chapter 6 Test B

1. Solve the following system of equations by the graphical method:

$$\begin{cases} 3x - y = 3 \\ x + 2y = 8 \end{cases}$$

Solve each of the following systems of equations algebraically.

2. $\begin{cases} 5x - 6y = -9 \\ 4x + 3y = -15 \end{cases}$

3. $\begin{cases} 3x + 4y = 9 \\ 4x - 3y = 12 \end{cases}$

4. $\begin{cases} x - \dfrac{5}{2}y = 2 \\ \dfrac{2}{5}x - y = 1 \end{cases}$

Solve the following algebraically.

5. Three packages of veal and four packages of chicken cost $19. Five packages of veal and two packages of chicken cost $20. Find the cost of a package of veal and a package of chicken.

## Cumulative Review: Chapters 4−6

In exercises 1−4, reduce the fraction to lowest terms.

1. $\dfrac{56}{-35}$

2. $\dfrac{6x^3}{16x^8}$

3. $\dfrac{45a^3b^5}{25a^4b^4}$

4. $\dfrac{2c+3c+4c}{5c^2+6c^2+7c^2}$

In exercises 5−20, perform the indicated operations and simplify as completely as possible.

5. $\dfrac{8x^2}{27} \div \dfrac{x}{6}$

6. $\dfrac{8x^2}{27} \cdot \dfrac{x}{6}$

7. $\dfrac{8x^2}{27} - \dfrac{x}{6}$

8. $\dfrac{8}{3y^3} - \dfrac{2}{3y^3}$

9. $\dfrac{3u^2+1}{4u^2} + \dfrac{u^2-1}{4u^2}$

10. $\dfrac{2x+5}{7x} + \dfrac{x-3}{7x} - \dfrac{3x+2}{7x}$

11. $\dfrac{10a^4b^5}{21b^3c^2} \div \dfrac{15a^2c^4}{14b^4}$

12. $\dfrac{10a^4b^5}{21b^3c^2} \cdot \dfrac{15a^2c^4}{14b^4}$

13. $\dfrac{8}{5d} + \dfrac{5}{3d}$

14. $\dfrac{3}{4k} - \dfrac{5}{6l}$

15. $\dfrac{5}{8u^3v^2} + \dfrac{3}{10u^2v^3}$

16. $\dfrac{5}{8u^3v^2} \cdot \dfrac{3}{10u^2v^3}$

17. $\left(\dfrac{5}{2y} - \dfrac{1}{y}\right) \div \dfrac{4}{y}$

18. $\dfrac{5}{2y} - \left(\dfrac{1}{y} \div \dfrac{4}{y}\right)$

19. $3x - \dfrac{2}{x} + \dfrac{4}{x^3}$

20. $\dfrac{3}{2p^3q} + \dfrac{5}{4p^2q^2} - \dfrac{7}{6pq^3}$

In exercises 21−28, solve the given equation.

21. $\dfrac{x}{4} + \dfrac{x}{6} = \dfrac{2x+9}{12}$

22. $\dfrac{t+1}{5} - \dfrac{2t}{15} = \dfrac{t-1}{3}$

23. $\dfrac{w}{2} - \dfrac{w}{3} = \dfrac{w}{5}$

24. $\dfrac{z}{4} + \dfrac{z}{12} = \dfrac{z}{3}$

25. $\dfrac{1-3y}{4} + \dfrac{4-3y}{6} = \dfrac{9y+7}{3}$

26. $\dfrac{x}{6} - \dfrac{1}{5} = \dfrac{x}{8}$

27. $0.05(x+1) + 0.5(x-1) = 0.65$

28. $\dfrac{2}{3}(2u-3) - \dfrac{3}{4}(3u+2) = 2$

In exercises 29−36, sketch the graph of the given equation in a rectangular coordinate system. Label the intercepts.

29.  $y = -x + 4$

30.  $2x - 7y = 14$

31.  $5y + 4x = 10$

32.  $6y - 9x = 18$

33.  $y - 4 = 0$

34.  $x + 4 = 0$

35.  $y + 2x = 0$

36.  $x - 2y = 0$

In exercises 37−40, find the slope of the line passing through the given pair of points.

37.  $(5, -3)$ and $(3, -7)$

38.  $(1, -8)$ and $(-5, 7)$

39.  $(3, 11)$ and $(3, -11)$

40.  $(3, 11)$ and $(-3, 11)$

In exercises 41−44, write an equation of the line with the given slope which passes through the given point.

41.  $m = -5, \ (1, -4)$

42.  $m = \dfrac{2}{3}, \ (0, -6)$

43.  $m = 0, \ (-8, 2)$

44.  $m$ is undefined, $(6, 7)$

45.  Write an equation of the line passing through the points $(4, -2)$ and $(-3, 5)$.

46.  Write an equation of the line whose $x$-intercept is 6 and whose $y$-intercept is 5.

In exercises 47−48, solve the given system of equations graphically.

47.  $\begin{cases} 3x - 2y = 6 \\ 2x + y = -10 \end{cases}$

48.  $\begin{cases} 3x + 2y = 10 \\ 2x + 3y = 10 \end{cases}$

In exercises 49−56, solve each system of equations algebraically.

49.  $\begin{cases} 3x - 2y = 6 \\ 2x + y = -10 \end{cases}$

50.  $\begin{cases} 3x + 2y = 10 \\ 2x + 3y = 10 \end{cases}$

51.  $\begin{cases} 6x + 4y = 1 \\ 3x - 8y = 8 \end{cases}$

52.  $\begin{cases} 8x + 3y = 19 \\ 9x - 8y = 10 \end{cases}$

53.  $\begin{cases} 4x = 3 - y \\ 3y = -4 + x \end{cases}$

54.  $\begin{cases} 5x - y = 5 \\ 4x + \dfrac{y}{5} = 1 \end{cases}$

55.  $\begin{cases} 2x - 5y = 10 \\ \quad 10y = 4x - 20 \end{cases}$

56.  $\begin{cases} 3x - y = 9 \\ x - \dfrac{1}{3}y = 1 \end{cases}$

Solve each of the following algebraically. Be sure to label what the variables represent.

57. The sum of four consecutive odd numbers is 14 more than twice the largest one. Find the four numbers.

58. The ratio of married couples to single people in an apartment house is 9 to 5. If there are 135 married couples living in the apartment house, how many single people live there?

59. A typist charges $1.50 more for typing a page of technical material than she does for typing a page of regular material. If a report contains 18 pages of regular material and 12 pages of technical material and if the total bill is $108, what is the charge for typing a page of regular material?

60. 100 liters of a 30% chlorine solution is to be formed by mixing a 20% chlorine solution, a 25% chlorine solution, and a 40% chlorine solution. If there is to be twice as much of the 25% solution in the mixture as the 20% solution, how many liters of each solution must be mixed?

61. A wallet contains only five and ten dollar bills. If there are 23 bill altogether and their total value is $200, how many of each type of bill are in the wallet?

## Cumulative Test: Chapters 4−6

In problems 1−6, perform the indicated operations and simplify as completely as possible. Final answers should be reduced to lowest terms.

1. $\dfrac{8p^4q^3}{3r^2} \cdot \dfrac{9r}{4pq^3}$

2. $\dfrac{12a^2b^3}{5c^5} \div 30a^3b^2c$

3. $\dfrac{4y}{3ab} + \dfrac{7y}{3ab} - \dfrac{2y}{3ab}$

4. $\dfrac{2w+5}{5w} - \dfrac{5-3w}{5w}$

5. $\dfrac{3}{4c^3d} - \dfrac{5}{6cd^3}$

6. $\left(8 \cdot \dfrac{2}{x^2}\right) \div \dfrac{12}{x^4}$

7. Solve for $h$: $\dfrac{h}{4} + \dfrac{h}{10} = 7$

8. Solve for $v$: $\dfrac{4v+5}{2} - \dfrac{2v+1}{6} = v+1$

In problems 9−10, sketch the graph of the given equation in a rectangular coordinate system. Label the intercepts.

9. $3x - 4y = 12$

10. $4x + 3y = 12$

Problems 11−13 refer to $\ell$, the straight line that passes through the points $(1, -3)$ and $(-3, 5)$.

11. Find the slope of $\ell$.

12. Write an equation of $\ell$.

13. Find the $y$-intercept of $\ell$.

In problems 14−15, solve the system of equations algebraically.

14. $\begin{cases} 5x - 6y = 13 \\ 3x + 2y = -9 \end{cases}$

15. $\begin{cases} x - \dfrac{5}{3}y = -3 \\ -6x + 10y = 15 \end{cases}$

Solve each of the following algebraically. Be sure to clearly label what the variables represent.

16. If there are 5,280 feet in one mile, how many miles are there in 20,000 feet?

17. Two consecutive odd numbers have the property that the sum of $\dfrac{2}{5}$ of the smaller and $\dfrac{2}{3}$ of the larger is equal to one more than the larger. Find the numbers.

18. There are 200 orchestra seats and 120 balcony seats in a theater. If an orchestra seat costs $6 more than a balcony seat, and if a sold-out performance brings in $4080, find the cost of a balcony seat.

19. Bill invests some money at 8% and four times as much money at 13%. The total amount he receives on these investments is $100 more than he would received if he had invested all of his money at 10%. How much did Bill invest at each rate?

## Chapter 7 Test A

1. Evaluate $2^{-3} - 3^{-2}4^0$

In problems 2−5, simplify as completely as possible. Express final answers with positive exponents only.

2. $x^{-2}x^{-3}$

3. $(x^{-2})^{-3}$

4. $\dfrac{(x^2y)^3(xy^2)^2}{x^4y^8}$

5. $\dfrac{(-2x^{-2}y^{-3})^3}{2(x^{-1}y^{-2})^5}$

6. Given the polynomial $3x^6 + 5x^3 - x^2 + 7x + 1$:

    (a) How many terms are there?

    (b) What is the coefficient of the second degree term? the first degree term?

    (c) What is the degree of the polynomial?

In problems 7−13, perform the indicated operations and simplify as completely as possible.

7. $2x^3y^2(5xy)(-3x^2y^3)$

8. $2x^3y^2(5xy - 3x^2y^3)$

9. $3y^2(x - y) - xy(3y + 2x) + y(2x^2 + 5y^2)$

10. $(2x + 1)(5x^2 - 3x + 4)$

11. Add $x^3 + 8x^2 - 6x + 1$ and $2x - 5x^2 + 3$

12. Subtract $x^2 - 8x + 9$ from $x^2 + 8x - 9$

13. $(z - 2)^2 - (z - 3)^2$

14. Write in scientific notation:

   (a) $258{,}000$

   (b) $0.0258$

15. Compute using scientific notation:
$$\frac{(0.0032)(600{,}000)}{0.08}$$

## Chapter 7 Test B

1. Evaluate $3^0 4^{-1} + 5^{-2}$

In problems $2-5$, simplify as completely as possible. Express final answers with positive exponents only.

2. $x^{-1}x^{-5}$

3. $(x^{-1})^{-5}$

4. $\dfrac{(xy^3)^3(x^3y^2)^2}{x^{10}y^{11}}$

5. $\dfrac{(-3x^{-3}y^{-4})^2}{-3(x^{-2}y^{-1})^4}$

6. Given the polynomial $4x^7 + 3x^4 - 5x^3 - x + 2$:

   (a) How many terms are there?

   (b) What is the coefficient of the fourth degree term? the third degree term?

   (c) What is the degree of the polynomial?

In problems $7-13$, perform the indicated operations and simplify as completely as possible.

7. $4x^2y^4(2x^3y)(-3xy^3)$

8. $4x^2y^4(2x^3y - 3xy^3)$

9. $2x^2y(1 - xy) - y(x^3 + 2x^2) + 2x^3(y + y^2)$

10. $(2x - 1)(3x^2 + 5x - 2)$

11. Add $2x^3 - 6x^2 + 5x - 3$ and $4 - 8x + 3x^2$

12. Subtract $x^2 - 6x - 4$ from $x^2 + 6x + 4$

13. $(1 - y)^2 - (2 - y)^2$

14. Write in scientific notation:

   (a) $7{,}950{,}000$

   (b) $0.000795$

15. Compute using scientific notation:
$$\frac{(0.00075)}{(0.03)(5,000,000)}$$

## Chapter 8 Test A

In problems $1-11$, factor as completely as possible. If not factorable, say so.

1. $2x^4 - 8x^3 + 6x^2$

2. $3x^3y^2 + 12x^2y^3 - 9x^2y^2$

3. $x^2 + 13x + 36$

4. $x^2 + xy - 12y^2$

5. $4x^5 - 16x^3$

6. $7x^2 + 14x + 14$

7. $x^2 - 16x + 64$

8. $2x^2 - 9x + 4$

9. $2x^2 - 4x + 9$

10. $8x^2 + 13x - 6$

11. $x^2y^2 + 6xy + 9$

In problems $12-14$, solve the given equation.

12. $(x - 7)(x + 1) = 0$

13. $z^2 = 10z$

14. $a(a + 1) = (a + 2)(a + 3)$

15. Divide: $\dfrac{40a^2b^3 + 30a^3b^4 - 24a^2b^2}{20a^3b^3}$

16. Divide: $\dfrac{3x^3 + 5x^2 - 7}{x + 2}$

## Chapter 8 Test B

In problems $1-11$, factor as completely as possible. If not factorable, say so.

1. $5x^4 + 15x^3 + 10x^2$

2. $7x^3y^4 - 28x^4y^3 - 14x^3y^3$

3. $x^2 - 11x + 24$

4. $x^2 - 7xy - 18y^2$

5. $3x^4 - 27x^2$

6. $4x^2 + 12x + 16$

7. $x^2 + 18x + 81$

8. $3x^2 + 4x - 15$

9. $3x^2 - 4x + 15$

10. $9x^2 - 14x - 8$

11. $x^2y^2 - 10xy + 25$

In problems 12−14, solve the given equation.

12. $(x + 8)(x − 2) = 0$

13. $z^2 = −12z$

14. $a(a − 1) = (a − 2)(a − 3)$

15. Divide: $\dfrac{54p^4q^3 − 24p^5q^4 + 27p^3q^3}{18p^4q^4}$

16. Divide: $\dfrac{2x^3 − 3x^2 + 1}{x − 3}$

## Chapter 9 Test A

1. Reduce to lowest terms: $\dfrac{3x^2 + 9x}{x^2 − 9x}$

2. Reduce to lowest terms: $\dfrac{3x^2 + 9x}{x^2 − 9}$

In problems 3−7, perform the indicated operations and simplify as completely as possible.

3. $\dfrac{4}{x − 2} − \dfrac{2}{x − 4}$

4. $\dfrac{x^2}{4x^2 − 8x} \cdot \dfrac{2x^2 − 8}{(x + 2)^2}$

5. $\dfrac{6}{x^2 − 4x} + \dfrac{3}{2x}$

6. $\dfrac{x^2 − 3x − 4}{x^2 − x − 6} \div \dfrac{x^2 − 4x}{x^2 + 2x}$

7. $\dfrac{2x − 3}{x^2 − 2x − 3} + \dfrac{5x − 4}{x^2 − 2x − 3} − \dfrac{3x + 5}{x^2 − 2x − 3}$

8. Solve for $x$: $\dfrac{4}{x + 2} + \dfrac{2}{x} = \dfrac{14}{3x}$

9. Solve for $y$: $2xy + z = 1 − \dfrac{2y}{3}$

10. Solve for $b$: $\dfrac{2b + 8}{b + 3} + 3 = \dfrac{2}{b + 3}$

Solve each of the following algebraically.

11. What number must be added to the numerator and the denominator of the fraction $\dfrac{3}{5}$ so that the resulting fraction is equal to $\dfrac{5}{3}$?

12. George can build a bookcase in 6 hours. If he works together with his assistant Frank, the two men can build the bookcase in 4 hours. How long would it take Frank to build the bookcase if he works alone?

## Chapter 9 Test B

1. Reduce to lowest terms: $\dfrac{x^2 - 16x}{2x^2 - 8x}$

2. Reduce to lowest terms: $\dfrac{x^2 - 16}{2x^2 - 8x}$

In problems 3−7, perform the indicated operations and simplify as completely as possible.

3. $\dfrac{2}{x+5} + \dfrac{1}{x+2}$

4. $\dfrac{3x^3 + 9x^2}{3x} \cdot \dfrac{(x-3)^2}{3x^2 - 27}$

5. $\dfrac{6}{x^2 + 9x} - \dfrac{2}{3x}$

6. $\dfrac{x^2 + 4x - 5}{x^2 + x - 12} \div \dfrac{x^2 + 5x}{x^2 - 3x}$

7. $\dfrac{6x - 5}{x^2 - 5x + 4} - \dfrac{2x + 3}{x^2 - 5x + 4} - \dfrac{x + 4}{x^2 - 5x + 4}$

8. Solve for $x$: $\dfrac{1}{x} + \dfrac{3}{x-3} = \dfrac{17}{2x}$

9. Solve for $q$: $3pq - 2r = 4 + \dfrac{1}{4}q$

10. Solve for $a$: $\dfrac{4a - 7}{4 - a} + 1 = \dfrac{5 + a}{4 - a}$

Solve each of the following algebraically.

11. The denominator of a fraction is one more than three times the numerator. If 2 is added to both the numerator and denominator, the resulting fraction is equal to $\dfrac{2}{5}$. Find the original fraction.

12. Julia can run 4 times as fast as she can walk. If it takes her $3\dfrac{1}{4}$ hours to complete a trip in which she walks 5 miles and runs 6 miles, find her walking speed.

## Cumulative Review: Chapters 7−9

In exercises 1−14, perform the indicated operations and simplify as completely as possible.

1. $(x + 2)(x - 7)$

2. $(x + 2)(x^2 + x - 7)$

3. $(x^2 + x + 2)(x^2 + x - 7)$

4. $5t(t + 1)(t - 4)$

5. $(6r - 5s)(3r + 2s)$

6. $3m(m - 2) + 2m(m - 3)$

7. $(x + 4)(x + 16) - (x - 8)^2$

8. $(2x + 1)(x - 8) + 2(x + 2)^2$

9. $(m+3)(3m-1)(3m+1)$          10. $(x-2)(x+5)(x+2)(x-5)$

11. Subtract $5z^2 + 4z - 8$ from $3z^2 + 6z - 1$

12. Add $y^2 - 8xy + 2x^2$ to the product of $x + 3y$ and $2x - y$

13. (a) What is the degree of the polynomial $7x^6 - 5x^4 + 3x^2 - 1$?

     (b) What is the coefficient of the fourth degree term?

14. Write the polynomial $2x - 3x^2 - 1 + 4x^4$ in complete standard form.

In exercises 15−20, divide the given polynomials. Use long division where necessary.

15. $\dfrac{2x^3 - 18x^4}{6x^2}$          16. $\dfrac{18s^3t^2 + 24s^2t^2 - 12s^2t^3}{8s^3t^3}$

17. $\dfrac{a^2 - 4a + 7}{a + 3}$          18. $\dfrac{3y^3 + y^2 - 12y - 4}{y + 2}$

19. $\dfrac{12t^3 + 10t^2 - 9}{2t - 1}$          20. $\dfrac{b^4 + 3b + 2}{b - 1}$

In exercises 21−28, simplify the expression as completely as possible. Final answers should be expressed with positive exponents only.

21. $2^1 + 2^0 + 2^{-1}$          22. $(3-1)^{-2} - (4-1)^{-2}$

23. $\dfrac{x^4 x^5}{(x^4)^5}$          24. $\dfrac{y^{-10}}{y^{-5}}$

25. $\dfrac{(-3x^4)^3}{-9(x^7)^2}$          26. $\dfrac{(a^{-1}b^{-2})^{-3}}{(b^{-3}a^{-2})^{-1}}$

27. $\dfrac{(2u^3v^{-3})^{-2}}{(3u^{-2}v^2)^{-3}}$          28. $\left(\dfrac{4mn^{-2}}{3m^{-3}n}\right)^{-1}$

29. Write in scientific notation: $123,000,000$

30. Write in scientific notation: $0.00000975$

31. Evaluate: $\dfrac{(3 \times 10^{-5})(2.4 \times 10^4)}{0.6 \times 10^2}$

32. Evaluate: $\dfrac{(0.00015)(48,000,000)}{(0.144)(80,000)}$

In exercises 33−52, factor the polynomial as completely as possible. If the polynomial is not factorable, say so.

33. $x^2 - 8x$          34. $x^2 - 8x + 12$

35. $x^2 - 8x - 9$

36. $x^2 - 8x + 9$

37. $4a^3b^2 - 6a^2b + 8ab^3$

38. $8r^4s^6 - 2r^6s^4$

39. $18v^2 - 32$

40. $18v^2 + 32$

41. $t^5 - 49t^3$

42. $3s^2 - 7s - 10$

43. $6t^2 - 13t + 6$

44. $6t^2 - 37t + 6$

45. $6t^2 - 12t + 6$

46. $6t^2 - 13t - 6$

47. $56 - x - x^2$

48. $x^2 - 29xy + 100y^2$

49. $x^2 + 100y^2$

50. $x^2 + ax + bx + ab$

51. $x^2 + 3x - 2xy - 6y$

52. $81x^4 - 16y^4$

In exercises 53−54, simplify the fraction.

53. $\dfrac{y^2 + 2y}{y^2 - 4}$

54. $\dfrac{s^2 - 3s - 4}{s^2 + 5s + 4}$

In exercises 55−63, perform the indicated operations and simplify as completely as possible.

55. $\dfrac{2}{5x} - \dfrac{1}{3x + 2}$

56. $\dfrac{4}{3cd^2} + \dfrac{3}{4c^2d}$

57. $\dfrac{x^2 - 1}{x - 1} \cdot \dfrac{x + 1}{x^2 + 1}$

58. $\dfrac{m^2n - mn^2}{m^2 - n^2} \div \dfrac{2mn}{m^2 + 2mn + n^2}$

59. $\dfrac{4}{x^2 + 4x + 3} - \dfrac{2}{x^2 + 3x + 2}$

60. $\dfrac{1}{a - 2} - \dfrac{1}{2a} - \dfrac{2}{3a - 6}$

61. $\dfrac{\dfrac{s}{3} - \dfrac{3}{s}}{\dfrac{s^2 + 6s + 9}{6s}}$

62. $\dfrac{3t}{t - 1} - \dfrac{t^2 + 2t}{t^2 - t}$

63. $\dfrac{5r + 4}{2r + 8} + \dfrac{3r + 7}{2r + 8} - \dfrac{8r + 3}{2r + 8}$

In exercises 64−73, solve the given equation. If it contains more than one variable, solve for the indicated variable.

64. $\dfrac{5}{x} - \dfrac{1}{4} = \dfrac{7}{2x}$

65. $\dfrac{2}{4a + 1} + \dfrac{5}{12a + 3} = \dfrac{1}{9}$

66. $3r + 2s = 8 - r - 2s$ for $s$

67. $\dfrac{3}{4}u - 8v + 2 = \dfrac{2}{3}u + v + 1$ for $u$

68. $\dfrac{3z-2}{z+5} - 2 = \dfrac{2z-7}{z+5}$

69. $\dfrac{1}{2}(x-1) + \dfrac{1}{6}(x+1) = \dfrac{1}{3}(x-1) + \dfrac{1}{4}(x+1)$

70. $\dfrac{6p+1}{3p+2} = \dfrac{2p-3}{p}$

71. $\dfrac{6p+1}{3p-1} = \dfrac{2p+1}{p}$

72. $c^2 - 4 = 0$

73. $c^2 - 4c = 0$

Solve each of the following algebraically. Be sure to label what the variable represents.

74. Tom can complete an assignment in 3 hours, Dick in 4 hours, and Harry in 12 hours. If the three work together, how long will it take for them to complete the assignment?

75. A man walks a distance of 4 kilometers at a certain rate and runs back to his starting point at three times his walking speed. If the round trip takes 2 hours, what is the man's running speed?

76. The denominator of a fraction is 5 more than twice its numerator. If 7 is subtracted from both the numerator and the denominator, the resulting fraction is equal to $-\dfrac{1}{2}$. Find the original fraction.

## Cumulative Test: Chapters 7–9

In problems 1–6, perform the indicated operations and simplify as completely as possible. Final answers should be reduced to lowest terms, and should be expressed with positive exponents only.

1. $(3s - 4t)(s^3 t + 2s^2 t^2 - 3st^3)$

2. $\dfrac{12a^3 b^{-4}}{9a^{-2} b^{-1}}$

3. $(2x+1)(x-2) + (x-3)(3x+5)$

4. $(2n-1)^2 - 2(n-1)^2$

5. $\dfrac{(4x^{-2})^{-2}}{(2x^{-3})^{-3}}$

6. $\dfrac{16k^3 \ell^5 - 10k^5 \ell^3}{4k^4 \ell^4}$

7. Use long division to find the quotient and remainder: $\dfrac{x^3 + 3x^2 - x + 3}{x+1}$

8. Use long division to find the quotient and remainder: $\dfrac{18 - 11x^2 + x^4}{x+3}$

9. Write the following in scientific notation:

   (a)  0.000000531

   (b)  531,000,000

10. Compute using scientific notation:
$$\frac{(0.00009)(1,600,000)}{(0.0045)(320,000)}$$

11. Subtract $x^3 + 3x^2 - 4x + 1$ from the sum of $x^3 - 2x + 2$ and $5x^2 - x - 4$

In problems $12-19$, factor the given polynomial as completely as possible.

12. $x^2 - 5x + 6$

13. $x^2 - 5x - 6$

14. $2s^4t^6 - 8s^6t^4$

15. $10x^2 + 30x - 70$

16. $3y^2 - y - 14$

17. $12z^2 + 13z - 25$

18. $c(2c + 1) - 3(2c + 1)$

19. $x^3 + xy^2 - 2x^2 - 2y^2$

In problems $20-21$, reduce to lowest terms.

20. $\dfrac{x^3 - 4x}{x^2 + 4x + 4}$

21. $\dfrac{2a^2 - 7a + 6}{a^2 - 5a + 6}$

In problems $22-26$, perform the indicated operations and simplify as completely as possible.

22. $\dfrac{6}{x + 3} + \dfrac{4}{x - 2}$

23. $\dfrac{3u^3 - 12u}{u^2 + 3u - 10} \cdot \dfrac{u^2 + 6u + 5}{6u^3 + 12u^2 + 6u}$

24. $\dfrac{9}{2x^2 + 5x + 2} + \dfrac{3}{2x^2 + x}$

25. $\dfrac{r^2 + 8r + 15}{r^2 + 8r + 16} \div \dfrac{r^2 + 6r + 9}{r^2 + 6r + 8}$

26. $\dfrac{\dfrac{1}{a} + \dfrac{1}{b}}{\dfrac{4}{a^2} - \dfrac{4}{b^2}}$

In problems $27-30$, solve the given equation.

27. $\dfrac{8}{2t - 3} = \dfrac{5}{6t - 9} + \dfrac{19}{15}$

28. Solve for $w$: $\dfrac{3}{4}w + 5v = kw - \dfrac{1}{2}v + 4$

29. $\dfrac{2x + 3}{x + 7} - \dfrac{2}{3} = \dfrac{x - 4}{x + 7}$

30. $(3x + 1)(x - 4) = (2x + 1)(x - 5)$

31. The distance between towns $A$ and $B$ is 60 miles. Adam drives from $A$ to $B$ at a certain speed and drives 10 mph faster when he returns from $B$ to $A$. If the entire trip took $3\dfrac{1}{2}$ hours, find Adam's driving speed each way.

## Chapter 10 Test A

Perform the indicated operations. Make sure your final answer is in simplest radical form. Assume all variables appearing under radical signs are non-negative.

1. $\sqrt{16x^{10}y^{12}}$

2. $\sqrt{32} + 4\sqrt{72} - \sqrt{200}$

3. $\sqrt{45x^5} - x^2\sqrt{20x}$

4. $\dfrac{\sqrt{180x^8y^7}}{\sqrt{12x^2y^3}}$

5. $\sqrt{48x^5y^{11}} - 2x^2y^3\sqrt{12xy^5}$

6. $\dfrac{6y^3}{\sqrt{8y}}$

7. $\left(\sqrt{x} + 2\sqrt{3}\right)\left(\sqrt{4x} - \sqrt{27}\right)$

8. $\left(\sqrt{a} + \sqrt{6}\right)^2 - \left(\sqrt{a+6}\right)^2$

9. $\dfrac{7}{3 + \sqrt{2}}$

10. $\dfrac{2x - 8y}{\sqrt{x} - 2\sqrt{y}}$

11. Decide whether or not $1 + \sqrt{3}$ is a solution to the equation $x^2 - 2 = 2x$.

12. Solve for $x$: $\sqrt{4x} - 1 = 5$

## Chapter 10 Test B

Perform the indicated operations. Make sure your final answer is in simplest radical form. Assume all variables appearing under radical signs are non-negative.

1. $\sqrt{36x^{14}y^8}$

2. $\sqrt{50} - 3\sqrt{98} + \sqrt{128}$

3. $\sqrt{75x^5} - x\sqrt{27x^3}$

4. $\dfrac{\sqrt{120x^5y^{12}}}{\sqrt{12xy^4}}$

5. $\sqrt{72x^9y^5} - 2x^3y\sqrt{18x^3y^3}$

6. $\dfrac{8z^4}{\sqrt{12z}}$

7. $\left(\sqrt{x} - 3\sqrt{2}\right)\left(\sqrt{9x} + \sqrt{8}\right)$

8. $\left(\sqrt{b+7}\right)^2 - \left(\sqrt{b} + \sqrt{7}\right)^2$

9. $\dfrac{6}{4 - \sqrt{10}}$

10. $\dfrac{3x - 27y}{\sqrt{x} + 3\sqrt{y}}$

11. Decide whether or not $1 - \sqrt{3}$ is a solution to the equation $x^2 = 2x + 2$.

12. Solve for $x$: $\sqrt{3x} + 1 = 7$

## Chapter 11 Test A

Solve each of the following equations. Choose any method you like.

1. $(x+1)(x-1) = 8$

2. $2x^2 - 4x - 1 = 0$

3. $3x^2 = 8x + 3$

4. $(2x-1)^2 = 4$

5. $(2x-1)^2 = 4x$

6. $3x^2 - 6x + 1 = -x^2 + 2x + 1$

7. $\dfrac{x+3}{x} = \dfrac{x}{x-1}$

8. $(x^2+1)(x+1) = (x^2-2)(x-2)$

9. $\dfrac{1}{x} + \dfrac{1}{x-1} = \dfrac{3}{2}$

10. Solve the following by completing the square, and check your answer by using the quadratic formula: $2x^2 - 16x = -5$.

11. Sketch the graph of $y = x^2 + 4x - 5$. Label the intercepts.

## Chapter 11 Test B

Solve each of the following equations. Choose any method you like.

1. $(x-2)(x+2) = 21$

2. $2x^2 + 4x + 1 = 0$

3. $5x^2 + 3x = 2$

4. $(3x+1)^2 = 16$

5. $(3x+1)^2 = 16x$

6. $2x^2 + 5x - 2 = x^2 - x + 1$

7. $\dfrac{2x-3}{x} = \dfrac{4x}{2x+1}$

8. $(x^2+2)(x-1) = (x^2-2)(x+1)$

9. $\dfrac{1}{x+1} + \dfrac{1}{x} = \dfrac{3}{2}$

10. Solve the following by completing the square, and check your answer by using the quadratic formula: $3x^2 + 18x = -2$.

11. Sketch the graph of $y = x^2 - 4x - 5$. Label the intercepts.

## Cumulative Review: Chapters 10–11

In exercises 1–8, simplify the expression as completely as possible. Fractions should be reduced to lowest terms.

1. $\sqrt{4a^{20}b^{30}}$

2. $\sqrt{\dfrac{12w^8z^{11}}{108wz^7}}$

3. $\dfrac{8}{\sqrt{11}}$

4. $\dfrac{8}{\sqrt{11}-3}$

5. $\dfrac{12}{\sqrt{14}+\sqrt{10}}$

6. $\dfrac{21x}{\sqrt{7x}}$

7. $\sqrt{192}$

8. $\sqrt{84u^6v^7}$

In exercises 9–17, perform the indicated operations and simplify as completely as possible.

9. $\sqrt{24rs}+\sqrt{54rs}$

10. $\sqrt{3}\left(4\sqrt{5}+\sqrt{10}\right)-2\sqrt{3}\left(\sqrt{10}+\sqrt{20}\right)$

11. $4\sqrt{\dfrac{5}{2}}-\sqrt{40}$

12. $\left(3\sqrt{a}-\sqrt{4b}\right)\left(\sqrt{4a}+3\sqrt{b}\right)$

13. $2\sqrt{50p^8q^5}-p^2q^2\sqrt{98p^4q}$

14. $\dfrac{5}{\sqrt{6}}+\dfrac{7}{\sqrt{96}}$

15. $\left(2\sqrt{6}+\sqrt{7}\right)\left(2\sqrt{6}-\sqrt{7}\right)$

16. $\left(\sqrt{x+5}\right)^4$

17. $\left(\dfrac{8}{\sqrt{13}-3}-\dfrac{26}{\sqrt{13}}\right)^2$

In exercises 18–33, solve the given equation.

18. $x^2+4x=0$

19. $x^2+4x=21$

20. $1-\dfrac{5}{x}=\dfrac{24}{x^2}$

21. $\dfrac{x+4}{x+1}=\dfrac{x-3}{x-2}$

22. $x^2+5x=4$

23. $4x^2-7x+5=x^2-2x+3$

24. $(x-2)(x+3)=0$

25. $(x-2)(x+3)=6$

26. $(2x+1)(x-2)=18$

27. $(2x+1)(x-2)=(x+2)(2x-3)$

28. $(a-3)^2=4$

29. $(a-4)^2=3$

30. $\sqrt{x+2}=5$

31. $\sqrt{x}+2=5$

32. $7 - \sqrt{4s} = 11$

33. $2\sqrt{y - 3} + 5 = 9$

34. Solve by completing the square: $x^2 + 4x + 2 = 0$

35. Solve by completing the square: $2x^2 - 6x + 1 = 0$

In exercises 36–39, sketch the graph of the given equation. Label the intercepts. Round to the nearest tenth where necessary.

36. $y = x^2 - 8x + 7$

37. $y = (3x + 4)^2$

38. $y = 16 - 9x^2$

39. $y = -x^2 + x + 3$

40. The length of a rectangle is 2 less than three times its width. If the area of the rectangle is 40, find its dimensions.

41. Consider the right triangle shown below:

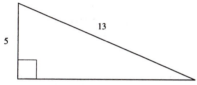

(a) Find its perimeter.

(b) Find its area.

42. Find the length of the side of a square if its diagonal is 8 inches.

43. A man jogs for 10 miles and then walks back to his starting point along the same route. His jogging speed is 5 mph faster than his walking speed. If the round trip takes him 3 hours, find his walking speed.

44. The sum of three times a number and twice the reciprocal of the number is 7. Find the number(s).

## Cumulative Test: Chapters 10–11

In problems 1–10, perform the indicated operations and simplify as completely as possible. Assume all variables appearing under radical signs are non-negative.

1. $\sqrt{128}$

2. $\sqrt{27r^3 s^5 t^8}$

3. $3\sqrt{28} + 2\sqrt{63}$

4. $5\sqrt{40b^3 c^8} - 3bc^2\sqrt{90bc^4}$

5. $\dfrac{45}{\sqrt{10}}$

6. $\sqrt{16x^{16}}$

7. $\dfrac{45}{\sqrt{10}-1}$

8. $\left(3\sqrt{2}+1\right)^2$

9. $\left(\sqrt{16w}-\sqrt{24}\right)\left(\sqrt{9w}+\sqrt{96}\right)$

10. $\dfrac{\sqrt{11}}{\sqrt{7}-\sqrt{3}}-\dfrac{\sqrt{3}}{\sqrt{11}+\sqrt{7}}$

11. $(x-1)(x+4)=50$

12. $(x+5)^2=5$

13. $\dfrac{2x}{4x-1}=\dfrac{x+2}{2x}$

14. $\sqrt{3x+1}+1=3$

15. $3(x^2+1)=10x$

16. $\dfrac{r^2}{4}+\dfrac{1}{3}=2$

17. $(2x+1)^2=4x(x+1)$

18. $(2x+1)^2=4(x+1)$

19. Find the area of a square with a diagonal of 6 cm.

20. A motorboat travels 28 miles at a certain speed before the motor fails. The driver then has to row an additional 6 miles to reach the shore, at a speed that is 12 mph less than the motorboat's speed. Find the rowing speed if the entire trip takes 5 hours.

In problems 21–22, sketch the graph of the given equation. Label the intercepts. Round to the nearest tenth where necessary.

21. $y=-x^2-6x-8$

22. $y=x^2-3$

## Practice Final Exam

In problems 1–4, evaluate the given expression.

1. $\dfrac{(-1)(-2)-(-3)(-4)}{-5}$

2. $|1-2[3-4(5-6)]|$

3. $x^2-xy-|x-y-z|$ when $x=-3$, $y=3$, and $z=2$

4. $(2^{-1}+3^{-1})^{-1}$

In problems 5–12, perform the indicated operations and simplify as completely as possible. (Where applicable, express your answers with positive exponents only in simplest radical form.)

5. $x^2(2x-3y^2)+2x(xy^2)-3(x^3-2x(xy))$

6. Subtract the sum of $x-1$ and $2x^2-x+2$ from the product of $x-1$ and $2x^2-x+2$.

7. $\dfrac{1}{8y}-\dfrac{3x^2}{5y^2}\cdot\dfrac{y}{6x}$

8. $\dfrac{(x^3y^2)^{-1}(x^{-2}y^3)^{-2}}{(xy^{-2})^{-3}}$

9. $\dfrac{x^2 - 4x - 5}{x^2 - 5x + 4} \div \dfrac{x^2 + 5x + 4}{x^2 + 4x - 5}$

10. $\dfrac{1}{x^2 - x} + \dfrac{1}{x^2 + x}$

11. $4x\sqrt{24x} - \dfrac{2}{5}\sqrt{150x^3}$

12. $\dfrac{3}{3 - \sqrt{6}} - \dfrac{3}{\sqrt{6} + \sqrt{3}}$

In problems 13−14, solve the given inequality and sketch its solution set on a number line.

13. $-11 \le 2(4 - x) - 5(1 + x) < 24$

14. $\dfrac{x + 1}{2} - \dfrac{x + 2}{3} + \dfrac{x + 3}{4} > 1$

In problems 15−20, solve the given equation.

15. $3(p + 2) + p(p + 3) = p(p - 2) + 2(p - 3)$

16. $\dfrac{3x - 2}{4} + 2 = \dfrac{2x - 3}{5}$

17. $\dfrac{3}{2x + 1} + \dfrac{2}{3x} = \dfrac{2}{x}$

18. $(3a + 1)(a - 1) = (2a - 1)(a + 1)$

19. $y^2 - 1 = \dfrac{5}{6}y$

20. $5x^2 + 6x - 1 = 0$

21. Compute using scientific notation: $\dfrac{(7,200,000)(0.0048)}{(64,000)(0.000003)}$

In problems 22−23, factor as completely as possible.

22. $4x^4 + 18x^3 - 36x^2$

23. $x^3 - xy^2 + 2x^2y - 2y^3$

24. Find the quotient and the remainder: $\dfrac{2x^3 - 1}{x + 1}$

25. Use the intercept method to sketch the graph of $5x - 6y = 15$.

26. Write an equation of the line whose $x$-intercept is 2 and whose $y$-intercept is $-8$.

27. Solve the following system algebraically: $\begin{cases} 7x + 12y = 3 \\ 5x - 3y = 6 \end{cases}$

Solve each of the following problems algebraically.

28. My father's age is seven more than three times my age. If the sum of our ages is five less than five times my age, how old am I?

29. Red poker chips are worth 2 cents, blue poker chips are worth 3 cents, and white poker chips are worth 5 cents. There are 150 chips in a pot, with twice the number of white chips as blue chips. If the total value of the pot is $5.80, how many red chips are contained in the pot?

30. Mrs. Brown invests some money at 6% and the rest at 9%, yielding her an annual interest of $306. If she had invested the same amounts in the opposite way, her annual interest would have increased to $324. How much money has Mrs. Brown invested in all?

31. Sam needs three more hours to plow a field than Hugo needs. Together, they can plow the field in two hours. How long would it take Hugo to plow the field if he works alone?

32. Find the two numbers whose sum is 1 and whose product is $-\dfrac{3}{4}$.

# Answers to Chapter Tests, Cumulative Reviews and Tests, and Practice Final

## Chapter 1 Test A

1. (a) F      (b) T      (c) F      (d) F      (e) $\{20, 30\}$
2. $-8$      3. $10$      4. $-4$      5. $3$
6. $-5$      7. $-20$      8. $4$      9. $45$
10. $-27$      11. $2 \cdot 2 \cdot 2 \cdot 3 \cdot 3$

## Chapter 1 Test B

1. (a) T      (b) F      (c) T      (d) F      (e) $\{3\}$
2. $-5$      3. $-6$      4. $0$      5. $4$
6. $-2$      7. $-16$      8. $-4$      9. $30$
10. $-32$      11. $2 \cdot 2 \cdot 3 \cdot 5$

## Chapter 2 Test A

1. $64$      2. $-64$      3. $-1$      4. $43$
5. $5$      6. $-6$      7. $-4$      8. $0$
9. $5x^4 - 5xy^3 - 5x^3y$      10. $10x^5y^7$
11. $2xy$      12. $4x^3y^2$
13. $-x + 1$
14. (a) $4n - 3$      (b) $3n + 7 = n - 1$
15. (a) $n$      (b) 7 dollars      (c) $3n - 1$      (d) 4 dollars
     (e) $7(3n - 1)$ dollars      (f) $4n$ dollars      (g) $25n - 7$ dollars

## Chapter 2 Test B

1. $-64$      2. $-64$      3. $16$      4. $3$
5. $6$      6. $9$      7. $14$      8. $-5$
9. $6x^2y^2 + 4xy - 4x^3$      10. $6x^5y^9$
11. $x^2y$      12. $5x^3y^4$
13. $-4x + 10$
14. (a) $4(n - 3)$      (b) $5n + 6 = n + 10$

15. (a) 6 dollars     (b) $n$     (c) 5 dollars     (d) $4n+1$

    (e) $5n$ dollars     (f) $6(4n+1)$ dollars     (g) $29n+6$ dollars

## Chapter 3 Test A

1. (a) identity     (b) conditional     (c) contradiction

2. (a) no     (b) yes     (c) no

3. (a) 3     (b) $-2$     (c) 4

    (d) $x \geq 2$

    (e) $-3 < x < 2$

4. 5 and 27     5. 10 dimes and 6 quarters

6. 120 miles

## Chapter 3 Test B

1. (a) conditional     (b) contradiction     (c) identity

2. (a) yes     (b) no     (c) yes

3. (a) 1     (b) $-3$     (c) 5

    (d) $x > 4$

    (e) $1 \leq x \leq 2$

4. 60 children's tickets     5. 7 inches, 7 inches, 11 inches

6. 6

## Cumulative Review: Chapters 1−3

1. $-9$     2. $-24$     3. 1     4. 3

5. 81     6. $-81$     7. $a^{12}$     8. $5b^8$

9. $2r^3s^5 - 4r^5s^3$     10. $y^2 - 3y - 2$

11. $6x^2y - 15x^2z$     12. $-30x^2yz$

13. $6t^4u^2$     14. $-t^3u^2 + 6t^4$

15. $4p + 32q$     16. $10c^5$

17. $v^3w^4 + 9v^5w^5$     18. $x^3yz^3$

19. $8a^2b^2 - 8ab$     20. $-2rt + 2t$

21. $-23h - 120$     22. $n - 2n^2 + n^3$

23. 0     24. $-d^3$     25. 0     26. 36

27. 12     28. $-14$     29. 9     30. $-9$

31. 3      32. 0      33. 2      34. 3

35. $-2$      36. 0      37. $-5$

38. $y < 3$      39. $z \leq 1$

40. $-5$      41. $0 \leq c < 3$

42. $1 < y < 2$      43. 6      44. 5

45. 8 and 19      46. width $= 4$, length $= 18$

47. 2 p.m.      48. 13 white geraniums

## Cumulative Test: Chapters 1–3

1. (a) 19    (b) $-64$      2. (a) 7    (b) 5

3. (a) $2x^3 - 4x^2 + 7x$    (b) $x + 8y$    (c) $-a^4b + 8a^2b^2$    (d) 0
   (e) $3u^2v^6 - 9u^5v^4$    (f) $3 - 6b + 12b^2 - 8b^3$

4. (a) 3    (b) 5    (c) $c < 8$    (d) identity

   (e) 4    (f) contradiction

5. (a) $x > 5$      (b) $-1 < x \leq 0$

       

4. (a) 46 and 10    (b) 3 p.m.    (c) 18 full-sized cars

## Chapter 4 Test A

1. (a) $-\dfrac{4}{7}$    (b) $y^5$    (c) $\dfrac{3x}{2}$    (d) $-\dfrac{7b^3}{2a^3}$

2. (a) $\dfrac{9x^2}{2y}$    (b) $\dfrac{1}{8y^3}$    (c) 1    (d) $\dfrac{b}{14a}$

   (e) $\dfrac{c}{2}$    (f) $\dfrac{c^2}{16}$    (g) $\dfrac{2}{x^2}$    (h) $\dfrac{u}{10}$

   (i) $\dfrac{17}{12y}$    (j) $\dfrac{4x - 25y}{10x^2y^3}$    (k) $-\dfrac{1}{x}$

3. (a) 12    (b) $x \leq -1$    (c) 2    (d) 12

4. 2.54 meters      5. 10

6. 5 nickels, 20 pennies, 60 dimes      7. $5000 at 7%, $6000 at 10%

8. width $= 12$ cm, length $= 25$ cm      9. 80 liters

## Chapter 4 Test B

1. (a) $\dfrac{3}{8}$    (b) $z^8$    (c) $-\dfrac{5x^3}{3}$    (d) $-\dfrac{5p}{2q}$

2. (a) $\dfrac{xy^2}{18}$    (b) $\dfrac{8}{x^3}$    (c) $1$    (d) $\dfrac{d}{12c}$

    (e) $\dfrac{m}{3}$    (f) $\dfrac{m^2}{36}$    (g) $\dfrac{2}{x^3}$    (h) $\dfrac{11v}{18}$

    (i) $\dfrac{11}{20r}$    (j) $\dfrac{9b^2 + 20a}{12a^2b^3}$    (k) $\dfrac{2}{x^2}$

3. (a) $24$    (b) $x < 3$    (c) $4$    (d) $60$

4. 5.66 cubic meters       5. 5

6. 10 quarters, 20 nickels, 30 dimes      7. 3 p.m.

8. $1500 at 8%, $3000 at 9%

9. 12 gallons of 25% solution, 8 gallons of 50% solution

## Chapter 5 Test A

1. yes      2. (a) $-3$      (b) $12$      (c) $-4$

3. (a)                                  (b)

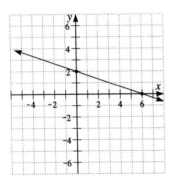
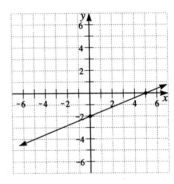

    (c)                                  (d)

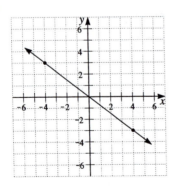
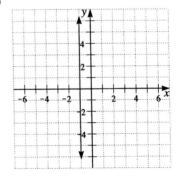

(e)

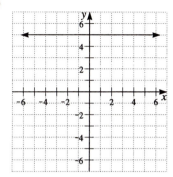

4.  $-1$

5.  $y + 1 = -\dfrac{2}{5}(x - 6)$ or $y = -\dfrac{2}{5}x + \dfrac{7}{5}$

6.  $y = -2x + 7$

7.  $y = 3; x = -2$

8.  $-\dfrac{2}{3}$

9.  $y = 3x + 5$

10.  $y = \dfrac{2}{3}x - 2$

## Chapter 5 Test B

1.  no

2.  (a)  2

(b)  $-3$

(c)  10

3.  (a)

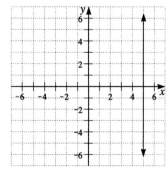

(b)

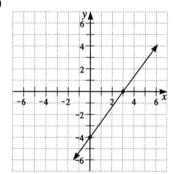

(c)

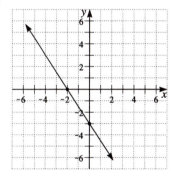

(d)

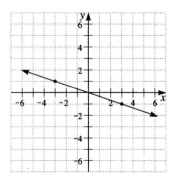

(e)

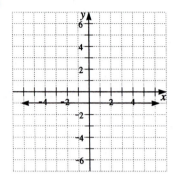

4. 1

5. $y + 4 = -\dfrac{3}{4}(x - 1)$ or $y = -\dfrac{3}{4}x - \dfrac{13}{4}$

6. $y = -4x + 10$

7. $y = -2; x = 4$

8. $\dfrac{3}{2}$

9. $y = 5x - 11$

10. $y = -\dfrac{1}{2}x + 2$

## Chapter 6 Test A

1.

2. $(4, -1)$

3. $(0, 2)$

4. no solution

5. plums: $0.60/pound; peaches: $0.50/pound

## Chapter 6 Test B

1.

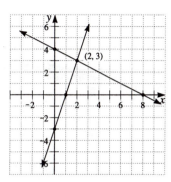

The point $(2, 3)$ is marked on the graph.

2. $(-3, -1)$

3. $(3, 0)$

4. no solution

5. veal: \$3/package; chicken: \$2.50/package

## Cumulative Review:  Chapters 4−6

1. $-\dfrac{8}{5}$

2. $\dfrac{3}{8x^5}$

3. $\dfrac{9b}{5a}$

4. $\dfrac{1}{2c}$

5. $\dfrac{16x}{9}$

6. $\dfrac{4x^3}{81}$

7. $\dfrac{16x^2 - 9x}{54}$

8. $\dfrac{2}{y^3}$

9. $1$

10. $0$

11. $\dfrac{4a^2b^6}{9c^6}$

12. $\dfrac{25a^6c^2}{49b^2}$

13. $\dfrac{49}{15d}$

14. $\dfrac{9\ell - 10k}{12k\ell}$

15. $\dfrac{25v + 12u}{40u^3v^3}$

16. $\dfrac{3}{16u^5v^5}$

17. $\dfrac{3}{8}$

18. $\dfrac{10 - y}{4y}$

19. $\dfrac{3x^4 - 2x^2 + 4}{x^3}$

20. $\dfrac{18q^2 + 15pq - 14p^2}{12p^3q^3}$

21. $3$

22. $2$

23. $0$

24. identity

25. $-\dfrac{1}{3}$

26. $\dfrac{24}{5}$

27. $2$

28. $-6$

314

29.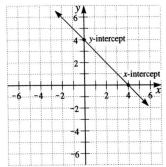

30.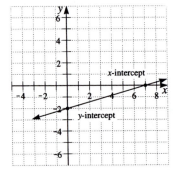

31.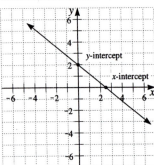

32.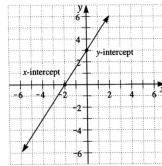

33.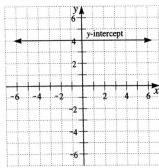

34.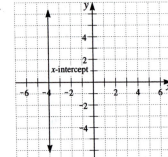

35.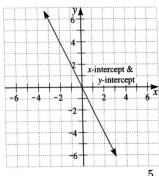

36.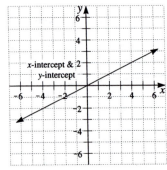

37.  2

38.  $-\dfrac{5}{2}$

39.  undefined

40.  0

41. $y + 4 = -5(x - 1)$ or $y = -5x + 1$      42. $y = \dfrac{2}{3}x - 6$      43. $y = 2$

44. $x = 6$      45. $y + 2 = -1(x - 4)$ or $y = -x + 2$

46. $y = -\dfrac{5}{6}x + 5$

47.
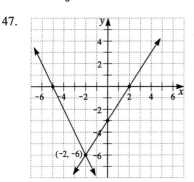

48.
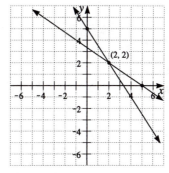

49. $(-2, -6)$      50. $(2, 2)$      51. $\left(\dfrac{2}{3}, -\dfrac{3}{4}\right)$      52. $(2, 1)$

53. $(1, -1)$      54. $\left(\dfrac{2}{5}, -3\right)$

55. infinitely many solutions: $\{(x, y) \mid 2x - 5y = 10\}$

56. no solution      57. 7, 9, 11, 13      58. 75 single people      59. $3

60. 20 liters of 20% solution, 40 liters of 25% solution, 40 liters of 40% solution

61. 6 $5 bills, 17 $10 bills

## Cumulative Test: Chapters 4−6

1. $\dfrac{6p^3}{r}$      2. $\dfrac{2b}{25ac^6}$      3. $\dfrac{3y}{ab}$      4. 1

5. $\dfrac{9d^2 - 10c^2}{12c^3 d^3}$      6. $\dfrac{4x^2}{3}$      7. 20      8. $-2$

9.
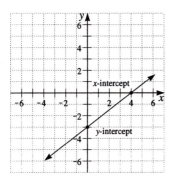

10.
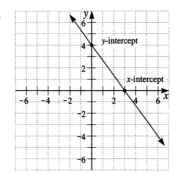

11. $-2$

12. $y - 5 = -2(x + 3)$ or $y = -2x - 1$

13. $-1$

14. $(-1, -3)$

15. no solution

16. 3.79 miles

17. 25 and 27

18. $9

19. $1000 at 8%, $4000 at 13%

## Chapter 7 Test A

1. $\dfrac{1}{72}$

2. $\dfrac{1}{x^5}$

3. $x^6$

4. $\dfrac{x^4}{y}$

5. $-\dfrac{4y}{x}$

6. (a) 5

(b) $-1$; 7

(c) 6

7. $-30x^6y^6$

8. $10x^4y^3 - 6x^5y^5$

9. $2y^3$

10. $10x^3 - x^2 + 5x + 4$

11. $x^3 + 3x^2 - 4x + 4$

12. $16x - 18$

13. $2z - 5$

14. (a) $2.58 \times 10^5$

(b) $2.58 \times 10^{-2}$

15. $2.4 \times 10^5$

## Chapter 7 Test B

1. $\dfrac{29}{100}$

2. $\dfrac{1}{x^6}$

3. $x^5$

4. $\dfrac{y^2}{x}$

5. $-\dfrac{3x^2}{y^4}$

6. (a) 5

(b) 3; $-5$

(c) 7

7. $-24x^6y^8$

8. $8x^5y^5 - 12x^3y^7$

9. $x^3y$

10. $6x^3 + 7x^2 - 9x + 2$

11. $2x^3 - 3x^2 - 3x + 1$

12. $12x + 8$

13. $2y - 3$

14. (a) $7.95 \times 10^6$

(b) $7.95 \times 10^{-4}$

15. $5 \times 10^{-9}$

## Chapter 8 Test A

1. $2x^2(x - 1)(x - 3)$

2. $3x^2y^2(x + 4y - 3)$

3. $(x + 4)(x + 9)$

4. $(x - 3y)(x + 4y)$

5. $4x^3(x - 2)(x + 2)$

6. $7(x^2 + 2x + 2)$

7. $(x - 8)(x - 8)$

8. $(2x - 1)(x - 4)$

9. not factorable

10. $(8x - 3)(x + 2)$

11. $(xy + 3)(xy + 3)$

12. $7, -1$

13. $0, 10$

14. $-\dfrac{3}{2}$

15. $\dfrac{2}{a} + \dfrac{3b}{2} - \dfrac{6}{5ab}$

16. quotient: $3x^2 - x + 2$; remainder: $-11$

## Chapter 8 Test B

1. $5x^2(x+1)(x+2)$
2. $7x^3y^3(y-4x-2)$
3. $(x-3)(x-8)$
4. $(x-9y)(x+2y)$
5. $3x^2(x-3)(x+3)$
6. $4(x^2+3x+4)$
7. $(x+9)(x+9)$
8. $(3x-5)(x+3)$
9. not factorable
10. $(9x+4)(x-2)$
11. $(xy-5)(xy-5)$
12. $-8, 2$
13. $0, -12$
14. $\dfrac{3}{2}$
15. $\dfrac{3}{q} - \dfrac{4p}{3} + \dfrac{3}{2pq}$
16. quotient: $2x^2+3x+9$;   remainder: 28

## Chapter 9 Test A

1. $\dfrac{3(x+3)}{x-9}$
2. $\dfrac{3x}{x-3}$
3. $\dfrac{2(x-6)}{(x-2)(x-4)}$
4. $\dfrac{x}{2(x+2)}$
5. $\dfrac{3}{2(x-4)}$
6. $\dfrac{x+1}{x-3}$
7. $\dfrac{4}{x+1}$
8. $4$
9. $y = \dfrac{3-z}{6x+2}$
10. no solution
11. $-8$
12. 12 hours

## Chapter 9 Test B

1. $\dfrac{x-16}{2(x-4)}$
2. $\dfrac{x+4}{2x}$
3. $\dfrac{3(x+3)}{(x+5)(x+2)}$
4. $\dfrac{x(x-3)}{3}$
5. $-\dfrac{2}{3(x+9)}$
6. $\dfrac{x-1}{x+4}$
7. $\dfrac{3}{x-1}$
8. $5$
9. $q = \dfrac{8r+16}{12p-1}$
10. no solution
11. $\dfrac{4}{13}$
12. 2 mph

## Cumulative Review:  Chapters 7−9

1. $x^2-5x-14$
2. $x^3+3x^2-5x-14$
3. $x^4+2x^3-4x^2-5x-14$
4. $5t^3-15t^2-20t$
5. $18r^2-3rs-10s^2$
6. $5m^2-12m$
7. $36x$
8. $4x^2-7x$
9. $9m^3+27m^2-m-3$
10. $x^4-29x^2+100$
11. $-2z^2+2z+7$
12. $4x^2-3xy-2y^2$

13. (a) 6      (b) $-5$

14. $4x^4 + 0x^3 - 3x^2 + 2x - 1$

15. $\dfrac{x}{3} - 3x^2$

16. $\dfrac{9}{4t} + \dfrac{3}{st} - \dfrac{3}{2s}$

17. quotient: $a - 7$;   remainder: 28

18. quotient: $3y^2 - 5y - 2$;   remainder: 0

19. quotient: $6t^2 + 8t + 4$;   remainder: $-5$

20. quotient: $b^3 + b^2 + b + 4$;   remainder: 6

21. $\dfrac{7}{2}$

22. $\dfrac{5}{36}$

23. $\dfrac{1}{x^{11}}$

24. $\dfrac{1}{y^5}$

25. $\dfrac{3}{x^2}$

26. $ab^3$

27. $\dfrac{27v^{12}}{4u^{12}}$

28. $\dfrac{3n^3}{4m^4}$

29. $1.23 \times 10^8$

30. $9.75 \times 10^{-6}$

31. $1.2 \times 10^{-2}$ or $0.012$

32. $6.25 \times 10^{-1}$ or $0.625$

33. $x(x - 8)$

34. $(x - 2)(x - 6)$

35. $(x - 9)(x + 1)$

36. not factorable

37. $2ab(2a^2b - 3a + 4b^2)$

38. $2r^4s^4(2s - r)(2s + r)$

39. $2(3v - 4)(3v + 4)$

40. $2(9v^2 + 16)$

41. $t^3(t - 7)(t + 7)$

42. $(3s - 10)(s + 1)$

43. $(2t - 3)(3t - 2)$

44. $(6t - 1)(t - 6)$

45. $6(t - 1)(t - 1)$

46. not factorable

47. $-(x + 8)(x - 7)$

48. $(x - 25y)(x - 4y)$

49. not factorable

50. $(x + a)(x + b)$

51. $(x - 2y)(x + 3)$

52. $(9x^2 + 4y^2)(3x + 2y)(3x - 2y)$

53. $\dfrac{y}{y - 2}$

54. $\dfrac{s - 4}{s + 4}$

55. $\dfrac{x + 4}{5x(3x + 2)}$

56. $\dfrac{16c + 9d}{12c^2d^2}$

57. $\dfrac{x^2 + 2x + 1}{x^2 + 1}$

58. $\dfrac{m + n}{2}$

59. $\dfrac{2}{(x + 2)(x + 3)}$

60. $\dfrac{-a + 6}{6a(a - 2)}$

61. $\dfrac{2(s - 3)}{s + 3}$

62. 2

63. $\dfrac{4}{r + 4}$

64. 6

65. 8

66. $s = 2 - r$

67. $u = 12(9v - 1)$

68. no solution

69. 3

70. $-1$

71. no solution

72. $2, -2$

73. $0, 4$

74. $1\dfrac{1}{2}$ hours

75. 8 kph

76. $\dfrac{4}{13}$

## Cumulative Test: Chapters 7–9

1. $3s^4t + 2s^3t^2 - 17s^2t^3 + 12st^4$

2. $\dfrac{4a^5}{3b^3}$

3. $5x^2 - 7x - 17$

4. $2n^2 - 1$

5.  $\dfrac{1}{2x^5}$

6.  $\dfrac{4\ell}{k} - \dfrac{5k}{2\ell}$

7.  quotient: $x^2 + 2x - 3$;  remainder: 6

8.  quotient: $x^3 - 3x^2 - 2x + 6$;  remainder: 0

9.  (a) $5.31 \times 10^{-7}$  (b) $5.31 \times 10^8$

10.  $1 \times 10^{-1}$ or $0.1$

11.  $2x^2 + x - 3$

12.  $(x - 2)(x - 3)$

13.  $(x - 6)(x + 1)$

14.  $2s^4 t^4 (t - 2s)(t + 2s)$

15.  $10(x^2 + 3x - 7)$

16.  $(3y - 7)(y + 2)$

17.  $(12z + 25)(z - 1)$

18.  $(2c + 1)(c - 3)$

19.  $(x^2 + y^2)(x - 2)$

20.  $\dfrac{x(x - 2)}{x + 2}$

21.  $\dfrac{2a - 3}{a - 3}$

22.  $\dfrac{10x}{(x + 3)(x - 2)}$

23.  $\dfrac{u + 2}{2(u + 1)}$

24.  $\dfrac{6}{x(x + 2)}$

25.  $\dfrac{(r + 5)(r + 2)}{(r + 4)(r + 3)}$

26.  $\dfrac{ab}{4(b - a)}$

27.  4

28.  $w = \dfrac{22v - 16}{4k - 3}$

29.  no solution

30.  1

31.  30 mph from $A$ to $B$; 40 mph from $B$ to $A$

## Chapter 10 Test A

1.  $4x^5 y^6$

2.  $18\sqrt{2}$

3.  $x^2\sqrt{5x}$

4.  $\sqrt{15}x^3 y^2$

5.  0

6.  $\dfrac{3y^2\sqrt{2y}}{2}$

7.  $2x + \sqrt{3x} - 18$

8.  $2\sqrt{6a}$

9.  $3 - \sqrt{2}$

10.  $2(\sqrt{x} + 2\sqrt{y})$

11.  yes

12.  9

## Chapter 10 Test B

1.  $6x^7 y^4$

2.  $-8\sqrt{2}$

3.  $2x^2\sqrt{3x}$

4.  $\sqrt{10}x^2 y^4$

5.  0

6.  $\dfrac{4z^3\sqrt{3z}}{3}$

7.  $3x - 7\sqrt{2x} - 12$

8.  $-2\sqrt{7b}$

9.  $4 + \sqrt{10}$

10.  $3(\sqrt{x} - 3\sqrt{y})$

11.  yes

12.  12

## Chapter 11 Test A

1.  $\pm 3$

2.  $\dfrac{2 \pm \sqrt{6}}{2}$

3.  $3, -\dfrac{1}{3}$

4.  $\dfrac{3}{2}, -\dfrac{1}{2}$

5.  $\dfrac{2 \pm \sqrt{3}}{2}$

6.  $0, 2$

7.  $\dfrac{3}{2}$

8.  $\dfrac{-1 \pm \sqrt{5}}{2}$

9.  $2, \dfrac{1}{3}$

10.  $\dfrac{8 \pm 3\sqrt{6}}{2}$

11.  The $x$-intercepts are $-5$ and $1$.
     The $y$-intercept is $-5$.

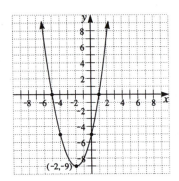

(-2,-9)

## Chapter 11 Test B

1.  $\pm 5$

2.  $\dfrac{-2 \pm \sqrt{2}}{2}$

3.  $\dfrac{2}{5}, -1$

4.  $1, -\dfrac{5}{3}$

5.  $\dfrac{1}{9}, 1$

6.  $-3 \pm 2\sqrt{3}$

7.  $-\dfrac{3}{4}$

8.  $0, 2$

9.  $1, -\dfrac{2}{3}$

10. $\dfrac{-9 \pm 5\sqrt{3}}{3}$

11. The $x$-intercepts are $-1$ and $5$.
    The $y$-intercept is $-5$.

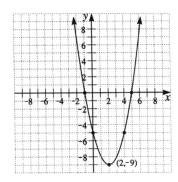

(2,-9)

## Cumulative Review:  Chapters 10−11

1.  $2a^{10}b^{15}$

2.  $\dfrac{w^3 z^2}{3}\sqrt{w}$

3.  $\dfrac{8\sqrt{11}}{11}$

4.  $4\left(\sqrt{11} + 3\right)$

5.  $3\left(\sqrt{14} - \sqrt{10}\right)$

6.  $3\sqrt{7x}$

7.  $8\sqrt{3}$

8.  $2u^3 v^3 \sqrt{21v}$

9.  $5\sqrt{6rs}$

10. $-\sqrt{30}$

11. $0$

12. $6a + 5\sqrt{ab} - 6b$

13. $3p^4 q^2 \sqrt{2q}$

14. $\dfrac{9\sqrt{6}}{8}$

15. $17$

16. $x^2 + 10x + 25$

17. $36$

18. $0, -4$

19. $-7, 3$

20. $-3, 8$

21. $\dfrac{5}{4}$

22. $\dfrac{-5 \pm \sqrt{41}}{2}$

23. $\dfrac{2}{3}, 1$

24. $-3, 2$

25. $-4, 3$

26. $-\dfrac{5}{2}, 4$

27. 1

28. 1, 5

29. $4 \pm \sqrt{3}$

30. 23

31. 9

32. no solution

33. 7

34. $-2 \pm \sqrt{2}$

35. $\dfrac{3 \pm \sqrt{7}}{2}$

36. The $x$-intercepts are 1 and 7.
    The $y$-intercept is 7.

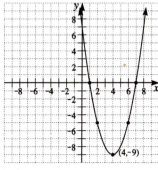

37. The $x$-intercept is $-\dfrac{4}{3}$.
    The $y$-intercept is 16.

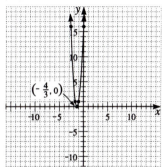

38. The $x$-intercepts are $-\dfrac{4}{3}$ and $\dfrac{4}{3}$.
    The $y$-intercept is 16.

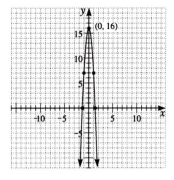

39. The $x$-intercepts are $\dfrac{1-\sqrt{13}}{2} \approx -1.3$ and

$\dfrac{1+\sqrt{13}}{2} \approx 2.3$.

The $y$-intercept is 3.

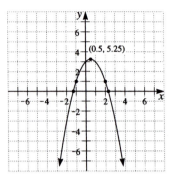

40. width $= 4$, length $= 10$      41.   (a)   30           (b)   30

42. $4\sqrt{2}$ inches     43.   5 mph     44.   $\dfrac{1}{3}, 2$

## Cumulative Test: Chapters 10−11

1. $8\sqrt{2}$          2.   $3rs^2t^4\sqrt{3rs}$      3.   $12\sqrt{7}$         4.   $bc^4\sqrt{10b}$

5. $\dfrac{9\sqrt{10}}{2}$         6.   $4x^8$        7.   $5\left(\sqrt{10}+1\right)$     8.   $19+6\sqrt{2}$

9. $12w + 10\sqrt{6w} - 48$      10.   $\dfrac{\sqrt{7}\left(\sqrt{11}+\sqrt{3}\right)}{4}$

11. $-9, 6$      12.   $-5 \pm \sqrt{5}$     13.   $\dfrac{2}{7}$      14.   1

15. $\dfrac{1}{3}, 3$      16.   $\dfrac{\pm 2\sqrt{15}}{3}$     17.   no solution     18.   $\pm \dfrac{\sqrt{3}}{2}$

19. 18 sq. cm      20.   2 mph

21. The $x$-intercepts are $-4$ and $-2$.
The $y$-intercept is $-8$.

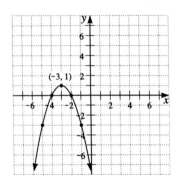

22. The $x$-intercepts are $\pm\sqrt{3} \approx \pm 1.7$.
    The $y$-intercept is $-3$.

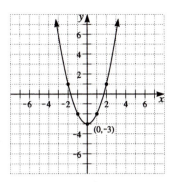

## Practice Final Exam

1. 2

2. 13

3. 10

4. $\dfrac{6}{5}$

5. $-x^3 - x^2y^2 + 6x^2y$

6. $2x^3 - 5x^2 + 3x - 3$

7. $\dfrac{5 - 4x}{40y}$

8. $\dfrac{x^4}{y^{14}}$

9. $\dfrac{(x-5)(x+5)}{(x-4)(x+4)}$

10. $\dfrac{2}{x^2 - 1}$

11. $6x\sqrt{6x}$

12. $3 + \sqrt{3}$

13. $-3 < x \le 2$

14. $x > 1$

15. $-2$

16. $-6$

17. $4$

18. $0, 3$

19. $-\dfrac{2}{3}, \dfrac{3}{2}$

20. $\dfrac{-3 \pm \sqrt{14}}{5}$

21. $1.8 \times 10^5$

22. $2x^2(2x - 3)(x + 6)$

23. $(x + 2y)(x - y)(x + y)$

24. quotient: $2x^2 - 2x + 2$; remainder: $-3$

25.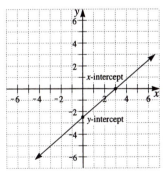

26. $y = 4x - 8$

27. $\left(1, -\dfrac{1}{3}\right)$

28. 12 years old

29. 30 red chips

31. $4200

31. 3 hours

32. $-\dfrac{1}{2}$ and $\dfrac{3}{2}$